Geometry and Computing

Series Editors

Herbert Edelsbrunner
Leif Kobbelt
Konrad Polthier

Editorial Advisory Board

Jean-Daniel Boissonnat
Gunnar Carlsson
Bernard Chazelle
Xiao-Shan Gao
Craig Gotsman
Leo Guibas
Myung-Soo Kim
Takao Nishizeki
Helmut Pottmann
Roberto Scopigno
Hans-Peter Seidel
Steve Smale
Peter Schröder
Dietrich Stoyan

For further volumes:
http://www.springer.com/series/7580

Rémi Ronfard
Gabriel Taubin

Editors

Image and Geometry Processing for 3-D Cinematography

With 165 Figures

Editors
Rémi Ronfard
LEAR Team
INRIA
avenue de l'Europe
38 330 Montbonnot
France
remi.ronfard@inria.fr

Gabriel Taubin
Brown University
Division of Engineering
Providence, RI 02912
USA
taubin@brown.edu

ISBN 978-3-642-12391-7   e-ISBN 978-3-642-12392-4
DOI 10.1007/978-3-642-12392-4
Springer Heidelberg Dordrecht London New York

Library of Congress Control Number: 2010930516

© Springer-Verlag Berlin Heidelberg 2010
This work is subject to copyright. All rights are reserved, whether the whole or part of the material is concerned, specifically the rights of translation, reprinting, reuse of illustrations, recitation, broadcasting, reproduction on microfilm or in any other way, and storage in data banks. Duplication of this publication or parts thereof is permitted only under the provisions of the German Copyright Law of September 9, 1965, in its current version, and permission for use must always be obtained from Springer. Violations are liable to prosecution under the German Copyright Law.
The use of general descriptive names, registered names, trademarks, etc. in this publication does not imply, even in the absence of a specific statement, that such names are exempt from the relevant protective laws and regulations and therefore free for general use.

Cover design: deblik, Berlin

Printed on acid-free paper

Springer is part of Springer Science+Business Media (www.springer.com)

The cinema is truth 24 frames-per-second.
-Jean-Luc Godard

Acknowledgements

The preparation of this book was made possible by INRIA's Associate Team VAMP (Video and Mesh Processing for 3-D Cinematography).

Contents

Image and Geometry Processing for 3-D Cinematography:
An Introduction 1
Rémi Ronfard and Gabriel Taubin

Part I 3-D Cinematography and Applications

Stereoscopic Cinema 11
Frédéric Devernay and Paul Beardsley

Free-Viewpoint Television 53
Masayuki Tanimoto

Free-Viewpoint Video for TV Sport Production 77
Adrian Hilton, Jean-Yves Guillemaut, Joe Kilner, Oliver Grau,
and Graham Thomas

Challenges for Multi-View Video Capture 107
Bennett Wilburn

Part II Recent Developments in Geometry

Performance Capture from Multi-View Video 127
Christian Theobalt, Edilson de Aguiar, Carsten Stoll,
Hans-Peter Seidel, and Sebastian Thrun

Combining Multi-view Stereo and Bundle Adjustment
for Accurate Camera Calibration 151
Yasutaka Furukawa and Jean Ponce

Cell-Based 3D Video Capture Method with Active Cameras 171
Tatsuhisa Yamaguchi, Hiromasa Yoshimoto, and Takashi
Matsuyama

ix

Dense 3D Motion Capture from Synchronized Video Streams 193
Yasutaka Furukawa and Jean Ponce

Part III Recent Developments in Image Processing

**Wavelet-Based Inverse Light and Reflectance from Images
of a Known Object** 215
Dana Cobzas, Cameron Upright, and Martin Jagersand

**3-D Lighting Environment Estimation with Shading and
Shadows** 239
Takeshi Takai, Susumu Iino, Atsuto Maki, and Takashi Matsuyama

3-D Cinematography with Approximate or No Geometry 259
Martin Eisemann, Timo Stich, and Marcus Magnor

View Dependent Texturing Using a Linear Basis 285
Martin Jagersand, Neil Birkbeck, and Dana Cobzas

Image and Geometry Processing for 3-D Cinematography: An Introduction

Rémi Ronfard and Gabriel Taubin

By 3-D cinematography we refer to techniques to generate 3-D models of dynamic scenes from multiple cameras at video frame rates. Recent developments in computer vision and computer graphics, especially in such areas as multiple-view geometry and image-based rendering have made 3-D cinematography possible. Important applications areas include production of stereoscopic movies, full 3-D animation from multiple videos, special effects for more traditional movies, and broadcasting of multiple-viewpoint television, among others. Drawing from two recent workshop on 3-D Cinematography, the 12 chapters in this book are original contributions by scientists who have contributed to the mathematical foundations of the field and practitioners who have developed working systems.

1 Overview of the Field

The name 3-D cinematography is motivated by the fact that it extends traditional cinematography from 2-D (images) to 3-D (solid objects that can be rendered with photorealistic textures from arbitrary viewpoints) at the same frame rate.

As the name implies, 3-D cinematography focuses on the inter-relations between cinematography (the art of placing cameras and lights to produce good motion pictures) with 3-D modeling. This book summarizes the main contributions from two recent workshops which brought together researchers and practitioners from diverse disciplines, including computer vision, computer graphics, electrical and optical engineering.

R. Ronfard (✉)
INRIA, Grenoble, France,
e-mail: remi.ronfard@inria.fr

G. Taubin
Brown University, Providence, RI, USA,
e-mail: taubin@brown.edu

R. Ronfard and G. Taubin (eds.), *Image and Geometry Processing for 3-D Cinematography*, Geometry and Computing 5, DOI 10.1007/978-3-642-12392-4_1,
© Springer-Verlag Berlin Heidelberg 2010

A first workshop on 3-D Cinematography took place in New York City in June 2006 jointly with the IEEE Conference on Computer Vision and Pattern Recognition.[1] Selected speakers from this workshop were invited to write extended papers, which after review were published as a special section in IEEE Computer Graphics and Applications [7]. At the time some prototypes had demonstrated the ability to reconstruct dynamic 3-D scenes in various forms and resolutions [5, 8]. Various names were used to refer to these systems, such as virtualized reality, free-viewpoint video, and 3-D video. All of these efforts were multi-disciplinary. These advances had clearly shown the promises of 3-D cinematography systems, such as allowing real-time, multiple-camera capture, processing, transmission, and rendering of 3-D models of real dynamic scenes. Yet, many research problems remained to be solved before such systems could be transposed from blue screen studios to the real world.

A second workshop on 3-D Cinematography took place at the BANFF Center in 2008.[2] This workshop focused on summarizing the progress made in the field during the two years subsequent to the first workshop, and in particular on real-time working systems and applications, with an emphasis on recent, realistic models for lights, cameras and actions. Indeed, 3-D cinematography can be regarded as the geometric investigation of lights (how to represent complex lighting, how to relight, etc.); cameras (how to recover the true camera parameters, how to simulate and control virtual cameras, etc.); and actions (how to represent complex movements in a scene, how to edit, etc.).

2   The Geometry of Lights, Cameras and Actions

The theory of 3-D cinematography can be traced back to the discovery of holography by Dennis Gabor in 1946. Gabor firmly believed that holography was destined to replace cinematography as we know it [3]. But to this day, dynamic holograms present overwhelming technological challenges in bandwidth, storage and computing power. Some of today's 3-D cinematography systems follow the more realistic goal of "sampling" the light rays observed from around the scene, as described in Tanimoto's chapter. Most other systems are based on photogrammetry and a partial reconstruction of the 3-D scene. A very good history of those later efforts in 3-D cinematography can be found in the review paper by Kanade and Narayanan [5].

From a geometric viewpoint, it is a hard problem to represent complex, time-varying scenes and their interactions with lights and cameras. One important question explored in this book is – what is the dimensionality of such scenes? Space decomposition methods are popular because they provide approximate answers, although not of very good quality. It has become increasingly evident that better representations are needed. Several partial solutions are proposed in the workshop

---

[1] http://perception.inrialpes.fr/3-Dcine/.

[2] http://www.birs.ca/birspages.php?task=displayevent\&event_id=08w5070.

papers, illustrated with examples. They include wavelet bases, implicit functions defined on a space grid, etc. It appears that a common pattern is the recovery of a controllable model of the scene, such that the resulting images can be edited (interaction).

Changing the viewpoint is only one (important) aspect, but changing the lighting and action is equally important [2]. Recording and representing three-dimensional scenes is an emerging technology made possible by the convergence of optics, geometry and computer science, with many applications in the movie industry, and more generally in entertainment. Note that the invention of cinema (camera and projector) was also primarily a scientific invention that evolved into an art form. We suspect the same thing will probably happen with 3-D movies.

3 Book Contents

The book is composed of 12 chapters, which elaborate on the content of talks given at the BANFF workshop. The chapters are organized into three sections. The first section presents an overview of the inter-relations between the art of cinematography and the science of image and geometry processing; the second section is devoted to recent developments in geometry; and the third section is devoted to recent developments in image processing.

3.1 3-D Cinematography and Applications

The first section of the book presents an overview of the inter-relations between the art of cinematography and the science of image and geometry processing.

The chapter *Stereoscopic Cinema* by Frédéric Devernay and Paul Beardsley is an introduction to stereoscopic 3-D cinematography, focusing on the main sources of visual fatigue which are specific to viewing binocular movies, and on techniques that can be used to produce comfortable 3-D movies. The causes of visual fatigue can be identified and classified into three main categories: geometric differences between both images which cause vertical disparity in some areas of the images, inconsistencies between the 3-D scene being viewed and the proscenium arch (the 3-D screen edges), and discrepancy between the accommodating and the convergence stimuli that are included in the images. For each of these categories, the authors propose solutions to issue warnings during the shooting, or to correct the movies in the post-production phase. These warnings and corrections are made possible by the use of state-of-the-art computer vision algorithms. The chapter explains where to place the two cameras in a real scene to obtain a correct stereoscopic movie. This is what many people understand as 3-D cinematography [4, 6]. In the context of this book, it is worth noting that stereoscopic 3-D recreates the perception of depth from a single viewpoint. In the future, it will be possible for a real scene to be photographed from a variety of viewpoints, and then its geometry to be fully or partially reconstructed, so that it can later be filmed by two virtual cameras, producing a stereoscopic movie with a variable viewpoint.

One important point made by Devernay and Beardsley is that the success of 3-D cinematography should be measured in terms of *psychological and perceptual* qualities, as can be done for traditional cinema [1]. Their chapter can therefore be a guide for future work in *perceptually-guided* 3-D reconstruction. We can only hope that this leads to a better understanding of depth perception in future immersive and interactive 3-D movies.

The chapter *Free-Viewpoint Television* by Masayuki Tanimoto describes a new type of television, named Free viewpoint Television or FTV, where movies are recorded from a variety of viewpoints. FTV is an innovative visual media that enables users to view 3-D scenes with freedom to interactively change the viewpoint. Geometrically, FTV is based on a ray-space representation, where each ray in real space is assigned a color value. By using this method, Tanimoto and his team constructed the world's first real-time FTV system, which comprises a complete data pipeline, from capturing, to processing and display. They also developed new types of ray capture and display technologies, such as a 360° mirror-scan ray capturing system and a 360° ray-reproducing display. In his chapter, Tanimoto argues that FTV is a natural interface between the viewer and a 3-D environment, and an innovative tool to create new types of content and art.

The chapter *Free-Viewpoint Video for TV Sport Production* by A. Hilton, J.-Y. Guillemaut, J. Kilner, O. Grau and G. Thomas, presents a case study for free-viewpoint television, namely, sports broadcasting. Contrasting Tanimoto's purely ray-based approach, here the authors follow a model-based approach, with geometric models for the field, the players and the ball. More specifically, this chapter reviews the challenges of transferring techniques developed for multiple view reconstruction and free-viewpoint video in a controlled studio environment, to broadcast production for football and rugby. This is illustrated by examples taken from the ongoing development of the *iview* free-viewpoint video system for sports production by the University of Surrey and the BBC. Production requirements and constraints for use of free-viewpoint video technology in live events are identified. Challenges presented by transferring studio technologies to large scale sports stadium are reviewed together with solutions being developed to tackle these problems. This work highlights the need for robust multiple view reconstruction and rendering algorithms which achieve free-viewpoint video, with the quality of broadcast cameras. The advances required for broadcast production also coincide with those of other areas of 3-D cinematography for film and interactive media production.

The chapter *Challenges for Multi-view Video Capture* by Bennett Wilburn further discusses the challenges associated with implementing large scale multi-view video capture systems, with the capture of football matches as a motivating example. This chapter briefly reviews existing multiview video capture architectures, their advantages and disadvantages, and issues in scaling them to large environments. Then it explains that today's viewers are accustomed to a level of realism and resolution which is not feasibly achieved by simply scaling up the performance of existing systems. The chapter surveys some methods for extending the effective resolution and frame rate of multiview capture systems. It explores the implications of real-time applications for smart camera design and camera array architectures, keeping in mind that real-time performance is a key goal for covering live sporting events. Finally, it comments briefly on some of the remaining challenges for photo-realistic view interpolation of multi-view video for live, unconstrained sporting events.

3.2 Recent Developments in Geometry

The second section of the book presents original contributions to the geometric modeling of large-scale, realistic live-action performances, with an emphasis on the modeling of cameras and actors.

The chapter *Performance Capture from Multi-view Video* by Christian Theobalt, Edilson de Aguiar, Carsten Stoll, Hans-Peter Seidel and Sebastian Thrun, presents an original method to capture performance from a handful of synchronized video streams. The method, which is based on a mesh representation of the human body, captures performance from mesh deformation, and without a kinematic skeleton. In contrast to traditional marker-based capturing methods, this approach does not require optical markings, and it is even able to reconstruct detailed geometry and motion of a dancer wearing a wide skirt. Another important feature of the method is that it reconstructs spatio-temporally coherent geometry, with surface correspondences over time. This is an important prerequisite for post-processing of the captured animations. All of this, the authors argue, can be obtained by capturing a flexible and precise geometrical model of the performers (actors). Their "performance capture" approach has a variety of potential applications in visual media production and the entertainment industry. It enables the creation of high quality 3-D video, a new type of media where the viewer has control over the camera's viewpoint. The captured detailed animations can also be used for visual effects in movies and games.

Most, if not all, 3-D cinematography techniques rely on precise multi-camera calibration methods. The multi-camera calibration problem has been resolved in a laboratory setting, but remains a challenging task in uncontrolled environments such as a theater stage, a sports field, or the set of a live-action movie. The chapter *Combining Multi-view Stereo and Bundle Adjustment for Accurate Camera Calibration* by Yasutaka Furukawa and Jean Ponce, presents a novel approach to camera calibration where top-down information from rough camera parameter estimates and multi-view stereo are used to effectively guide the search for additional image correspondences, and to significantly improve camera calibration parameters using a standard bundle adjustment algorithm.

The chapter *Cell-Based 3-D Video Capture Method with Active Cameras* by Tatsuhisa Yamaguchi, Hiromasa Yoshimoto, and Takashi Matsuyama, deals with the important problem of planning a 3-D cinematographic experiment, in such a way that the best use can be made of a limited number of cameras in a vast area such as a theater stage or a sports field. 3-D video is usually generated from multi-view videos taken by a group of cameras surrounding an object in action. To generate nice-looking 3-D video, several simultaneous constraints should be satisfied: (1) the cameras should be well calibrated, (2) for each video frame, the 3-D object surface should be well covered by a set of 2D multi-view video frames, and (3) the resolution of the video frames should be high enough to record the object surface texture. From a mathematical point of view, it is almost impossible to find such camera arrangement over a large performance area such as a stadium or a concert stage, where the performers move across the stage. Active motion-controlled cameras can be used in those cases. In this chapter, Matsuyama and co-workers describe a *cellular method* for planning the robotic movements of a set of active cameras, such that the above constraints can be met at all times. Their method is suitable for scripted performances such as music, dance or theater.

The chapter *Dense 3-D Motion Capture from Synchronized Video Streams* by Yasutaka Furukawa and Jean Ponce, describes a novel approach for recovering the deformable motion of a free-form object from synchronized video streams acquired by calibrated cameras. Contrary to most previous work, the instantaneous geometry of the observed scene is represented by a polyhedral mesh with a fixed topology. This represent a very significant step towards applications of 3-D cinematography in computer animation, virtual worlds and video games. The initial mesh is constructed in the first frame using multi-view stereo. Deformable motion is then captured by tracking vertices of the mesh over time, using two optimization processes per frame: a local one using a rigid motion model in the neighborhood of each vertex, and a global one using a regularized nonrigid model for the whole mesh. Qualitative and quantitative experiments using realistic data sets show that this algorithm effectively handles complex nonrigid motions and severe occlusions.

3.3   Recent Developments in Image Processing

The third section of the book presents original contributions in image processing of large-scale, realistic live-action performances, with an emphasis on the modeling, capture and rendering of lighting and texture.

Indirectly estimating light sources from scene images and modeling the light distribution is an important, but difficult problem in computer vision. A practical solution is of value both as input to other computer vision algorithms and in graphics rendering. For instance, photometric stereo and shape from shading requires known light sources. With estimated light such techniques could be applied in everyday environments, outside of controlled laboratory conditions. Light estimated from images is also helpful in augmented reality, to consistently relight artificially introduced objects. Simpler light models use individual point light sources but only work for simple illumination environments. The chapter *Wavelet-Based Inverse Light and Reflectance from Images of a Known Object* by Dana Cobzas, Cameron Upright and Martin Jagersand describes a novel light model using Daubechies wavelets and a method for recovering light from cast shadows and specular highlights in images. Their model is suitable for complex environments and illuminations. Experiments are presented with both uniform and textured objects and under complex geometry and light conditions. The chapter evaluates the estimation process stability, and the quality of scene relighting. The approach is based on a smooth wavelet representation compared to a non-smooth Haar basis, and on two other popular light representations (a discrete set of infinite light sources and a global spherical harmonics basis).

The chapter *3-D Lighting Environment Estimation with Shading and Shadows* by Takeshi Takai, Susumu Iino, Atsuto Maki, and Takashi Matsuyama, propose the Skeleton Cube to estimate time-varying lighting environments: e.g., lighting by candles and fireworks. A skeleton cube is a hollow cubic object placed in the scene to estimate its surrounding light sources. For the estimation, video of the cube is taken by a calibrated camera and the observed self-shadows and shading patterns are analyzed to compute 3-D distribution of time-varying point light sources. An iterative search algorithm is presented for computing the 3-D light source distribution and several simulation and real world experiments illustrate the effectiveness of the method.

As illustrated in the second section of this book, estimating the full 3-D geometry of a dynamic scene is an essential 3-D cinematography operation for sparse recording setups. When the geometric model and/or the camera calibration are imprecise, however, traditional methods based on multi-view texturing lead to blurring and ghosting artifacts during rendering. The chapter *3-D Cinematography with Approximate or No Geometry* by Martin Eisemann, Timo Stich and Marcus Magnor, presents original image-based strategies to alleviate, and even eliminate, rendering artifacts in the presence of geometry and/or calibration inaccuracies. By keeping the methods general, they can be used in conjunction with many different image-based rendering methods and projective texturing applications.

3-D cinematography of complex live-action scenes with transparencies and multiple levels-of-details requires advances in multi-view image representations. The chapter *View-Dependent Texturing Using a Linear Basis* by Martin Jagersand, Neil Birkbeck and Dana Cobzas, describes a three-scale hierarchical representation of scenes and objects, and explain how it is suitable for both computer vision capture of models from images and efficient photo-realistic graphics rendering. Their model consists of three different scales. The macro-scale is represented by conventional triangulated geometry. The meso-scale is represented as a displacement map. The micro-scale is represented by an appearance basis spanning viewpoint variation in texture space. To demonstrate their model, Jagersand et al. implemented a capture and rendering system based entirely on budget cameras and PC's. For efficient rendering the meso and micro level routines are both coded in graphics hardware using pixel shader code. This maps well to regular consumer PC graphics cards, where capacity for pixel processing is much higher than geometry processing. Thus photo-realistic rendering of complex scenes is possible on mid-grade graphics cards. Their chapter is illustrated with experimental results of capturing and rendering models from regular images of humans and objects.

4 Final Remarks

This book presents an overview of the current research in 3-D cinematography, with an emphasis on the geometry of lights, cameras and actions. Together, the 12 chapters make a convincing point that a clever combination of geometric modeling and image processing is making it possible to capture and render moderately complex dynamic scenes in full 3-D independently of viewpoint.

The next frontier is the synthesis of virtual camera movements along arbitrary paths extrapolating cameras arranged on a plane, a sphere, or even an entire volume. This problem raises difficult issues. What are the dimensions of the allowable space of cinematographic cameras that professional cinematographers would want to synthesize? In other words, what are the independent parameters of the virtualized cameras that can be interpolated from the set of existing views? Further, what is the range of those parameters that can be achieved using a given physical camera setup? Among the theoretically feasible parameter values, which are the ones that will produce sufficient resolution, photorealism, and subjective image quality? These questions remain open for future research in this new world of 3-D cinematography.

Acknowledgements Rémi Ronfard and Gabriel Taubin were supported by the VAMP Associate Team program (Video and Mesh Processing for 3-D Cinematography) at INRIA. Gabriel Taubin has also been supported by the National Science Foundation under Grants No. CCF-0915661, CCF-0729126, CNS-0721703, and IIS-0808718.

References

1. Cutting, J.: Perceiving scenes in film and in the world. In: J.D. Anderson, B.F. Anderson (eds.) Moving Image Theory: Ecological Considerations, pp. 9–17. University of Southern Illinois Press, Carbondale, IL (2007)
2. Debevec, P.: Virtual cinematography: relighting through computation. Computer **39**(8), 57–65 (2006). doi:http://dx.doi.org/10.1109/MC.2006.285
3. Gabor, D.: Three-dimensional cinema. New Sci. **8**(191), 141–145 (1960)
4. Hummel, R.: 3-D cinematography. American Cinematographer, April (2008)
5. Kanade, T., Narayanan, P.J.: Virtualized reality: perspectives on 4D digitization of dynamic events. IEEE Comput. Graph. Appl. **27**(3), 32–40 (2007). doi:http://dx.doi.org/10.1109/MCG.2007.72
6. Kozachik, P.: 2 worlds in 3 dimensions. American Cinematographer, February (2009)
7. Ronfard, R., Taubin, G.: Introducing 3D cinematography. IEEE Comput. Graph. Appl. **27**(3), 18–20 (2007). doi:http://dx.doi.org/10.1109/MCG.2007.64
8. Starck, J., Hilton, A.: Surface capture for performance-based animation. IEEE Comput. Graph. Appl. **27**, 21–31 (2007). doi:http://doi.ieeecomputersociety.org/10.1109/MCG.2007.68

Part I
3-D Cinematography and Applications

Stereoscopic Cinema

Frédéric Devernay and Paul Beardsley

Abstract Stereoscopic cinema has seen a surge of activity in recent years, and for the first time all of the major Hollywood studios released 3-D movies in 2009. This is happening alongside the adoption of 3-D technology for sports broadcasting, and the arrival of 3-D TVs for the home. Two previous attempts to introduce 3-D cinema in the 1950s and the 1980s failed because the contemporary technology was immature and resulted in viewer discomfort. But current technologies – such as accurately-adjustable 3-D camera rigs with onboard computers to automatically inform a camera operator of inappropriate stereoscopic shots, digital processing for post-shooting rectification of the 3-D imagery, digital projectors for accurate positioning of the two stereo projections on the cinema screen, and polarized silver screens to reduce cross-talk between the viewers left- and right-eyes – mean that the viewer experience is at a much higher level of quality than in the past. Even so, creation of stereoscopic cinema is an open, active research area, and there are many challenges from acquisition to post-production to automatic adaptation for different-sized display. This chapter describes the current state-of-the-art in stereoscopic cinema, and directions of future work.

1 Introduction

1.1 Stereoscopic Cinema, 3-D Cinema, and Others

Stereoscopic cinema is the art of making stereoscopic films or motion pictures. In stereoscopic films, depth perception is enhanced by having different images for the left and right human eye, so that objects present in the film are perceived by the

---

F. Devernay (✉)
INRIA Grenoble – Rhóne-Alpes, Montbonnot Saint Martin, France
e-mail: frederic.devernay@inria.fr

P. Beardsley
Disney Research, Zürich, Switzerland
e-mail: pab@disneyresearch.com

R. Ronfard and G. Taubin (eds.), *Image and Geometry Processing for 3-D Cinematography*, Geometry and Computing 5, DOI 10.1007/978-3-642-12392-4_2, © Springer-Verlag Berlin Heidelberg 2010

spectator at different depths. The visual perception process that reconstructs the 3-D depth and shape of objects from the left and right images is called stereopsis.

Stereoscopic cinema is also frequently called 3-D cinema,[1] but in this work we prefer using the term stereoscopic, as 3-D cinema may also refer to other novel methods for producing moving pictures:

- Free-viewpoint video, where videos are captured from a large number of cameras and combined into 4-D data that can be used to render a new video as seen from an arbitrary viewpoint in space
- 3-D geometry reconstructed from multiple viewpoints and rendered from an arbitrary viewpoint

Smolic et al. [54] did a large review on the subjects of 3-D and free-viewpoint video, and other chapters in the present book also deal with building and using these moving pictures representations.

In this chapter, we will limit ourselves to movies that are made using exactly two cameras placed in a stereoscopic configuration. These movies are usually seen through a 3-D display device which can present two different images to the two human eyes (see Sexton and Surman [51] for a review on 3-D displays). These display devices sometimes take as input more than two images, especially glasses-free 3-D displays where a number of viewpoints are displayed simultaneously but only two are seen by a human spectator. These viewpoints may be generated from stereoscopic film, using techniques that will be described later in this chapter, or from other 3-D representations of a scene.

Many scientific disciplines are related to stereoscopic cinema, and they will sometimes be referred to during this chapter. These include computer science (especially computer vision and computer graphics), signal processing, optics, optometry, ophthalmology, psychophysics, neurosciences, human factors, and display technologies.

1.2 A Brief History of Stereoscopic Cinema

The interest in stereoscopic cinema appeared at the same time as cinema itself, and the "prehistory" (from 1838 to 1922 when the first public projection was made [22,83]) as well as "history" (from 1922 to 1952, when the first feature-length commercial stereoscopic movie, *Bwana Devil*, was released to the public [83]) follow closely the development of cinematography. The history of the period from 1952 to 2004 can be retraced from the thoughtful conversations found in Zone [82].

In the early 1950s, the television was causing a large decrease in the movie theater attendance, and stereoscopic cinema was seen as a method to get back this audience. This explains the first wave of commercial stereoscopic movies in the 1950s, when Cinemascope also appeared as another television-killer. Unfortunately

---

[1] Stereoscopic cinema is even sometimes redundantly called Stereoscopic 3-D (or S3D) cinema.

Stereoscopic Cinema

the quality of stereoscopic movies, both in terms of visual quality and in terms of cinematographic content, was not on par with 2-D movies in general, especially Cinemascope. Viewing a stereoscopic film at that time meant the promise of a headache, because stereoscopic filming techniques were not fully mastered, and the spectacular effects were overly used in these movies, reducing the importance of screenplay to close to nothing. One notable counter-example to the typical weak screenplay of stereoscopic movies was *Dial M for Murder* by Alfred Hitchcock, which was shot in 3-D, but incidentally had a much bigger success as a 2-D movie. Finally, Cinemascope won the war against TV, and the production of stereoscopic cinema declined.

The second stereoscopic cinema "wave", in the 1980s, was formed both by large format (IMAX 3-D) stereoscopic movies, and by standard stereoscopic movies which tried again the recipe from the 1950s (spectacular content but weak screenplay), with no more success than the first wave. There were a few high-quality movies, such as *Wings of Courage* by Jean-Jacques Annaud in 1995, which was the first IMAX 3-D fiction movie, made with the highest standards, but at a considerable cost. The IMAX-3D film camera rig is extremely heavy and difficult to operate [82]. Stereoscopic cinema wouldn't take off until digital processing finally brought the tools that were necessary to make stereoscopic cinema both easier to shoot and to watch.

As a matter of fact, the rebirth of stereoscopic cinema came from animation, which produced movies like *Chicken Little*, *Open Season* and *Meet the Robinsons*, which were digitally produced, and thus could be finely tuned to lower the strain on the spectator's visual system, and experiment new rules for making stereoscopic movies. Live-action movies produced using digital stereoscopic camera systems came afterwards, with movies such as *U2 3D*, and of course *Avatar*, which held promises of a stereoscopic cinema revival.

Stereoscopic cinema brought many new problems that were not addressed by traditional movie making. Many of these problems deal with geometric considerations: how to place the two cameras with respect to each other, where to place the actors, what camera parameters (focal length, depth of field, ...) should be used .... As a matter of fact, many of these problems were somehow solved by experience, but opinions often diverged on the right solution to film in stereoscopy. The first theoretical essay on stereoscopic cinema was written by Spottiswoode et al. [56]. The Spottiswoodes made a great effort to formalize the influence of the camera geometry on the 3-D perception by the audience, and also proposed a solution on the difficult problem of transitions between shots. Their scientific approach of stereoscopic cinema imposed very strict rules on moviemaking, and most of the stereoscopic moviemakers didn't think it was the right way to go [82].

Many years later, Lenny Lipton, who founded StereoGraphics and invented the CrystalEyes electronic shutter glasses, tried again to describe the scientific foundations of stereoscopic cinema [36]. His approach was more viewer-centric, and he focused more on how the human visual system perceives 3-D, and how it reacts to stereoscopic films projected on a movie screen. Although the resulting book contains many mathematical errors, and even forgets most of the previous finding by the Spottiswoodes [53], it remains one of the very few efforts in this domain.

The last notable effort at trying to formalize stereoscopic movie making was this of Mendiburu [41], who omitted maths, but explained with clear and simple drawings the complicated geometric effects that are involved in stereoscopic cinema. This book was instantaneously adopted as a reference by the moviemaking community, and is probably the best technical introduction to the domain.

1.3 Computer Vision, Computer Graphics, and Stereoscopic Cinema

The discussions in this chapter straddle Computer Vision, Computer Graphics, and Stereoscopic Cinema: Computer Vision techniques will be used to compute and locate the defects in the images taken by the stereoscopic camera, and Computer Graphics techniques will be used to correct these defects.

1.3.1 A Few Definitions

Each discipline has its own language and words. Before proceeding, let us define a few geometric elements that are useful in stereoscopic cinema [28]:

Interocular (also called Interaxial): (The term baseline is also widely used in computer vision, but not in stereoscopic cinema.) The distance between the two eyes/cameras, or rather their optical centers. It is also used sometimes used to designate the segment joining the two optical centers. The average human interocular is 65 mm, with large variations around this value.

Hyperstereo (or miniaturization): The process of filming with an interocular larger than 65 mm (it can be up to a few dozen meters), with the consequence that the scene appears smaller when the stereoscopic movie is viewed by a human subject.

Hypostereo (or gigantism): The process of filming with an interocular smaller than 65 mm, resulting in a "bigger than life" appearance when viewing the stereoscopic movie. It can be as small as 0 mm.

Roundness factor: Suppose a sphere is filmed by a stereoscopic camera. When displaying it, the roundness factor is the ratio between its apparent depth and its apparent width. If it is lower than 1, the sphere appears as a flattened disc parallel to the image plane. If it is bigger than 1, the sphere appears as a spheroid elongated in the viewer's direction. We will see that the roundness factor depends on the object's position in space.

Disparity: The difference in position between the projections of a 3-D point in the left and right images, or the left and right retinas. In a standard stereo setup, the disparity is mostly horizontal (the corresponding points are aligned vertically), but vertical disparity may happen (and has to be corrected).

Screen plane: The position in space where the display projection surface is located (supposing the projection surface is planar).

Vergence, convergence, divergence: The angle formed by the optical axis of the two eyes in binocular vision. The optical axis is the half 3-D line corresponding

Stereoscopic Cinema

to the line-of-sight of the center of the fovea. It can be positive (convergence) or negative (divergence).

Plane of convergence: The vertical plane parallel to the screen plane containing the point that the two eyes are looking at. If it is in front of the screen plane, then the object being looked at appears in front of the screen. When using cameras, the plane of convergence is the zero-disparity plane (it is really a plane if images are rectified, as will be seen later).

Proscenium arch (also called stereoscopic window and floating window): (Fig. 5) The perceived depth of the screen borders. If the left and right borders of the left and right images do not coincide on the screen, the proscenium arch is not in the screen plane. It is also called the stereoscopic window, since the 3-D scene looks as if it were seen through that 3-D window. As will be explained later, objects closer than the proscenium arch should not touch the left or right side of the arch.

1.3.2 Stereo-Specific Processes

The stereoscopic movie production pipeline shares a lot with standard 2-D movie production [41]. Computer vision and computer graphics tools used in the process can sometimes be used with no or little modification. For example, matchmoving (which is more often called *Structure from Motion* or *SfM* in Computer Vision) can take into account the fact that two cameras are taking the same scene, and may use the additional constraint that the two camera positions are fixed with respect to each other.

Many processes, though, are specific to stereoscopic cinema production and post-production, and cannot be found in 2-D movie production:

- *Correcting geometric causes of visual fatigue* such as images misalignments, will be covered by Sects. 2.5 and 5.1.
- *Color-balancing left and right images* is especially necessary when a half-silvered mirror is used to separate images for the left and right cameras and wide angle lenses are used: the transmission and reflection spectrum response of these mirrors depend on the incidence angle and have to be calibrated. This can be done using color calibration devices and will not be covered in this chapter.
- *Adapting the movie to the screen size (and distance)* is not as simple as scaling the left and right views: the stereoscopic display is not easily scalable like a 2-D display, because the human interocular is fixed and therefore does not "scale" with the screen size or distance. A consequence is that the same stereoscopic movie displayed on screens of different sizes will probably give quite different 3-D effects (at least quantitatively). The adaptation can be done at the shooting stage (Sect. 3) or in post-production (Sect. 5). These processes can either be used to give the stereoscopic scene the most natural look possible (which usually means a roundness factor close to 1), or to "play" with 3-D effects, for example by changing the interocular or the position where the infinity plane appears.
- *Local 3-D changes (or 3-D touchup)* consist in editing the 3-D content of the stereoscopic scene. This usually means providing new interactive editing tools

that work both on the images and on the disparity map. These tools usually share a lot with colorizing tools, since they involve cutting-up objects, tracking them, and changing their depth (instead of color). This is beyond the scope of this chapter.

- *Playing with the depth of field* is sometimes necessary, especially when adapting the 3-D scene to a given screen distance: the depth of field should be consistent with the distance to the screen, in order to minimize vergence-accommodation conflicts (Sect. 5.4).
- *Changing the proscenium arch* is sometimes necessary, because objects in front of the screen may cross the screen borders and become inconsistent with the stereoscopic window (Sect. 5.5).
- *3-D compositing (with real or CG scenes)* should be easier in stereoscopic cinema, since it has 3-D content already. However, there are some additional difficulties: 2-D movies mainly have to deal with positioning the composited objects within the scene and dealing with occlusion masks. In 3-D, the composited scene must also be consistent between the two eyes, and its 3-D shape must be consistent with the original stereoscopic scene (Sect. 5.6). Relighting the scene also brings out similar problems.

2 Three-Dimensional Perception and Visual Fatigue

In traditional 2-D cinema, since the result is always a 2-D moving picture, almost anything can be filmed and displayed, without any effect on the spectator's health (except maybe for light flashes and stroboscopic effects), and the artist has a total freedom on what can be shown to the spectator. The result will appear as a moving picture drawn on a plane placed at some distance from the spectator, and will always be physically plausible.

In stereoscopic cinema, the two images need to be mutually consistent, so that a 3-D scene (real or virtual) can be reconstructed by the human brain. This implies strong geometric and photometric constraints between the images displayed to the left and the right eye, so that the 3-D shape of the scene is perceived correctly, and there is not too much strain on the human visual system which would result in visual fatigue.

Our goal in this chapter is to deal with the issues related to geometry and visual fatigue in stereoscopic cinema, but without any artistic considerations. We will just try to define bounds within which the artistic creativity can freely wander. These bounds were almost non-existent in 2-D movies, but as we will see they are crucial in stereoscopic movies.

Besides, since 3-D perception is naturally more important in stereoscopic movies, we have to understand what are the different visual features that produce depth perception. Those features will be called *depth cues*, and surprisingly most are monoscopic and can be experienced by viewing a 2-D image. Stereoscopy is only one depth cue amongst many others, although it may be the most complicated to deal with. The stereographer Phil Streather, quoting Lenny Lipton, said: "Good 3D is not just about setting a good background. You need to pay good attention to the seven monocular cues – aerial perspective, inter position, light and shade, relative size, texture gradients, perspective and motion parallax. Artists have used the first five of those cues for centuries. The final stage is depth balancing".

2.1 Monoscopic Depth Cues

The perception of 3-D shape is caused by the co-occurrence of a number of consistent visual artifacts called depth cues. These depth cues can be split into monoscopic cues, and stereoscopy (or stereopsis). For a review of 3-D shape perception from the cognitive science perspective, see Todd [64]. The basic seven monoscopic depth cues, as described in Lipton [36, Chap. 2] (see also [60]), and illustrated Fig. 1, are:

- *Light and shade*
- *Relative size*, or retinal image size (smaller objects are farther)
- *Interposition*, or overlapping (overlapped objects lie behind)
- *Textural gradient* (increase in density of a projected texture as a function of distance and slant)
- *Aerial perspective* (usually caused by haze)
- *Motion parallax* (2-D motion of closer objects is faster)
- *Perspective*, or linear perspective

As can bee seen in a famous drawing by Hogarth (Fig. 2), their importance can be easily demonstrated by using contradictory depth cues.

Lipton [36] also refers to what he calls a "physiological cue": *Accommodation* (the monoscopic focus response of the eye, or how much the ciliary muscles contract to maintain a clear image of an object as its distance changes). However, it is not

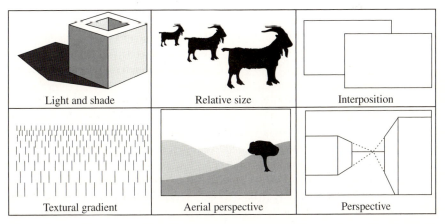

Fig. 1 Six monoscopic depth cues (from Stereographics [60]). The seventh is motion parallax, which is hard to illustrate, and depth of field can also be considered as a depth cue (see Fig. 3)

Fig. 2 "Whoever makes a DESIGN without the Knowledge of PERSPECTIVE will be liable to such Absurdities as are shown in this Frontispiece" (engraving by William Hogarth, 1754), a proof by contradiction of the importance of many monoscopic depth cues

clear from psychophysics experiments whether this should be considered as depth cue, i.e. if it gives an indication of depth in the absence of any other depth cue.

Although it is usually forgotten in the list of depth cues, we should also add *depth of field*, or retinal image blur [27, 70] (it is different from the accommodation cue cited before, which refers to the accommodation *distance* only, not to the depth of field), the importance of which is well illustrated by Fig. 3. The depth of field of the Human eye is around 0.3 Diopters (D) in normal situations, although finer studies [39] claim that it also slightly depends on parameters such as the pupil size, wavelength, and spectral composition. Diopters are inverse of meters: at a focus distance of 3 m, a depth of field of ± 0.3 D, means that the in-focus range is from $1/(\frac{1}{3}+0.3) \approx 1.6$ m to $1/(\frac{1}{3}-0.3) = 30$ m, whereas at a focus distance of 30 cm, the in-focus depth range is only from 27.5 to 33 cm (it is easy to understand from this formula why we prefer using diopters rather than a distance range to measure the depth of field: diopters are independent of the focus distance, and can easily be converted to a distance range). This explains why the photograph in Fig. 3 looks like

Fig. 3 Focus matters! This photo is of a real scene, but the depth of field was reduced by a tilt-shift effect to make it look like a model [26] (photo by oseillo)

a model rather than an actual-size scene [26]: the in-focus parts of the scene seem to be only about 30 cm away from the spectator. The depth of field range is not much affected by age, so this depth cue may be learned from observations over a long period, whereas the accommodation range goes from 12 D for children, to 8 D for young adults, to ... below 1 D for presbyopes.

2.2 Stereoscopy and Stereopsis

Stereoscopy, i.e. the fact that we are looking at a scene using our two eyes, brings two additional physiological cues [36]:

- *Vergence* (the angle between the line-of-sight of both eyes)
- *Disparity* (the positional difference between the two retinal images of a scene point, which is non-zero for objects behind or in front of the convergence point)

These cues are used by the perception process called stereopsis, which gives a sensation of depth from two different viewpoints, mainly from the horizontal disparity.

Although stereoscopy and motion parallax are very powerful 3-D depth cues, it should be noted that human observers asked to make judgments about the 3-D metric structure of a scene from these cues are usually subject to large systematic errors [62, 65].

2.3 Conflicting Cues

All these cues (the eight monoscopic cues and stereopsis) may be conflicting, i.e. giving opposite indications on the scene geometry. Many optical illusions make heavy use of these conflicting cues, i.e. when an object seems smaller because lines in the image suggest a vanishing point. Two famous examples are the Ames room and the pseudoscope.

The Ames room (invented by Adelbert Ames in 1934) is an example where monocular cues are conflicting. Ames room is contained in a large box, and the spectator can look at it though a single viewpoint, which is a hole in one of the walls. From this viewpoint, this room seems to be cubic-shaped because converging lines in the scene suggest the three standard lines directions (the vertical, and two orthogonal horizontal directions), but it is really trapezoidal (Fig. 4). Perspective cues are influenced by prior knowledge of what a room should look like, so that persons standing in each far corner of the room will appear to be either very small or very big. The room itself can be seen from a peep hole at the front, thus forbidding binocular vision. In this precise case, binocular vision would easily disambiguate the conflicting cues by concluding that the room is not cubic. The Ames room was used in movies such as *Eternal Sunshine of the Spotless Minds* by Michel Gondry or the *Lord of the Rings* trilogy.

Another example of conflicting cues, which is more related to stereoscopic cinema, is illustrated by the pseudoscope. The pseudoscope (invented by Charles Wheatstone) is an binocular device which switches the viewpoints from the left and right eyes, so that all stereoscopic cues are reversed, but the monoscopic cues still remain and usually dominate the stereoscopic cues. The viewer still has the impression of "seeing in 3-D", and the closer objects in the scene actually seem bigger than they are, because the binocular disparity indicates that these big objects (in the image) are far away. This situation happened quite often during the projection of stereoscopic movies in the past [82], where the filters in front of the projectors or the film reels were accidentally reversed, but the audience usually did not notice what was wrong, and still had the impression of having seen a 3-D movie, though they thought it probably was a bad one because of the resulting headache.

Conflicting perspective and stereoscopic cues were actually used heavily by Pete Kozachik, Director of Photography on Henry Selick's stereoscopic film Coraline, to give the audience different sensations [11, 33]:

> Henry wanted to create a sense of confinement to suggest Coraline's feelings of loneliness and boredom in her new home. His idea had interiors built with a strong forced perspective

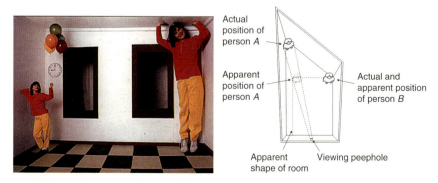

Fig. 4 Ames room: an example of conflicting perspective and relative size cues

Stereoscopic Cinema

and shot in 3-D to give conflicting cues on how deep the rooms really were. Later, we see establishing shots of the more appealing Other World rooms shot from the same position but built with normal perspective. The compositions match in 2-D, but the 3-D depth cues evoke a different feel for each room.

2.4 Inconsistent Cues

Inconsistent cues are usually less disturbing for the spectator than conflicting cues. They are defined as cues that indicate different amounts of depth in the same direction. They have been used for ages in bas-relief, where the lighting cue enhances the depth perceived by the binocular system, and as a matter of fact bas-relief is usually better appreciated from a far distance, where the stereoscopic cues have less importance.

An effect that is often observed when looking at stereoscopic photographs is called the *cardboard effect* [40, 76]: some depth is clearly perceived, but the amount of depth is too small with respect to the expected depth from the image size of the objects, resulting in objects appearing as flat, or drawn on cardboard of billboards. We will explain later how to predict this effect, and most importantly how to avoid it.

Another well-known stereoscopic effect is called the *puppet-theater effect* (also called *pinching*): background objects do not appear as small as expected, so that foreground objects appear proportionately smaller.

These inconsistent cues can easily be avoided if there is total control on the shooting geometry, including camera placement. If there are some unavoidable constraints on the shooting geometry, we will explain in Sect. 5 how *some* of these inconsistent cues related to stereoscopy can be corrected in post-production (Sect. 5).

2.5 Sources of Visual Fatigue

Visual fatigue is probably the most important point to be considered in stereoscopic cinema. Stereoscopic movies in the past often resulted in a bad viewing experience, and this reduced a lot the acceptance of stereoscopic cinema by a large public. Ukai and Howarth [66] produced a reference study on visual fatigue caused by viewing stereoscopic films, and is a good introduction to this field.

The symptoms of visual fatigue may be conscious (headache, tiredness, soreness of the eyes) or unconscious (perturbation of the oculomotor system). It should actually be considered as a public health concern [68], just as the critical fusion frequency on CRT screens 50 years ago, as it may actually lead to difficulties in judging distances (which is very important in such tasks as driving). Ukai and Howarth [66, Sect. 6] even report the case of an infant whose oculomotor system was permanently disturbed by viewing a stereoscopic movie. Although the long-term effects of viewing stereoscopic cinema were not studied due to the fact that this medium is not yet widespread, many studies exist on the effects on health of using virtual reality displays [34, 68, 69]. Virtual reality displays are widely used in the industry (from desktop displays to immersive displays), and are sometimes used daily by people working in industrial design, data visualization, or simulation.

The sources of visual fatigue that are specific to stereoscopic motion pictures are mainly due to binocular asymmetry, i.e. photometric or geometric differences between the left and right retinal images. Kooi and Toet [32] experimentally measured thresholds on the various asymmetries that will lead to visual incomfort (incomfort is the lowest grade of conscious visual fatigue). For example, they measured, in agreement with Pastoor's rule of thumb [46] that a 35 arcmin horizontal disparity range is quite acceptable for binocular perception of 3-D and 70 arcmin disparity is too much to be viewed. They also found out that the human visual system is most sensitive to vertical binocular disparities. The various quantitative binocular thresholds computed from their experiments can be found in Kooi and Toet [32, Table 4].

The main sources of visual fatigue can be listed as:

- *Crosstalk* (sometimes called *crossover* or *ghosting*), which is usually due to a stereoscopic viewing system with a single screen: a small fraction of the intensity from the left image can be seen in the right eye, and vice-versa. The typical values for crosstalk [32] are 0.1–0.3% with polarization-based systems, and 4–10% with LCD shutter glasses. Preprocessing can be applied to the images before displaying in order to reduce crosstalk by subtracting a fraction of the left image from the right image [41] – a process sometimes called *ghost-busting*.
- *Breaking the proscenium rule* (or breaking the stereoscopic window) happens when there are interposition errors between the stereoscopic imagery and the edges of the display (see Fig. 5 and Mendiburu [41, Chap. 5]). A simple way to correct this is to move the proscenium closer to the spectator by adding borders to the image, but this is not always as easy as it seems (see Sect. 5.5).
- *Horizontal disparity limits* are the minimum and maximum disparity values that can be accepted without producing visual fatigue. An obvious bound for disparity

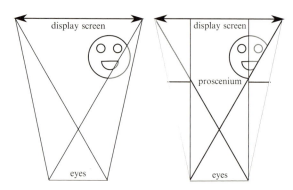

Fig. 5 Breaking the proscenium rule: (*left*) part of the object in front of the proscenium arch is not visible in one eye, which breaks the proscenium rule; (*right*) masking part of the image in each eye moves the proscenium closer than the object, and the proscenium rule is re-established

is that the eyes should not diverge. Another limit concerns the range of acceptable disparities within a stereoscopic scene that can be fused simultaneously by the human visual system.

- *Vertical disparity* causes torsion motion of the ocular globes, and is only tolerable for short time intervals. Our oculomotor system learned the epipolar geometry of our eyes over a lifetime of real-world experience, and any deviation from the learned motion causes strain.
- *Vergence-accommodation conflicts* occur when the focus distance of the eyes is not consistent with their vergence angle. It happens quite often when viewing stereoscopic cinema, since the display usually consists of a planar surface placed at a fixed distance. Strictly speaking, any 3-D point that is not in the convergence plane will have an accommodation distance, which is exactly the screen distance, different from the vergence distance. However, this constraint can be somewhat relaxed by using the depth of field of the visual system, as will be seen later.

Geometric asymmetries come very often either from a misalignment or from a difference between the optics in the camera system or in the projection system, as seen in Fig. 6. In the following, we will only discuss *horizontal disparity limits*, *vertical disparity*, and *vergence-accommodation conflicts*, since the other sources of visual fatigue are easier to deal with.

2.5.1 Horizontal Disparity Limits

The most simple and obvious disparity limit is eye divergence. In their early work on stereoscopic cinema, the Spottiswoodes said: "It is found that divergence is likely to cause eyestrain, and therefore screen parallaxes in excess of the eye separation should be avoided". But they also went on to say, in listing future development requirements, that "Much experimental work must be carried out to determine limiting values of divergence at different viewing distances which are acceptable without eyestrain".

These limiting values are the maximum disparities acceptable around the convergence point, usually expressed as angular values, such that the binocular fusion

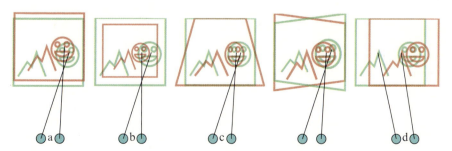

Fig. 6 A few examples of geometric asymmetries: **a** vertical shift, **b** size or magnification difference, **c** distortion difference, **d** keystone distortion due to toed-in cameras, **e** horizontal shift – leading to eye divergence in this case (adapted from Ukai and Howarth [66])

of the 3-D scene is performed without any form of eyestrain or visual fatigue. Many publications dealt with the subject of finding the horizontal disparity limits [29, 46, 78].

The horizontal disparity limits are actually closely related to the depth of field, as noted by Lambooij et al. [34]:

> An accepted limit for DOF in optical power for a 3 mm pupil diameter (common under normal daylight conditions) and the eyes focusing at infinity, is one-third of a diopter. With respect to the revisited Panum's fusion area,[2] disparities beyond one degree (a conservative application of the 60–70 arcmin recommendation), are assumed to cause visual discomfort, which actually results from the human eye's aperture and depth of field. Though this nowadays serves as a rule-of-thumb, it is acknowledged as a limit, because lower recommendations have been reported as well. If both the limits of disparity and DOF are calculated in distances, they show very high resemblance.

Yano et al. [77] also showed that images containing disparities beyond the depth of field (± 0.2 D depth of field, which means $\pm 0.82°$ in disparity) cause visual fatigue.

2.5.2 Vertical Disparity

Let us suppose that the line joining both eyes is horizontal, and that the stereoscopic display screen is vertical and parallel to this line. The images of any 3-D point projected onto the display screen using each eye optical center as the centers of projection are two points which are aligned horizontally, i.e. have no vertical disparity. Thus all the scene points that are displayed on the screen should have no vertical disparity. Vertical disparity (see Fig. 6 for some examples) may come from a misalignment of the cameras or of the display devices, from a focal length difference between the optics of the cameras, from keystone distortion, due to a toed-in camera configuration, or from nonlinear (e.g. radial) distortions.

However, it is to be noted that vertical disparities exist in the visual system: remember that the eye is not a linear perspective camera, but a spheric sensor, so that an object which is not in the median plane between both eyes will be closer to one eye than to the other, and thus its image will be bigger in one eye than in the other (a spheric sensor basically measures angles). The size ratio between the two images is called the vertical size ratio, or VSR. VSR is naturally present when rectified images (i.e. with no vertical disparity) are projected on a flat display screen: a vertical rod, though it's displayed with the same size on the left and right images on the flat display, subtends a larger angle in the nearest eye.

Psychophysical experiments showed that vertical disparity gradients have a strong influence on the perception of stereoscopic shape, depth and size [5, 43, 48]. For example, the so-called induced-size effect [43] is caused by a vertical gradient of vertical disparity (vertical-size parallax transformation) between the half images of an isolated surface, which creates an impression of a surface slanted in depth. Ogle

---

[2] In the human visual system, the space around the current fixation point which can be fused is called Panum's area or fusion area. It is usually measured in minutes of arc (arcmin).

[43, 44] called it the induced-size effect because it is as though the vertical magnification of the image in one eye induces an equivalent horizontal magnification of the image in the other eye.[3]

Allison [3] also notes that vertical disparities can be used to fool the visual system: "Images on the retinae of a fronto-parallel plane placed at some distance actually have some keystone distortion, which may be used as a depth cue. Displaying keystone-distorted images on that fronto-parallel screen actually exaggerates that keystone distortion when the viewer is focusing on the center of the screen, and would thus giving a cue that the surface is nearer than the physical screen".

By displaying keystoned images, the VSR is distorted in a complicated way which may be inconsistent with the horizontal disparities. Besides, that distortion depends on the viewer position with respect to the screen. When the images displayed on the screen are rectified, although depth perception may be distorted depending on the viewer position, horizontal and vertical disparities will always be consistent, as long as the viewer's interocular (the line joining the two optical centers) is kept parallel to the screen and horizontal (this viewing position may be hard to obtain in the side rows of wide movie theaters).

Woods et al. [72] discuss sources of distortion in stereo camera arrangements as well as the human factors considerations required when creating stereo images. These experiments show that there is a limit in the screen disparity which it is comfortable to show on stereoscopic displays. A limit of 10 mm screen disparity on a 16″ display at a viewing distance of 800 mm was found to be the maximum that all 10 subjects of the experiment could view. Their main recommendation is to use a parallel camera configuration in preference to converged cameras, in order to avoid keystone distortion and depth plane curvature.

Is Vertical Disparity Really a Source of Visual Fatigue? From the fact that vertical disparities are actually a depth cue, there has been a debate on whether vertical disparities can be a source of visual fatigue, since they are naturally present in retinal images. Stelmach et al. [59] and [55] claim that keystone and depth plane curvature cause minimal discomfort: images plane shift (equivalent to rectified images) and toed-in cameras are equally comfortable in their opinion. However, we must distinguish between visual discomfort, which is conscious, and visual fatigue, where the viewer may not be conscious of the problem during the experiment, but headache, eyestrain, or long-term effects can happen.

Even in the case where the vertical disparities are not due to a uniform transform of the images, such as a rotation, scaling or homography, Stevenson and Schor [61] demonstrated that human stereo matching does not actually follow the epipolar lines, and human subjects can still make accurate near/far depth discrimination when the vertical disparity range is as high as 45 arcmin.

Allison [4] concludes that, although keystone-distorted images coming from a toed-in camera configuration can be displayed with their vertical disparities without

---

[3] However, Read and Cumming [47] show that non-zero vertical disparity sensors are in fact not necessary to explain the induced size effect.

discomfort, the images should preferably be rectified, because the additional depth cues caused by keystone distortion perturb the actual depth perception process.

2.5.3 Vergence-Accommodation Conflicts

When looking at a real 3-D scene, the distance of accommodation (i.e. the focus distance) is equal to the distance of convergence, which is the distance to the perceived object. This relation between these two oculomotor functions, called Donder's line [77], is learned through the first years of life, and is used by the visual system to quickly focus-and-converge on objects surrounding us. The relation between vergence and accommodation does not have to follow exactly Donder's line: there is an area around it where vergence and accommodation agree, which is called Percival's zone of comfort [27, 77].

When viewing a stereoscopic movie, the distance of accommodation differs from the distance of convergence, which is the distance to the perceived object (Fig. 7). This discrepancy causes a perturbation of the oculomotor system [17, 77], which causes visual fatigue, and may even damage the visual acuity, which is reported by Ukai and Howarth [66] to have plasticity until the age of 8 or later. This problem has been largely studied for virtual reality (VR) displays, and lately for 3-D television (3DTV), but has been largely overlooked in movie theater situations. In fact, many stereoscopic movies, especially IMAX-3D movies, make heavy use of spectacular effects by presenting perceived objects which are very close to the spectator, when the screen distance is about 20 m.

Wann and Mon-Williams [68] cite it as one of the main sources of stress in VR displays. They observed that, in a situation where other sources of visual discomfort were eliminated, prolonged use of a stereoscopic display caused short-term modifications in the normal accommodation-vergence relationship.

Hoffman et al. [27] designed a special 3-D display where vergence and focus cues are consistent [2]. The display is designed so that its depth of field approximately corresponds to the Human depth of field. By using this display, they showed that "when focus cues are correct or nearly correct, (1) the time required to

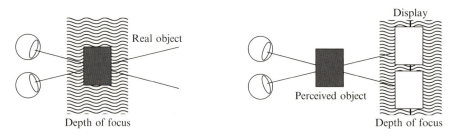

Fig. 7 Vergence and accommodation: they are consistent when viewing a real object (*left*), but may be conflicting when viewing a stereoscopic display, since the perceived object may not lie within the depth of field range (*right*). Adapted from Emoto et al. [17]

Stereoscopic Cinema 27

identify a stereoscopic stimulus is reduced, (2) stereo-acuity in a time-limited task is increased, (3) distortions in perceived depth are reduced, and (4) viewer fatigue and discomfort are reduced". However, this screen is merely experimental, and it is impractical for home or movie theater use.

The depth of field may be converted to disparities in degrees using simple geometric reasoning. The depth of field of the human visual system is, depending on the authors, between ± 0.2 D or $\pm 0.82°$ [77] and ± 0.3 D [12]. The limit to binocular fusion is from $2°$ to $3°$ at the front or back of the stereoscopic display, and Percival's zone of comfort is about one-third of this, i.e. $0.67°$ to $1°$. We note that the in-focus range almost corresponds to Percival's zone of comfort for binocular fusion, which probably comes from the fact that the visual system only learned to fuse non-blurred objects within the in-focus range.

Let us take for example a conservative value of ± 0.2 D for the depth of field. For a movie theater screen placed at 16 m, the in-focus range is from $1/(1/16 + 0.2) \approx 3.8$ m to infinity, whereas for a 3DTV screen placed at 3.5 m, it is from $1/(1/3.5 + 0.2) \approx 2$ m to $1/(1/3.5 + 0.2) \approx 11.7$ m. As we will see later, this means that the camera focus range should theoretically be different when shooting movies for a movie theater or a 3-D television.

3 Picking the Right Shooting Geometry

3.1 The Spottiswoode Point of View

Spottiswoode et al. [56] wrote the first essay on the perceived geometry in stereoscopic cinema. They devised how depth is distorted by "stereoscopic transmission" (i.e. recording and reproduction of a stereoscopic movie), and how to achieve "continuity in space", or making sure that there is a smooth transition in depth when switching from one stereoscopic shot to another.

They did a strong criticism of the "human vision" systems (i.e. systems that were trying to mimic the human eyes, with a 6.5 cm interocular, and a $0.3°$ convergence). They claimed that all stereoscopic parameters can be and should be adapted, either at shooting time or as post-corrections, depending on screen size, to get the desirable effects (depth magnification or reduction, and continuity in space). According to them, the main stereoscopic parameter is the *nearness factor* N, defined as the ratio between the viewing distance from screen and the distance to the fused image (with our notations, $N = H'/Z'$): "for any pair of optical image points, the ratio of the spectator's viewing distance from the screen to his distance to the fused image point is a constant, no matter whereabouts in the theater he may be sitting". Continuity in space is achieved by slowly shifting over a few seconds the images in the horizontal direction before and after the cuts. The spectator does not notice that the images are slowly shifting: the vergence angle of the eyes is adapted by the human visual system, and depth perception is almost the same. Due to the technical tools available at this time, this is the only post-correction method they propose.

They list three classes of stereoscopic transmission:

- *Ortho-infinite*, where infinity points are correctly represented at infinity
- *Hyper-infinite*, where objects short of infinity are represented at infinity (which means that infinity points cause divergence)
- *Hypo-infinite*, where objects at infinity are represented closer than infinity (which causes the cardboard effect, see Sect. 2.4)

In order to help the stereographer picking up the right shooting parameters, they invented a calculator, called the *Stereomeasure*, which computes the relation between the various parameters of 3-D recording and reproduction. They claim that a stereoscopic movie has to be made for given projection conditions (screen size and distance to screen), which is a fact sometimes overlooked either by stereographers, or by the 3DTV industry (however, as discussed in Sect. 5, the stereoscopic movie may be adapted to other screen sizes and distances).

In the Spottiswoode setup, the standard distance from spectator to screen should be from $2W'$ to $2.5W'$ (W' is the screen width). They place the proscenium arch at $N = 2$ (half distance from screen), and almost everything in the scene happens between $N = 0$ (infinity) and $N = 2$ (half distance) ($N = 1$ is the screen plane).

Their work has been strongly criticized by many stereographers, sometimes with wrong arguments [36, 53], but the main problem is probably that the artists do not want to be constrained by mathematics when they are creating: cinematography has always been an art of freedom, with a few rules of thumb that could always be ignored. But the reality is here: the constraints on stereoscopic cinema are much stronger than on 2-D cinema, and bypassing the rules results in bad movies causing eyestrain or headache to the spectator. A bad stereoscopic movie can be a very good 2-D movie, but adding the stereoscopic dimension will always modify the perceived quality of the movie, either by adding a feeling of "being there", or by obfuscating the intrinsic qualities of the movie with ill-managed stereoscopy.

There are some problems with the Spottiswoode's theorisation of stereoscopic transmission, though, and this section will try to shed some light on some of these:

- The parametrization by the nearness factor hides the fact that strong nonlinear depth and size distortions may occur in some cases, especially on far points.
- Divergence at infinity will happen quite often when images are shifted in order to achieve continuity in space.
- Shifting the images may break the vergence-accommodation constraints, in particular Percival's zone of comfort, and cause visual fatigue.

Woods et al. [72] extended this work and also computed spatial distortion of the perceived geometry. Although their study is more focused on determining horizontal and vertical disparity limits, they also studied depth plane curvature effects, where a fronto-parallel plane appears to be curved. This situation arises when non-rectified images are used and the camera configuration is toed-in (i.e. with a non-zero vergence angle).

More recently, Masaoka et al. [40] from the NHK labs also did a similar study, and presented a software tool which is able to predict spatial distortions that happen when using given shooting and viewing parameters.

3.2 Shooting and Viewing Geometries

As was shown by Spottiswoode et al. [56] in the early days of stereoscopic cinema, projecting a stereoscopic movie on different screen sizes and distances will produce different perceptions of depth.

One obvious solution was adopted by the large format stereoscopic cinema (IMAX 3-D): if the film is shot with parallel cameras (i.e. vergence is zero), and is projected with parallel projectors that have a human-like interocular (usually 6 cm for IMAX-3D), then infinity points will always be perceived exactly at infinity, and divergence will never occur [82]. Large-format stereoscopic cinema has less constraints than stereoscopic cinema targeted at standard movie theaters, since the screen is practically borderless, and the audience is located near the center of the hemispherical screen. The camera interocular is usually close to the human interocular, but it may be played with easily, depending on the scene to be shot (as in *Bugs! 3-D* by Phil Streather, where hypostereo or gigantism is heavily used). If the camera interocular is the same as the human interocular, then everything in the scene will appear at the same depth and size as if seen with normal vision, which is a very pleasant experience.

However, shooting parallel is not always advisable or even possible for standard stereoscopic movies. Close scenes for example, where the subject is closer than the movie theater screen, require camera convergence, or the film subject will appear too close to the spectator and will break the proscenium rule most of the time. But using camera convergence has many disadvantages, and we will see that it may cause heavy distortions on the 3-D scene.

Let us study the distortions caused by given shooting and viewing geometries. The geometric parameters we use (Fig. 8) are very simple but describe fully the stereoscopic setup. Compared to camera-based parameters, as used by Yamanoue et al. [76], we can easily attach a simple meaning to each parameter and understand its effect on space distortions. We assume that the stereoscopic movie is rectified and thus contains no vertical disparity (see Sect. 5.1), so that the convergence plane, where the disparity is zero, is vertical and parallel to the line joining the optical centers of the cameras.

The 3-D distortions in the perceived scene essentially come from different scene magnifications in the fronto-parallel (or width and height) directions, and in the depth direction. Spottiswoode et al. [56] defined the *shape ratio* as the ratio between depth magnification and width magnification, Yamanoue et al. [76] call it E_p or *depth reduction*, and Mendiburu [41] uses the term *roundness factor*. We will use *roundness factor* in the remaining of our study. A low roundness factor will result in the cardboard effect, and a rule of thumb used by stereographers is that it should never be below 0.2, or 20%.

Let b, W, H, Z be the stereoscopic camera parameters, and b', W', H', Z' be the viewing parameters, as described in Fig. 8. Let d be the disparity in the images. The disparity on the display screen is $d' = d + d_0$, taking into account an optional shift d_0 between the images (shifting can be done at the shooting stage, in post-production, or at the display stage).

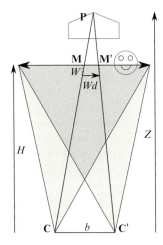

Symbol	Camera	Display
C, C′	Camera optical center	Eye optical center
P	Physical scene point	Perceived 3-D point
M, M′	Image points of **P**	Screen points
b	Camera interocular	Eye interocular
H	Convergence distance	Screen distance
W	Width of convergence plane	Screen size
Z	Real depth	Perceived depth
d	Disparity (as a fraction of W)	

Fig. 8 Shooting and viewing geometries can be described using the same small set of parameters

Triangles **MPM′** and **CPC′** are homothetic, consequently:

$$\frac{Z-H}{Z} = \frac{W}{b}d. \tag{1}$$

It can easily be rewritten as:

$$d = \frac{b}{W}\frac{Z-H}{Z}, \text{ or } Z = \frac{H}{1-\frac{W}{b}d}, \tag{2}$$

which holds both for the shooting and viewing geometries (i.e. with and without primes). From these, we can compute the image disparity d from the real depth Z:

$$d = \frac{b}{W}\frac{Z-H}{Z}, \tag{3}$$

and then compute the perceived depth Z' from the disparity d':

$$Z' = \frac{H'}{1-\frac{W'}{b'}(d+d_0)}. \tag{4}$$

Finally, eliminating the disparity d from both equations gives the relation between real depth Z and perceived depth Z':

$$Z' = \frac{H'}{1-\frac{W'}{b'}(\frac{b}{W}\frac{Z-H}{Z}+d_0)} \text{ or } Z = \frac{H}{1-\frac{W}{b}(\frac{b'}{W'}\frac{Z'-H'}{Z'}-d_0)} \tag{5}$$

3.3 Depth Distortions

Let us now compute the depth distortion from the perceived depth. In the general case, points at infinity ($Z \to +\infty$) are perceived at

$$Z' = \frac{H'}{1 - \frac{W'}{b'}(\frac{b}{W} + d_0)}. \tag{6}$$

Eye divergence happens when Z' becomes negative. The eyes diverge when looking at scene points at infinity ($Z \to +\infty$) if and only if:

$$\frac{b'}{W'} < \frac{b}{W} + d_0. \tag{7}$$

In this case, the real depth which is mapped to $Z' = \infty$ can be computed as

$$Z = \frac{1}{1 - \frac{W}{b}\left(\frac{b'}{W'} - d_0\right)}, \tag{8}$$

and any object at a depth beyond this one will cause divergence.

The relation between Z and Z' is nonlinear, except if $\frac{W}{b} = \frac{W'}{b'}$ and $d_0 = 0$, which we call the *canonical setup*. In this case, the relation between Z and Z' simplifies to:

$$Z' = Z\frac{H'}{H}. \quad \left(\text{Canonical setup:} \quad \frac{W}{b} = \frac{W'}{b'}, d_0 = 0\right) \tag{9}$$

Let us now study how depth distortion behaves when we are not in the canonical setup anymore. For our case study, we start from a purely canonical setup, with $b = b' = 6.5$ cm, $W = W' = 10$ m, $H = H' = 15$ m and $d_0 = 0$. In the following charts, depth is measured from the convergence plane or the screen plane (depth increases away from the cameras/eyes).

The first chart (Fig. 9) shows the effect of changing only the camera interocular b (the distance to the convergence plane H remains unchanged, which implies that the vergence angle changes accordingly). This shows the well-known effects called *hyperstereo* and *hypostereo*: the roundness factor of in-screen object varies from high values (hyperstereo) to low values (hypostereo).

Let us now suppose that the subject being filmed is moving away from the camera, and we want to keep its image size constant by adjusting the zoom and vergence accordingly, i.e. W remains constant. To keep the object's roundness factor constant, we also keep the interocular b proportional to the object distance H ($b = \alpha H$). The effect on perceived depth is shown in Fig. 10. We see that the depth magnification or roundness factor close to the screen plane remains equal to 1, as expected, but the depth of out-of-screen objects is distorted, and a closer analysis would show that the wide-interocular-zoomed-in configuration creates a *puppet-theater effect*, since farther objects have a larger image size than expected. This is one configuration

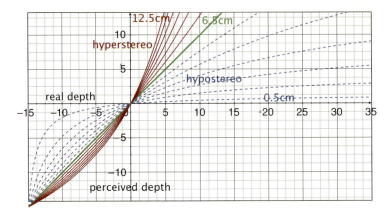

Fig. 9 Perceived depth as a function of real depth for different values of the camera interocular b. This graph demonstrates the well-known hyperstereo and hypostereo phenomenon

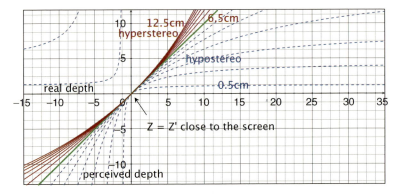

Fig. 10 When the object moves away, but we keep the object image size constant by zooming in (i.e. H varies but W is constant), if we keep the camera baseline b proportional to convergence distance H, the depth magnification close to the screen plane remains equal to 1. But be careful: divergence may happen!

where the roundness factor of the in-screen objects can be kept constant while changing the stereoscopic camera parameters, and in the next section we will show how to compute all the changes in camera parameters that give a similar result.

3.4 Shape Distortions and the Depth Consistency Rule

If we want the camera setup and the display setup to preserve the shape of all observed objects up to a global 3-D scale factor, a first constraint is that there must be a linear relation between depths, so we must be in the canonical setup described by (9). A second constraint is that the ratio between depth magnification and image magnification, called the *roundness factor*, must be equal to 1. This

Stereoscopic Cinema

means that the *only* configuration with faithful depth reproduction is when the shooting and viewing geometries are *homothetic* (i.e. there is a scale factor between the two geometries):

$$\frac{W'}{W} = \frac{H'}{H} = \frac{b'}{b}. \qquad \text{(Homothetic configuration)} \qquad (10)$$

In general, the depth and space distortion is nonlinear: it can easily be shown that the perceived space is a homographic transform of the real space. Shooting from farther away while zooming in with a bigger interocular doesn't distort (much) depth, as we showed in the previous section, and that's probably the right way to zoom in – the baseline should be proportional to the convergence distance. But one has to take care about the fact that the infinity plane in scene space may cause divergence.

More generally, we can introduce the *depth consistency rule*: Objects which are close to the convergence plane in the real scene or close to the screen in projection space ($Z = H$ or $Z' = H'$) should have a depth which is consistent with their apparent size, i.e. their roundness factor should be equal to 1.

The depth ration between scene space and projection space close to the convergence plane can be computed from (5) as:

$$\frac{\partial Z'}{\partial Z}(Z = H) = \frac{b}{HW} \frac{H'W'}{b'}, \qquad (11)$$

and the apparent size ratio is simply $\frac{W'}{W}$. Setting the ratio between both, i.e. the roundness factor, equal to 1 leads to the *depth consistency rule*:

$$\frac{b}{H} = \frac{b'}{H'}. \qquad \text{(Depth consistency rule)} \qquad (12)$$

One important fact arises from the depth consistency rule: for objects that are close to the convergence plane or the screen plane, screen size (W') does not matter, whereas most spectators expect to get exaggerated 3-D effects when looking at a movie on a bigger screen. Be careful, though, that for objects farther than the screen, especially at $Z \to +\infty$, divergence may occur on bigger screens. Since neither b, H, or b' can be changed when projecting a 3-D movie, the key parameter for the depth consistency rule is in fact screen distance, which dramatically influences the perceived depth: the bigger the screen distance, the higher the roundness factor will be. This means that if a stereoscopic movie, which was made to be seen in a movie theater from a distance of 16 m, is viewed on a 3DTV from a distance of 4 m, the roundness factor will be divided by 4 and the spectator will experience the classical *cardboard effect* where objects look flat.

What can we do to enforce the depth consistency rule, i.e. to produce the same 3-D experience in different environments, given the fact that b is fixed and H' is usually constrained by the viewing conditions (movie theater vs. home cinema vs. TV)? The only possible solution consists in artificially changing the shooting parameters in post production using techniques that will be described in Sect. 5.

3.5 Shooting with the Right Depth of Field

When disparities outside of the Percival zone of comfort are present, Ukai and Howarth [66] showed that vergence-accommodation conflicts that arise can be attenuated by reducing the depth of field: the ideal focus distance should be on the plane of convergence (i.e. the screen), and the depth of field should match the expected depth of field of the viewing conditions. They even showed that objects that are out of this focus range are surprisingly better perceived in 3-D if they are blurred than if they are in focus.

We saw (Sect. 2.1) that the human eye depth of field in normal conditions is between 0.2 and 0.3 D (diopters) – let us say 0.2 D to be conservative. This depth of field should be converted to a depth range in the targeted viewing conditions: $Z'_{min} = 1/\left(\frac{1}{H'} - 0.2\right)$, $Z'_{max} = 1/\left(\frac{1}{H} + 0.2\right)$ (if Z'_{max} is negative, then it is considered to be at infinity). Then, the perceived depth range should be converted to real depth range $[Z_{min}, Z_{max}]$, using the inverse of formula (5), and the camera aperture should be computed from this depth range.

Unfortunately, the viewing conditions are usually not known at shooting time, or the movie has to be made for a various range of screen distances and sizes. We will describe a possible solution to this general case in Sect. 5.4.

3.6 Remaining Issues

Even when the targeted viewing conditions are known precisely, the shooting conditions are sometimes constrained: for example, when filming wild animals, a minimum distance may be necessary, resulting in using a wide baseline and a long focal length (i.e. a large $\frac{b}{W}$). The resulting stereoscopic movie, although it will have a correct roundness factor around the screen plane, will probably have strong divergence at infinity, since $\frac{b'}{W'} < \frac{b}{W}$, which may look strange, even if the right depth of field is used.

Another kind of problem may happen, even with objects which are close to the screen plane: Psychophysics experiments showed that specular reflections may also be used as a shape cue, and more specifically as a curvature cue [8,9]. Except when using a purely homothetic configuration, these depths cues will be inconsistent with other cues, even near the screen plane, and this effect will be particularly visible when using long focal lengths. This problem can probably not be overcome by changing the shooting geometry, and the best solution is probably to edit manually the specular reflections in post-production to make them look more natural, or to use the right makeup on the actors ....

Inconsistent specular reflections can also be due to the use of a mirror rig (a stereoscopic camera rig using a half-silvered mirror to separate images for the left and right cameras): specular reflections are usually polarized by nature, and the transmission and reflection coefficients of the mirror depend on the polarization of incoming light. As a result, specular reflections have a different aspect in the left and right images.

4 Lessons for Live-Action Stereoscopic Cinema from Animated 3-D

Shooting live-action stereoscopic cinema has progressed over the last decade with the arrival of more versatile camera rigs. Older stereo rigs had two fixed cameras, and the stereo extrinsic parameters – baseline and vergence – were changed by manual adjustment between shots. Newer rigs allow dynamic change of the stereo parameters during a shot. This still falls short of a final stage of versatility, which is the modification of the left- and right-eye images during post-production, to effectively change the stereo extrinsic parameters – in other words, this ultimate goal is to use the shot footage as a basis for synthesizing left- and right-eye images with any required stereo baseline and vergence.

New technologies are bringing us closer to the goal of synthesizing left- and right-eye views with different stereo parameters during post-production. But what are the benefits, and how will this create a better viewer experience? To answer this question, in this section we look at current practices in animated stereoscopic cinema. In animation, the creative team has complete control over camera position, camera motion, camera intrinsics, stereo extrinsics, and the 3-D structure of the scene. This allows scope to experiment with the stereoscopic experience, including doing 3-D manipulations and distortions that do not correspond to any physical reality but which enhance viewer experience. We describe four core techniques of animated stereoscopic cinema. The challenge for live-action 3-D is to create new technologies so that these techniques can be applied to live-action footage as easily as they are currently applied to animation.

4.1 Proscenium Arch or Floating Window

The proscenium arch or floating window was introduced earlier. The simplest way to project stereoscopic imagery is to capture images from a left- and right- camera and then put those images directly onto the cinema screen. This imposes a specific epipolar geometry on the viewer, and our eyes adjust to it. Now consider the four-sided boundary of the physical cinema screen. It also imposes epipolar constraints on the eyes (in fact these are epipolar constraints which are consistent with the whole rest of the physical world). But note that there is no reason why the epipolar geometry imposed by the stereoscopic images will be consistent with the epipolar constraints associated with the screen boundary. The result is conflicting visual cues.

The solution is to black-mask the two stereoscopic images. Consider the basic 3-D animation setup, and two cameras observing a 3-D scene. Now imagine between the camera and the scene a rectangular window, in 3-D, so that the cameras view the scene through the window, but the window is surrounded by a black wall where nothing is visible. This is the proscenium arch or floating window, a virtual 3-D entity interjected between the cameras and the scene. The visual effect of the window is achieved by black-masking the boundaries of the left- and right- eye images. Since this floating window is a 3-D entity that is consistent with the cameras and the rest of the 3-D scene, it does not give rise to conflicting visual cues.

4.2 Floating Windows and Audience Experience

Section 4.1 described a basic motivation for the floating window, motivated by comfort in the viewing experience. There is a further use for the technique. Note that the window can be placed anywhere in the 3-D scene, It can lie between the camera and the scene, part-way through the scene, or behind the scene. This is not perceived explicitly but placing the 3-D scene behind the floating window relative to the audience produces a more subdued passive feeling, according to accepted opinion in the creative community. Placing the scene in front of the floating window relative to the audience produces a more engaged active feeling. Also note that the window does not need to be fronto-parallel to the viewer. Instead it can be tilted in 3-D space. Again the viewer is typically unaware of this consciously, but orientation of the floating window can produce subliminal effects, e.g. a forward tilt of the upper part of the window can produce a looming feeling.

4.3 Window Violations

The topic of floating windows leads naturally to another technique. Consider again the 3-D setup - cameras, floating window, and 3-D scene. First consider objects that are on the far side of the window from the cameras. When they are center-stage, they are visible through the window. As they move off-stage and are obscured by the surrounding wall of the virtual window. This is all consistent from the viewpoint of the two stereoscopic cameras that are viewing the scene. Now consider an object that lies between the cameras and the window. While it's center-stage and visible in the two eyes, everything is fine. As it moves off-stage, it intersects the view-frustums created by each camera and the floating window, and nothing is visible outside the frustum. But this is inconsistent to the eye – it's as though we are looking at someone in front of a doorway and their silhouette disappears as it passes over the view frustums in the left- and right-eyes of the doorway which is to the rear of them.

The solution to this problem is simply not to allow such violations. This is straightforward in animation but in live-action stereoscopic cinema, of course, it requires the non-trivial recovery of the stereo camera positions and the 3-D scene to detect when this is happening, and avoid it.

4.4 Multi-rigging

So far, we considered manipulations of the left- and right- eye images that are consistent with a physical 3-D reality. Multi-rigging, however, is a technique for creating stereoscopic images which produce a desired viewing experience, but could not have arisen from a physically correct 3-D situation. Consider a scene with objects A and B. The left camera is kept fixed but there are two right cameras, one for shooting object A and one for shooting object B. Two different right eye images are generated, and they are then composited so that objects A and B appear in a single composite right eye image.

Why do this? Consider the case where A is close to the camera and B is far away. In a normal setup, B will appear flat due to distance. But by using a large stereo baseline for shooting object B, it is possible to capture more information around the occluding contour of the object, and give it a greater feeling of roundedness, even though it is placed more distantly in the scene. Again this is straightforward for animated stereoscopic cinema, but requires 3-D capture of the scene, and the ability to modify stereo baseline at post-production time, to apply it to live-action stereoscopic cinema.

5 Post-production of Stereoscopic Movies

When filming with stereoscopic cameras, even if the left and right views are rectified in post-production (Sect. 5.1), there are many reasons for which the movie may not be adapted to given viewing conditions, among which:

- The screen distance and screen size are different from the ones the movie was filmed for, resulting in a different *roundness factor* for objects close to the screen.
- Because of a screen size larger than expected, the points at infinity cause eye divergence.
- There were constraints on camera placement (as those that happen when filming sports or wildlife), which cause large disparities, or even divergence, on far-away objects.
- The stereoscopic camera was not adjusted properly when filming.

As we will see during this section, changing the shooting geometry in post-production is theoretically possible, although the advanced techniques require high-quality computer vision and computer graphics algorithms to perform a process called view interpolation. Due to the fact that 3-D information can be extracted from the stereoscopic movies, there are also a few other stereo-specific post-production processes that may be improved by using computer vision techniques.

5.1 Eliminating Vertical Disparity: Rectification of Stereoscopic Movies

Vertical disparity is one of the sources of visual fatigue in stereoscopic cinema, and it may come from misaligned cameras or optics, or from toed-in camera configurations, where vertical disparity cannot be avoided.

In Computer Vision, 3-D reconstruction from stereoscopic images is usually preceded by a transformation of the original images [19, Chap. 12]. This transformation, called *rectification* is a 2-D warp of the images that aligns matching points on the same y coordinate in the two warped images. The rectification of a stereoscopic pair of images is usually preceded by the computation of the *epipolar geometry* of the stereoscopic camera system. Knowing the epipolar geometry, one can map a point in one of the images, to a line or curve in the other image which is the projection of the optical ray issued from that point onto that image. The rectification process transforms the epipolar lines or curves into horizontal lines, so that stereoscopic matching by computer vision algorithms is made easier.

It appears that this rectification process is exactly what we need to eliminate vertical disparity in stereoscopic movies, so that all the available results from Computer Vision on computing the epipolar geometry [6, 18, 24, 42, 57] or on the rectification of stereoscopic pairs [1, 14, 20, 23, 38, 73, 79] will be useful to accurately eliminate vertical disparity.

However, the rectification of a single image pair acquired in a laboratory with infinite depth of field for stereoscopic matching, and rectification of a stereoscopic movie that will be presented to spectators is not exactly the same task. The requirements of a rectification method designed for stereoscopic cinema are:

1. It should be able to work without knowing anything from the stereoscopic camera parameters, because this data is not always available, and may be lost in the film production pipeline, whereas images are usually not lost.
2. It should require no calibration pattern or grid, since the wide range of camera configurations and optics would require too many different calibration grid sizes: all the computation should be made from the images themselves (we call it *blind rectification*).
3. The aspect ratio of rectified images should be as close as possible to the aspect ratio of the original images.
4. The rectified image should fill completely the frame: no "unknown" or "black" area can be tolerated.
5. The rectification of the whole movie should be smooth (jitter or fast-varying rectification transforms will create shaky movies) so that the rectification parameters should either be computed for a whole shot, or it should be slowly varying over time.

Stereoscopic Cinema

6. The camera parameters (focal length, focus, ...) and the rig parameters (interocular, vergence, ...) may be fixed during a shot, but they could also be slowly varying.
7. The images may have artistic qualities that are difficult to handle for computer vision algorithms: lack of texture, blur, saturation (white or black), sudden change in illumination, noise, etc.

Needless to say, very few rectification methods fulfill these requirements, and the best solution will probably have to take the best out of the cited methods in order to give acceptable results. Cheng et al. [14] recently proposed a method to rectify stereo sequences with varying camera motions and zooming effects, while keeping the image aspect ratio, but there is still room for improvements, and there will probably be many other publications on this subject in the near future.

A proper solution would probably contain the following ingredients:

- A method for epipolar geometry computation and rectification that takes into account nonlinear distortion [1,6,57]: even if cinema optics are close-to-perfect, images have a very high resolution, and small nonlinear distortions with an amplitude of a few pixels may happen at short focal lengths.
- A multi-scale feature detection and matching method, which will be able to handle blurred or low-texture areas in the images.
- Proper parameterization of the rectification functions, so that only rectifications that conserve the aspect ratio are allowed. Methods for panorama stitching reached this goal by using the camera rotation around its optical center as a parameter – this may be a good direction.
- Temporal filtering, which is not an easy task since the filtered rectifications must also satisfy the above constraints.

5.2 Shifting and Scaling the Images

Image shifting is a process already used by Spottiswoode et al. [56], in particular to achieve *continuity in space* during shot transitions. The human brain is actually not very sensitive to stereoscopic shifting: if two fixed rectified images are shown in stereo to a subject, and the images are shifted slowly, the subject will not notice the change in shift, and surprisingly the scene will not appear to move closer or farther as would be expected. However, shifting the images modifies the eye vergence and not the accommodation, and thus may break the vergence-accommodation constraints and cause visual fatigue: when images are shifted, one must verify that the disparities are still within Percival's zone of comfort (which depends on viewing geometry).

Image shifting is mostly used for "softening the cuts": If the disparity of the main area of interest changes abruptly from one shot to the other, the human eyes may take some time to adjust the vergence, and it will cause visual fatigue or even alter the oculomotor system, as shown by Emoto et al. [17]. As explained by Spottiswoode et al. [56], shifting the images is one solution to this problem.

One way to make a smooth transition between shots is the following: let us say the main subject of shot 1 is at disparity d_1, and the main subject of shot 2 is at disparity d_2 (they do not have to be at the same position in the image). About 1 s before the transition, the images from shot 1 are slowly shifted from $d_0 = 0$ to $d_0 = \frac{d_2 - d_1}{2}$. When the transition (cut or fade) happens, the disparity of the main subject in shot 1 is $d = \frac{d_1 + d_2}{2}$. During the transition, the first image of shot 2 is presented with a shift of $d_0 = \frac{d_1 - d_2}{2}$, so that the disparity of the main subject is also $d = \frac{d_1 + d_2}{2}$. After the transition, the disparity continues to be slowly shifted, in order to arrive at a null shift ($d_0 = 0$) about 1 s after the transition.

During this process, care must be taken not only to stay inside Percival's zone of comfort, but also to avoid divergence in the areas in focus (the divergence threshold depends on screen size).

Another simple post-production process consists in scaling the images. Scaling is equivalent to changing the width W of the convergence plane, and allows reframing the scene by panning simultaneously both rescaled images.

There is no quality loss when shifting or scaling the images, except the one due to resampling the original images. In particular, no spatial or temporal artifact may appear in shifted or scaled sequences, whereas they may happen in the processes described hereafter.

5.3 View Interpolation, View Synthesis, and Disparity Remapping

5.3.1 Definitions and Existing Work

Image shifting and scaling cannot solve all the issues caused by having different (i.e. non-homothetic) shooting and viewing geometries. Ideally, we would like to be able to change two parameters of the shooting geometry in order to get homothetic setups: the camera interocular b and the distance to the convergence plane H. This section describes the computer vision and computer graphics techniques which can be used to achieve these effects in stereoscopic movie post-production. The results that these methods have obtained so far are not on par with the picture quality requirements of the stereoscopic cinema, but this research field is progressing constantly, and these requirements will probably be met within the next few years.

The first thing we would like to do is change the camera interocular b, in order to follow the depth consistency rule $\frac{b}{H} = \frac{b'}{H'}$. To achieve this, we use *view interpolation*, a technique that takes as input a set of synchronized images taken from different but close viewpoints, and generates a new view *as if* it were taken by a camera placed at a specified position between the original camera positions. It is different from a well-known post-production effect called *retiming*, which generates intermediate images between two consecutive images in time taken by the same camera, although some people have been successfully using retiming techniques to do view interpolation if the cameras are close enough. As shown in the first line

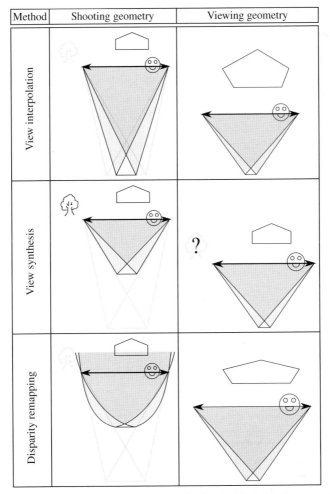

Fig. 11 Changing the shooting geometry in post-production (original shooting geometry is drawn in *light gray in the left column*). All methods can preserve the roundness of objects in the screen plane, at some cost for out-of-screen objects: view interpolation distorts dramatically their depth and size; view synthesis generates a geometrically correct scene, but it is missing a lot of scene information about objects that cannot be seen in the original images; disparity remapping may preserve the depth information of the whole scene, but the apparent width of objects is not preserved, resulting in puppet-theater effect

of Fig. 11, although the roundness factor of objects near the screen plane is well preserved, depth and size distortions are present and are what would be expected from the interpolated camera setup: far objects are heavily distorted both in size and depth, and divergence may happen at infinity.

If we also want to change the distance to screen, we have to use *view synthesis*. It is a similar technique, where the cameras may be farther apart (they can even surround the scene), and the synthesized viewpoint can be placed more freely. View synthesis usually requires a shooting setup with at least a dozen of cameras, and may even use hundreds of cameras. It is sometimes used in free-viewpoint video, when the scene is represented as a set of images, not as a textured 3-D mesh. The problem is that we usually only have two cameras, and for view synthesis to work, at least all parts of the scene that would be visible in the synthesized viewpoint must be also visible in the original viewpoints. Unfortunately, we easily notice (second line of Fig. 11) that many parts of the scene that were not visible in the original viewpoints become visible in the interpolated viewpoints (for example the tree in the figure). Since invisible parts of the scene cannot be invented, view synthesis is clearly not the right technique to solve this problem.

What we propose is a mixed technique between view interpolation and view synthesis, that preserves the global visibility of objects in the original viewpoints, but does not produce depth distortion or divergence: *disparity remapping*. In disparity remapping (third line of Fig. 11), we apply a nonlinear transfer function to the disparity function, so that perceived depth is proportional to real depth (the scale factor is computed at the convergence plane). In practice, it consists in shifting all the pixels in the image that have a given disparity by the disparity value that would be perceived in the viewing geometry for an object placed at the same depth. That way, divergence may not happen, since objects that were at a given distance from the convergence plane will be projected at the same distance, up to a fixed scale factor, and thus points at infinity are effectively projected at infinity. However, since the object image size is not changed by disparity remapping, there will still be some kind of puppet-theater effect, since far-away objects may appear bigger in the image than they should be. Of course, any disparity remapping function, even one that does not conserve depth, could be used to get special effects, to obtain effects similar to the multi-rigging techniques used in animation (Sect. 4.4).

There has been already a lot of work on view interpolation [10, 15, 25, 31, 45, 49, 63, 67, 71, 74, 75, 80, 81], and disparity remapping as we defined it uses the exact same tools. They have solved the problem with various levels of success, and the results are constantly improving, but none could definitely get rid of the two kinds of artifacts that plague view interpolation:

- Spatial artifacts, which are usually "phantom" objects or surfaces that appear due to specular reflections, occluded areas, blurry or non-textured areas, semi-transparent scene components (although many recent methods handle semi-transparency in some way).
- Temporal artifacts, which are mostly "blinking" effects that happen if the spatial artifact are not temporally consistent and seem to "pop up" at each frame.

5.3.2 Asymmetric Processing

The geometry changes leading to view interpolation shown in Fig. 11 are symmetric, i.e. the same changes are made to the left and right view. Interpolating both views may alter the quality and create artifacts and both views, resulting in a lower-quality stereoscopic movie. However, this may not be the best way to change the shooting geometry by interpolation.

Stelmach et al. [58] showed that the quality of the binocular percept is dominated by the sharpest image. Consequently, the artifacts could be reduced by interpolating only one of the two views, and perhaps smoothing the interpolated view if it contains artifacts. However, a more recent study by Seuntiens et al. [50] contradicts the work of Stelmach et al. [58], and shows that the perceived-quality of an asymmetrically-degraded image pair is about the average of both perceived qualities when one of the two views is degraded by a very strong JPEG compression. Still, they admit that "asymmetric coding is a valuable way to save bandwidth, but one view must be of high quality (preferably the original) and the compression level of the coded view must be within acceptable range", so that for any kind of asymmetric processing of a stereo pair (such as view interpolation), one has to determine what is this "acceptable range" to get a perceived quality which is better than with symmetric processing. Kooi and Toet [32] also note that interocular blur suppression, i.e. fusing an image with a blurred image, requires a few more seconds than with the original image pair. Gorley and Holliman [21] devised another metric for stereoscopic pair quality, which is based on comparing areas around SIFT features. This proved to be more consistent with perceived quality measured by human subjects than the classical PSNR (Peak Signal-to-Noise Ratio) measurement used in 2-D image quality measurements. Further experiments should be led on the perception of asymmetrically-coded stereoscopic sequences to get a better idea of the allowed amount of degradation in one of the images.

One could argue that the best-quality image in an asymmetrically-processed stereoscopic sequence should be the one corresponding to the dominant eye,[4] but this means that interpolation should depend on the viewer, which is not possible in a movie theater or in consumer 3DTV. Seuntiens et al. [50] also studied the effect of eye dominance on the perception of asymmetrically-compressed images, and their conclusion is that eye dominance has no effect on the quality perception, which means that it may be possible to do asymmetric processing while remaining user-independent.

5.4 Changing the Depth of Field

We have seen in Sects. 2.5.3 and 3.5 that not only the shooting geometry should be adapted to the display, but also the depth of field: the in-focus plane should be at the screen depth (i.e. at zero disparity), and the depth of field should follow the human depth of field (i.e. 0.2–0.3 D), so that there is no visual fatigue due to horizontal disparity limits or vergence-accommodation conflicts.

Having a limited depth of field may seem limited at first, but Ukai and Howarth [66] showed that 3-D is actually perceived even for blurred objects, and this also

---

[4] About 70% of people prefer the right eye, and 30% prefer the left, and this is correlated to right- or left-handedness, but many people are cross-lateral (e.g. right-handed with a left dominant eye).

results in less visual fatigue when these objects are out of the theoretical focus range. Another big issue with stereoscopic displays is the presence of crosstalk. Lenny Lipton claims [37]: "Working with a digital projector, which contributes no ghosting artifact, proves that the presence of even a faint ghost image may be as important as the accommodation/convergence breakdown". Actually, blurring the objects that have high disparities should also strongly reduce that ghosting effect, so that changing the depth of field may have the double advantage of diminishing visual fatigue and reducing crosstalk.

In animated 3-D movies, the depth of field can be artificially changed, since images with infinite depth of field can be generated, together with a depth map: [7, 30, 35]. In live-action stereoscopic cinema, the depth of field can be reduced using similar techniques, given a depth map extracted from the live stereoscopic sequence. Increasing the depth of field is theoretically possible, but would enhance dramatically the image noise.

5.5 Dealing with the Proscenium Rule

The proscenium rule states that the stereoscopic sequence should be seen at though it were happening behind a virtual arch formed by the stereoscopic borders of the screen, called the proscenium arch. When an object appears in front of the proscenium, and it touches either the left or the right edges of the screen, then the proscenium rule is violated, because part of the object becomes not visible in both eyes, whereas it should be (see Sect. 4.1 and Fig. 5). When the object touches the top or bottom borders, the stereoscopic scene is still consistent with the proscenium arch, and some even argue that in this case the proscenium appears bent towards the object [41, Chap. 5].

Since the proscenium arch is a virtual window formed by the 3-D reconstruction of the left and right stereoscopic image borders, it can be virtually moved forward or backward. Moving the proscenium forward is the most common solution, especially when using the screen borders as the proscenium would break the proscenium rule, both in live-action stereoscopic movies and in animated 3-D (Sect. 4.3). It can also be a narrative element by itself, as may be learned from animated 3-D (Sect. 4.2). The left and right proscenium edges may even have different depths, without the audience even noticing it (this effect is called "floating the stereoscopic window" by Mendiburu [41]).

The proscenium arch is usually moved towards the spectator by adding a black border on the left side of the right view and on the right side of the left view. However, on display devices with high rates of crosstalk, moving the proscenium may generate strong artifacts on the screen left and right borders, where one of the views is blackened out. One way to avoid the crosstalk issues is to use a semi-opaque mask instead of a solid black mask, as suggested and used by stereographer Phil Streather. Other alternative solutions may be tried, such as blurring the edges, or a combination of blurring and intensity attenuation.

5.6 Compositing Stereoscopic Scenes and Relighting

Compositing consists in combining several image sources, either from live action or computer-generated, to obtain one shot that is ready to be edited. It can be used to insert virtual objects in real scenes, real objects in virtual scenes, or real objects in real scenes. The different image sources should be acquired using the same camera parameters. For live-action footage, this does usually mean that either match-moving or motion-controlled cameras should be used to get the optical camera parameters (internal and external). Matting masks then have to be extracted for each image source: masks are usually automatically-generated for computer-generated images, and several solutions exist for live-action footage: automatic (blue screen / green screen / chroma key), or semi-automatic (rotoscoping / hand drawing). The image sources are then combined together with matting information to form the final sequence. In non-stereoscopic cinema, if there is no camera motion, the compositing operation is often 2-D only, and a simple transform (usually translation and scaling) is applied to each 2-D image source.

The 3-D nature of the stereoscopic cinema imposes a huge constraint on compositing: not only the image sources must be placed correctly in each view, but the 3-D geometry must be consistent across all image sources. In particular, this means that the shooting geometry (Sect. 3) must be the same, or must be adjusted in post-production to match a reference shooting geometry (Sect. 5.3). If there is a 3-D camera motion, it can be recovered using the same match-moving techniques as in 2-D cinema, which may take into account the rigid link between left and right cameras to recover a more robust camera motion.

The matting stage could also be simplified in stereoscopic cinema: the disparity map (which gives the horizontal disparity value at each pixel in the left and right rectified views) or the depth map (which can be easily computed from the disparity map) can be used as a Z-buffer to handle occlusions automatically between the composited scenes. The disparity map representation should be able to handle transparency, since the pixels at depth discontinuities should have two depths, two colors, and a foreground transparency value [80]. Recent algorithms for stereo-based 3-D reconstruction automatically produce both a disparity map and matting information, which can be used in the compositing and view interpolation stages [10,52,63,74,75]. When working on a sequence that has to be modified using view interpolation or disparity remapping, it should be noted that the interpolated disparity map can be obtained as a simple transform of the disparity map computed from the original rectified sequence, and this solution should be preferred to avoid the 3-D artifacts caused by running a stereo algorithm on interpolated images (more complicated transforms happen in the case of view synthesis).

Getting consistent lighting between the composited image sources can be difficult: even if the two scenes to be composed were captured or rendered using the same lighting conditions, the mutual lighting (or radiosity) will be missing from the composited scene, as well as the cast shadows from one scene to the other.

If lighting consistency cannot be achieved during the capture of real scenes (for example if one of the shots comes from archive footage), the problem becomes much more complicated: it is very difficult to change the lighting of a real scene, even when its approximate 3-D geometry can be recovered (by stereoscopic reconstruction in our case), because that 3-D reconstruction will miss the material properties of the 3-D surfaces, i.e. how each surface reacts to incident light, and how light is back-scattered from the surface, and the 3-D texture and normals of the surface are very difficult to estimate from stereo. These material properties are contained in the BRDF (Bi-Directional Reflectance Function), which has a complicated general form with many parameters, but can be simplified from some classes of materials (e.g. isotropic, non-specular, ...).

The brute force solution to this problem would be to acquire the real scene under all possible lighting conditions, and then select the right one from this huge data set. Hopefully, only a subset of all possible lighting conditions is necessary to be able to recompose any lighting condition: A simple set of lighting conditions (the light basis) is used to capture the scene, and more complicated lighting conditions can be computed by linear combination of the images taken under the light basis. This is the solution adopted by Debevec et al. for their *Light Stage* setup [13, 16].

5.7 Adding Titles or Subtitles

Whereas adding titles or subtitles in 2-D is rather straightforward (the titles should not overlap with important information in the image, such as faces or text), it becomes more problematic in stereoscopic movies.

Probably the most important constraint is that titles should never appear inside or behind objects in the scene: this would result in a conflict between depth (the object is in front of the title) and occlusion (the text occludes the object) which will cause visual fatigue in the long run.

Another constraint is that titles should appear at the same depth as the focus of attention, because re-adjusting the vergence may take as much as 1 s, and cause visual fatigue [17]. The position of the title within the image is less important, since the oculomotor system has a well-trained "scanning" function for looking at different places at the same depth. If the display device has too much crosstalk, then a better idea may be to find a place within the screen plane which is not occluded by objects in the scene (because they are highly contrasted, out-of-screen titles may cause double images in this case).

A gross disparity map produced by automatic 3-D reconstruction should be enough to help satisfy these two constraints, the only additional information that has to be provided is *where* is the person who is speaking. Then, the position and depth can be computed from the disparity map and the speaker location: The position can be at the top or the bottom of the image, or following the speaker's face like a speech balloon, and the depth should either be the screen depth (at zero disparity) or the depth of the speaker's face.

6 Conclusion

This chapter has reviewed the current state of understanding in stereoscopic cinema. It has discussed perceptual factors, choices of camera geometry at time of shooting, and a variety of post-production tools to manipulate the 3-D experience. We included lessons from animated stereoscopy to illustrate how the creative process proceeds when there is complete control over camera geometry and 3-D content, indicating useful goals for stereoscopic live-action work. Looking to the future, we anticipate that 3-D screens will become pervasive in cinemas, in the home as 3-D TV, and in hand-held 3-D displays. A new market in consumer stereoscopic photography has also made an appearance in 2009, as Fuji released the W1 Real 3-D binocular-stereoscopic camera plus auto-stereoscopic photo frame. All of this activity promises exciting developments in stereoscopic viewing and in the underlying technology of building sophisticated 3-D models of the world geared towards stereoscopic content.

References

1. Abraham, S., Förstner, W.: Fish-eye-stereo calibration and epipolar rectification. ISPRS J. Photogramm. Remote Sens. **59**(5), 278–288 (2005). doi:10.1016/j.isprsjprs.2005.03.001
2. Akeley, K., Watt, S.J., Girshick, A.R., Banks, M.S.: A stereo display prototype with multiple focal distances. ACM Trans. Graph. **23**(3), 804–813 (2004). doi:10.1145/1015706.1015804
3. Allison, R.S.: The camera convergence problem revisited. In: Proc. SPIE Stereoscopic Displays and Virtual Reality Systems XI, vol. 5291, pp. 167–178 (2004). doi:10.1117/12.526278. http://www.cse.yorku.ca/percept/papers/Allison-Cameraconvergenceproblemrevisited.pdf
4. Allison, R.S.: Analysis of the influence of vertical disparities arising in toed-in stereoscopic cameras. J. Imaging Sci. Technol. **51**(4), 317–327 (2007). http://www.cse.yorku.ca/percept/papers/jistpaper.pdf
5. Allison, R.S., Rogers, B.J., Bradshaw, M.F.: Geometric and induced effects in binocular stereopsis and motion parallax. Vision Res. **43**, 1879–1893 (2003). doi:10.1016/S0042-6989(03)00298-0. http://www.cse.yorku.ca/percept/papers/Allison-Geometric_and_induced_effects.pdf
6. Barreto, J.P., Daniilidis, K.: Fundamental matrix for cameras with radial distortion. In: Proc. ICCV (2005). doi:10.1109/ICCV.2005.103. http://www.cis.upenn.edu/~kostas/mypub.dir/barreto05iccv.pdf
7. Bertalmio, M., Fort, P., Sanchez-Crespo, D.: Real-time, accurate depth of field using anisotropic diffusion and programmable graphics cards. In: Proc. 2nd Intl. Symp. on 3D Data Processing, Visualization and Transmission (3DPVT), pp. 767–773 (2004). doi:10.1109/TDPVT.2004.1335393
8. Blake, A., Bülthoff, H.: Does the brain know the physics of specular reflection? Nature **343**(6254), 165–168 (1990). doi:10.1038/343165a0
9. Blake, A., Bulthoff, H.: Shape from specularities: computation and psychophysics. Philos Trans R Soc London B Biol Sci **331**(1260), 237–252 (1991). doi:10.1098/rstb.1991.0012
10. Bleyer, M., Gelautz, M., Rother, C., Rhemann, C.: A stereo approach that handles the matting problem via image warping. In: Proc. IEEE Conf. on Computer Vision and Pattern Recognition (CVPR) (2009). doi:10.1109/CVPRW.2009.5206656. http://research.microsoft.com/pubs/80301/CVPR09_StereoMatting.pdf
11. Bordwell, D.: Coraline, cornered (2009). http://www.davidbordwell.net/blog/?p=3789. Accessed 10 Jun 2009, archived at http://www.webcitation.org/5nFFuaq3f

12. Campbell, F.W.: The depth of field of the human eye. J. Mod. Opt. **4**(4), 157–164 (1957). doi:10.1080/713826091
13. Chabert, C.F., Einarsson, P., Jones, A., Lamond, B., Ma, W.C., Sylwan, S., Hawkins, T., Debevec, P.: Relighting human locomotion with flowed reflectance fields. In: SIGGRAPH '06: ACM SIGGRAPH 2006 Sketches, p. 76. ACM, New York, NY, USA (2006). doi:10.1145/1179849.1179944. http://gl.ict.usc.edu/research/RHL/SIGGRAPHsketch_RHL_0610.pdf
14. Cheng, C.M., Lai, S.H., Su, S.H.: Self image rectification for uncalibrated stereo video with varying camera motions and zooming effects. In: Proc. IAPR Conference on Machine Vision Applications (MVA). Yokohama, Japan (2009). http://www.mva-org.jp/Proceedings/2009CD/papers/02-03.pdf
15. Criminisi, A., Blake, A., Rother, C., Shotton, J., Torr, P.H.: Efficient dense stereo with occlusions for new view-synthesis by four-state dynamic programming. Int. J. Comput. Vis. **71**(1), 89–110 (2007). doi:10.1007/s11263-006-8525-1
16. Debevec, P., Wenger, A., Tchou, C., Gardner, A., Waese, J., Hawkins, T.: A lighting reproduction approach to live-action compositing. ACM Trans. Graph. (Proc. ACM SIGGRAPH 2002) **21**(3), 547–556 (2002). doi:http://doi.acm.org/10.1145/566654.566614
17. Emoto, M., Niida, T., Okano, F.: Repeated vergence adaptation causes the decline of visual functions in watching stereoscopic television. J. Disp. Technol. **1**(2), 328–340 (2005). doi:10.1109/JDT.2005.858938. http://www.nhk.or.jp/strl/publica/labnote/pdf/labnote501.pdf
18. Fitzgibbon, A.W.: Simultaneous linear estimation of multiple view geometry and lens distortion. In: Proceedings of the IEEE Conference on Computer Vision and Pattern Recognition, vol. 1, pp. 125–132 (2001). doi:10.1109/CVPR.2001.990465. http://www.robots.ox.ac.uk/~vgg/publications/papers/fitzgibbon01b.pdf
19. Forsyth, D., Ponce, J.: Computer Vision: A Modern Approach. Prentice-Hall, New York (2003)
20. Fusiello, A., Trucco, E., Verri, A.: A compact algorithm for rectification of stereo pairs. Mach. Vis. Appl. **12**, 16–22 (2000)
21. Gorley, P., Holliman, N.: Stereoscopic image quality metrics and compression. In: Woods, A.J., Holliman, N.S., Merritt, J.O. (eds.) Proc. SPIE Stereoscopic Displays and Applications XIX, vol. 6803, p. 680305. SPIE (2008). doi:10.1117/12.763530. http://www.dur.ac.uk/n.s.holliman/Presentations/SDA2008_6803-03.PDF
22. Gosser, H.M.: Selected Attempts at Stereoscopic Moving Pictures and Their Relationship to the Development of Motion Picture Technology, 1852–1903. Ayer, Salem, NH (1977)
23. Hartley, R.: Theory and practice of projective rectification. Int. J. Comput. Vis. **35**, 115–127 (1999)
24. Hartley, R., Zisserman, A.: Multiple-View Geometry in Computer Vision. Cambridge University Press, Cambridge (2000)
25. Hasinoff, S.W., Kang, S.B., Szeliski, R.: Boundary matting for view synthesis. Comput. Vis. Image Underst. **103**(1), 22–32 (2006). doi:10.1016/j.cviu.2006.02.005. http://www.cs.toronto.edu/~hasinoff/pubs/hasinoff-matting-2005.pdf
26. Held, R.T., Cooper, E.A., O'Brien, J.F., Banks, M.S.: Using blur to affect perceived distance and size. In: ACM Trans. Graph. **29**(2), 1–16. ACM, New York, USA (2010). 0730–0301. doi:acm.org/10.1145/1731047.1731057
27. Hoffman, D.M., Girshick, A.R., Akeley, K., Banks, M.S.: Vergence–accommodation conflicts hinder visual performance and cause visual fatigue. J. Vis. **8**(3), 1–30 (2008). doi:10.1167/8.3.33
28. Hummel, R.: 3-D cinematography. American Cinematographer Manual, pp. 52–63. American Society of Cinematographers, Hollywood, CA (2008)
29. Jones, G.R., Holliman, N.S., Lee, D.: Stereo images with comfortable perceived depth. US Patent 6798406 (2004). http://www.google.com/patents?vid=USPAT6798406
30. Kakimoto, M., Tatsukawa, T., Mukai, Y., Nishita, T.: Interactive simulation of the human eye depth of field and its correction by spectacle lenses. Comput. Graph. Forum **26**(3), 627–636 (2007). doi:10.1111/j.1467-8659.2007.01086.x. http://nis-lab.is.s.u-tokyo.ac.jp/~nis/abs_eg.html

Stereoscopic Cinema

31. Kilner, J., Starck, J., Hilton, A.: A comparative study of free-viewpoint video techniques for sports events. In: Proc. 3rd European Conference on Visual Media Production, pp. 87–96. London, UK (2006). http://www.ee.surrey.ac.uk/CVSSP/VMRG/Publications/kilner06cvmp.pdf
32. Kooi, F.L., Toet, A.: Visual comfort of binocular and 3D displays. Displays 25(2–3), 99–108 (2004). doi:10.1016/j.displa.2004.07.004
33. Kozachik, P.: 2 worlds in 3 dimensions. American Cinematographer 90(2), 26 (2009)
34. Lambooij, M.T.M., IJsselsteijn, W.A., Heynderickx, I.: Visual discomfort in stereoscopic displays: a review. In: Proc. SPIE Stereoscopic Displays and Virtual Reality Systems XIV, vol. 6490 (2007). doi:10.1117/12.705527
35. Lin, H.Y., Gu, K.D.: Photo-realistic depth-of-field effects synthesis based on real camera parameters. In: Advances in Visual Computing (ISVC 2007), Lecture Notes in Computer Science, vol. 4841, pp. 298–309. Springer, Berlin (2007). doi:10.1007/978-3-540-76858-6_30
36. Lipton, L.: Foundations of the Stereoscopic Cinema. Van Nostrand Reinhold, New York (1982)
37. Lipton, L.: The stereoscopic cinema: from film to digital projection. SMPTE J. 586–593 (2001)
38. Loop, C., Zhang, Z.: Computing rectifying homographies for stereo vision. In: Proc. IEEE Conf. on Computer Vision and Pattern Recognition, vol. 1, pp. –131 Vol. 1 (1999). doi:10.1109/CVPR.1999.786928. http://research.microsoft.com/~zhang/Papers/TR99-21.pdf
39. Marcos, S., Moreno, E., Navarro, R.: The depth-of-field of the human eye from objective and subjective measurements. Vision Res. 39(12), 2039–2049 (1999). doi:10.1016/S0042-6989(98)00317-4
40. Masaoka, K., Hanazato, A., Emoto, M., Yamanoue, H., Nojiri, Y., Okano, F.: Spatial distortion prediction system for stereoscopic images. J. Electron. Imaging 15(1) (2006). doi:10.1117/1.2181178. http://www.nhk.or.jp/strl/publica/labnote/pdf/labnote505.pdf
41. Mendiburu, B.: 3D Movie Making: Stereoscopic Digital Cinema from Script to Screen. Focal, Oxford (2009)
42. Micusik, B., Pajdla, T.: Estimation of omnidirectional camera model from epipolar geometry. In: Proc. IEEE Conf. on Computer Vision and Pattern Recognition (CVPR), vol. 1, pp. 485–490 (2003). doi:10.1109/CVPR.2003.1211393. ftp://cmp.felk.cvut.cz/pub/cmp/articles/micusik/Micusik-CVPR2003.pdf
43. Ogle, K.N.: Induced size effect: I. A new phenomenon in binocular space perception associated with the relative sizes of the images of the two eyes. Am. Med. Assoc. Arch. Ophthalmol. 20, 604–623 (1938). http://www.cns.nyu.edu/events/vjclub/classics/ogle-38.pdf
44. Ogle, K.N.: Researches in Binocular Vision. Hafner, New York (1964)
45. Park, J.I., Um, G.M., Ahn, C., Ahn, C.: Virtual control of optical axis of the 3DTV camera for reducing visual fatigue in stereoscopic 3DTV. ETRI J. 26(6), 597–604 (2004)
46. Pastoor, S.: Human factors of 3DTV: an overview of current research at Heinrich-Hertz-Institut Berlin. In: IEE Colloquiumon Stereoscopic Television, pp. 11/1–11/4. London (1992). http://ieeexplore.ieee.org/xpls/abs_all.jsp?arnumber=193706
47. Read, J.C.A., Cumming, B.G.: Does depth perception require vertical-disparity detectors? J. Vis. 6(12), 1323–1355 (2006). doi:10.1167/6.12.1. http://journalofvision.org/6/12/1
48. Rogers, B.J., Bradshaw, M.F.: Vertical disparities, differential perspective and binocular stereopsis. Nature 361, 253–255 (1993). doi:10.1038/361253a0. http://www.cns.nyu.edu/events/vjclub/classics/rogers-bradshaw-93.pdf
49. Rogmans, S., Lu, J., Bekaert, P., Lafruit, G.: Real-time stereo-based view synthesis algorithms: a unified framework and evaluation on commodity gpus. Signal Process. Image Commun. 24(1–2), 49–64 (2009). doi:10.1016/j.image.2008.10.005. http://www.sciencedirect.com/science/article/B6V08-4TT34BC-3/2/03a35e44fbc3c0f7ff19fcdbe474f73a. Special issue on advances in three-dimensional television and video
50. Seuntiens, P., Meesters, L., Ijsselsteijn, W.: Perceived quality of compressed stereoscopic images: effects of symmetric and asymmetric JPEG coding and camera separation. ACM Trans. Appl. Percept. 3(2), 95–109 (2009). doi:10.1145/1141897.1141899
51. Sexton, I., Surman, P.: Stereoscopic and autostereoscopic display systems. IEEE Signal Process. Mag. 16(3), 85–99 (1999). doi:10.1109/79.768575

52. Sizintsev, M., Wildes, R.P.: Coarse-to-fine stereo vision with accurate 3D boundaries. Image Vis. Comput. **28**(3), 352 – 366 (2010). doi:10.1016/j.imavis.2009.06.008. http://www.cse.yorku.ca/techreports/2006/CS-2006-07.pdf

53. Smith, C., Benton, S.: reviews of Foundations of the stereoscopic cinema by Lenny Lipton. Opt. Eng. **22**(2) (1983). http://www.3dmagic.com/Liptonreviews.htm

54. Smolic, A., Kimata, H., Vetro, A.: Development of MPEG standards for 3D and free viewpoint video. Tech. Rep. TR2005-116, MERL (2005). http://www.merl.com/papers/docs/TR2005-116.pdf

55. Speranza, F., Stelmach, L.B., Tam, W.J., Glabb, R.: Visual comfort and apparent depth in 3D systems: effects of camera convergence distance. In: Proc. SPIE Three-Dimensional TV, Video and Display, vol. 4864, pp. 146–156. SPIE (2002). doi:10.1117/12.454900

56. Spottiswoode, R., Spottiswoode, N.L., Smith, C.: Basic principles of the three-dimensional film. SMPTE J. **59**, 249–286 (1952). http://www.archive.org/details/journalofsociety59socirich

57. Steele, R.M., Jaynes, C.: Overconstrained linear estimation of radial distortion and multi-view geometry. In: Proc. ECCV (2006). doi:10.1007/11744023_20

58. Stelmach, L., Tam, W., Meegan, D., Vincent, A., Corriveau, P.: Human perception of mismatched stereoscopic 3D inputs. In: International Conference on Image Processing (ICIP), vol. 1, pp. 5–8 (2000). doi:10.1109/ICIP.2000.900878

59. Stelmach, L.B., Tam, W.J., Speranza, F., Renaud, R., Martin, T.: Improving the visual comfort of stereoscopic images. In: Proc. SPIE Stereoscopic Displays and Virtual Reality Systems X, vol. 5006, pp. 269–282 (2003). doi:10.1117/12.474093

60. The Stereographics Developer's Handbook – Background on Creating Images for CrystalEyes® and SimulEyes®. (1997). http://www.reald-corporate.com/scientific/downloads/handbook.pdf

61. Stevenson, S.B., Schor, C.M.: Human stereo matching is not restricted to epipolar lines. Vision Res. **37**(19), 2717–2723 (1997). doi:10.1016/S0042-6989(97)00097-7

62. Sun, G., Holliman, N.: Evaluating methods for controlling depth perception in stereoscopic cinematography. In: Proc. SPIE Stereoscopic Displays and Applications XX, vol. 7237 (2009). doi:10.1117/12.807136. http://www.dur.ac.uk/n.s.holliman/Presentations/SDA2009-Sun-Holliman.pdf

63. Taguchi, Y., Wilburn, B., Zitnick, C.: Stereo reconstruction with mixed pixels using adaptive over-segmentation. In: Proc. IEEE Conf. on Computer Vision and Pattern Recognition (CVPR), pp. 2720–2727 (2008). doi:10.1109/CVPR.2008.4587691. http://research.microsoft.com/users/larryz/StereoMixedPixels_CVPR2008.pdf

64. Todd, J.T.: The visual perception of 3D shape. Trends Cogn. Sci. **8**(3), 115–121 (2004). doi:10.1016/j.tics.2004.01.006

65. Todd, J.T., Norman, J.F.: The visual perception of 3-D shape from multiple cues: are observers capable of perceiving metric structure? Percept. Psychophys. **65**(1), 31–47 (2003). http://app.psychonomic-journals.org/content/65/1/31.abstract

66. Ukai, K., Howarth, P.A.: Visual fatigue caused by viewing stereoscopic motion images: background, theories, and observations. Displays **29**(2), 106–116 (2007). doi:10.1016/j.displa.2007.09.004

67. Wang, L., Jin, H., Yang, R., Gong, M.: Stereoscopic inpainting: joint color and depth completion from stereo images. In: CVPR 2008 – IEEE Computer Socitey conference on Computer Vision and Pattern Recognition, pp. 1–8 (2008). doi:10.1109/CVPR.2008.4587704

68. Wann, J.P., Mon-Williams, M.: Health issues with virtual reality displays: what we do know and what we don't. ACM SIGGRAPH Comput. Graph. **31**(2), 53–57 (1997)

69. Wann, J.P., Rushton, S., Mon-Williams, M.: Natural problems for stereoscopic depth perception in virtual environments. Vision Res. **35**(19), 2731–2736 (1995). doi:10.1016/0042-6989(95)00018-U

70. Watt, S.J., Akeley, K., Ernst, M.O., Banks, M.S.: Focus cues affect perceived depth. J. Vis. **5**(10), 834–862 (2005). doi:10.1167/5.10.7. http://journalofvision.org/5/10/7/

71. Woodford, O., Reid, I.D., Torr, P.H.S., Fitzgibbon, A.W.: On new view synthesis using multiview stereo. In: Proceedings of the 18th British Machine Vision Conference, vol. 2, pp. 1120–1129. Warwick (2007). http://www.robots.ox.ac.uk/~ojw/stereo4nvs/Woodford07a.pdf

72. Woods, A., Docherty, T., Koch, R.: Image distortions in stereoscopic video systems. In: Proc. SPIE Stereoscopic Displays and Applications IV, vol. 1915, pp. 36–48. San Jose, CA (1993). doi:10.1117/12.157041. http://www.cmst.curtin.edu.au/publicat/1993-01.pdf
73. Wu, H.H.P., Chen, C.C.: Scene reconstruction pose estimation and tracking. In: Projective Rectification with Minimal Geometric Distortion, chap. 13, pp. 221–242. I-Tech Education and Publishing, Vienna (2007). http://intechweb.org/book.php?id=10
74. Xiong, W., Chung, H., Jia, J.: Fractional stereo matching using expectation-maximization. IEEE Trans. Pattern Anal. Mach. Intell. **31**(3), 428–443 (2009). doi:10.1109/TPAMI.2008.98. http://www.cse.cuhk.edu.hk/~leojia/all_final_papers/pami_stereo08.pdf
75. Xiong, W., Jia, J.: Stereo matching on objects with fractional boundary. In: IEEE Conference on Computer Vision and Pattern Recognition (CVPR) (2007). doi:10.1109/CVPR.2007.383194. http://www.cse.cuhk.edu.hk/~leojia/all_final_papers/alpha_stereo_cvpr07.pdf
76. Yamanoue, H., Okui, M., Okano, F.: Geometrical analysis of puppet-theater and cardboard effects in stereoscopic HDTV images. IEEE Trans. Circuits Syst. Video Technol. **16**(6), 744–752 (2006). doi:10.1109/TCSVT.2006.875213
77. Yano, S., Emoto, M., Mitsuhashi, T.: Two factors in visual fatigue caused by stereoscopic HDTV images. Displays **25**, 141–150 (2004). doi:10.1016/j.displa.2004.09.002
78. Yeh, Y.Y., Silverstein, L.D.: Limits of fusion and depth judgment in stereoscopic color displays. Hum. Factors **32**(1), 45–60 (1990)
79. Zhou, J., Li, B.: Rectification with intersecting optical axes for stereoscopic visualization. Proc. ICPR **2**, 17–20 (2006). doi:10.1109/ICPR.2006.986
80. Zitnick, C.L., Kang, S.B., Uyttendaele, M., Winder, S., Szeliski, R.: High-quality video view interpolation using a layered representation. In: Proc. ACM SIGGRAPH, vol. 23, pp. 600–608. ACM, New York, NY, USA (2004). doi:10.1145/1015706.1015766. http://research.microsoft.com/~larryz/ZitnickSig04.pdf
81. Zitnick, C.L., Szeliski, R., Kang, S.B., Uyttendaele, M.T., Winder, S.: System and process for generating a two-layer, 3D representation of a scene. US Patent 7015926 (2006). http://www.google.com/patents?vid=USPAT7015926
82. Zone, R.: 3-D fimmakers : Conversations with Creators of Stereoscopic Motion Pictures. The Scarecrow Fimmakers Series, No. 119. Scarecrow, Lanham, MD (2005)
83. Zone, R.: Stereoscopic Cinema and the Origins of 3-D Film, 1838–1952. University Press of Kentucky, Lexington, KY (2007)

Free-Viewpoint Television

Masayuki Tanimoto

Abstract Free-viewpoint TV (FTV) is an ultimate 3DTV that enables users to view a 3D world by freely changing their viewpoints as if they were there. FTV will bring an epochal change in the history of television since this function has not yet been achieved by conventional TV technology. FTV will greatly contribute to the development of industry, society, life and culture because FTV has an infinite number of viewpoints and hence it has very high sensing and representation capabilities. Tanimoto Lab of Nagoya University proposed the concept of FTV and constructed the world's first real-time system including the complete chain of operation from image capture to display. We also realized FTV on a single PC and FTV with free listening-point audio. FTV is a ray-based system rather than a pixel-based system. We are creating ray-based image engineering through the development of FTV. MPEG regards FTV as the most challenging 3D media and has been conducting the international standardization activities of FTV. The first phase of FTV was MVC (Multi-view Video Coding) and the second phase of FTV is 3DV (3D Video). MVC was completed in March 2009. 3DV is a standard that targets serving a variety of 3D displays.

1 Introduction

Television realized the human dream of seeing a distant world in real time and has served as the most important visual information technology to date. Now, television can provide us scenes overcoming distance not only in space but also in time by introducing storage devices into television. However, TV shows us the same scene even if we change our viewpoint in front of the display. This is quite different from what we experience in the real world. With TV, users can get only a single view of a 3D world. The view is determined not by users but by a camera placed in the 3D world. Although many important technologies have been developed, this function of TV has never changed.

M. Tanimoto
Graduate School of Engineering, Nagoya University Furo-cho, Chikusa-ku, Nagoya, 464-8603 Japan
e-mail: tanimoto@nuee.nagoya-u.ac.jp

R. Ronfard and G. Taubin (eds.), *Image and Geometry Processing for 3-D Cinematography*, Geometry and Computing 5, DOI 10.1007/978-3-642-12392-4_3, © Springer-Verlag Berlin Heidelberg 2010

Tanimoto Lab of Nagoya University has developed a new type of television named FTV (Free-viewpoint TV) [1–7]. FTV is an innovative visual media that enables us to view a 3D scene by freely changing our viewpoints as if we were there. FTV will bring an epochal change in the history of visual media since such a function has not yet been achieved by conventional media technology. FTV is based on the ray-space method [8–11].

The most essential element of visual systems is not the pixel but the ray. FTV is not a conventional pixel-based system but a ray-based system. We have been developing ray capture, processing, and display technologies for FTV [4].

We proposed the concept of FTV and verified its feasibility with the world's first real-time experiment [12, 13], in which a real, moving object was viewed using a virtual bird's-eye camera whose position was moved freely in a space and was controlled by a 3D mouse.

We have been developing technologies for FTV and realized FTV on a single PC and a mobile player. We also realized FTV with free listening-point audio [14].

FTV opens a new frontier in the field of signal processing since its process such as coding and image generation is performed in the ray-space, a new domain with higher dimensionality than possible with conventional TV. A new user interface is also needed for FTV to make full use of 3D information.

As the ultimate 3DTV, FTV needs higher performance in all related devices, equipment and systems, as well as accelerated development of the electronics industry. As the next-generation TV, it will find applications in the fields of communication, entertainment, advertising, education, medicine, art, archives and so on. As part of the social information infrastructure, it can increase the security of public facilities, roads, vehicles, schools, factories, etc.

In MPEG (Moving Picture Experts Group), FTV has been a framework since 2001 that allows viewing of a 3D world by freely changing the viewpoint. MPEG regarded FTV as the most challenging 3D media and started the international standardization activities of FTV. The first phase of FTV was MVC (Multi-view Video Coding) [15], which enables the efficient coding of multiple camera views. MVC was completed in March 2009. The second phase of FTV is 3DV (3D Video) [16]. 3DV is a standard that targets serving a variety of 3D displays.

In this chapter, the technologies, system and advanced ray technologies of FTV are presented. The progress of international standardization of FTV is also described.

2 Ray-Space Representation

We developed FTV based on ray-space representation [8–11]. In ray-space representation, one ray in the 3D real space is represented by one point in the ray space. The ray space is a virtual space. However, it is directly connected to the real space. The ray space is generated easily by collecting multi-view images while giving consideration to the camera parameters.

Let (x, y, z) be three space coordinates and (θ, ϕ) be the parameters of direction. A ray going through space can be uniquely parameterized by its location (x, y, z) and its direction (θ, ϕ); in other words, a ray can be mapped to a point in this 5D, ray-parameter space. In this ray-parameter space, we introduce the function f, whose value corresponds to the intensity of a specified ray. Thus, all the intensity data of rays can be expressed by

$$f(x, y, z; \theta, \phi). \qquad (1)$$
$$-\pi \leq \theta < \pi, -\pi/2 \leq \phi < \pi/2.$$

We call this ray-parameter space the "ray-space."

Although the 5D ray-space mentioned above includes all information viewed from any viewpoint, it is highly redundant due to the straight traveling paths of the rays. Thus, when we treat rays that arrive at a reference plane, we can reduce the dimension of the parameter space to 4D.

We use two types of ray-space for FTV. One is the orthogonal ray-space, where a ray is expressed by the intersection of the ray and the reference plane and the ray's direction as shown in Fig. 1 Another is the spherical ray-space, where the reference plane is set to be normal to the ray as shown in Fig. 2 The orthogonal ray-space is used for FTV with a linear camera arrangement, whereas the spherical ray-space is used for FTV with a circular camera arrangement. The linear camera arrangement is used for parallel view and the circular camera arrangement is used for convergent view as shown in Fig. 3.

Both the orthogonal ray-space and the spherical ray-space are 4D and 5D, including time. If we place cameras within a limited region, the obtained rays are limited, and the ray-space constructed from these rays is a subspace of the ray-space. For example, if we place cameras in a line or in a circle, we have only one part of the data of the whole ray-space. In such cases, we define a smaller ray-space.

For the linear camera arrangement, the ray-space is constructed by placing many camera images upright and parallel, as shown in Fig. 4, forming the FTV signal in this case. The FTV signal consists of many camera images, and the horizontal

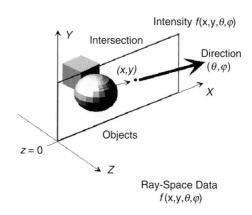

Fig. 1 Definition of orthogonal ray-space

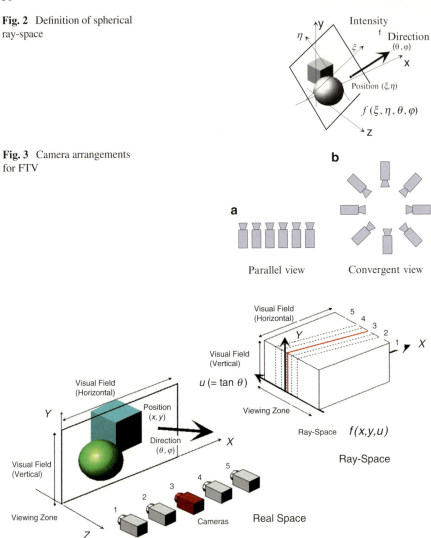

Fig. 2 Definition of spherical ray-space

Fig. 3 Camera arrangements for FTV

Fig. 4 Acquisition of orthogonal ray-space

cross-section has a line structure as shown in Fig. 5. The line structure of the ray-space is used for ray-space interpolation and compression. Vertical cross-sections of the ray-space give view images at the corresponding viewpoints as shown in Fig. 6.

For the circular camera arrangement, the spherical ray-space is constructed from many camera images, and its horizontal cross-section has a sinusoidal structure as shown in Fig. 7. The sinusoidal structure of the ray-space is also used for ray-space interpolation and compression.

Free-Viewpoint Television

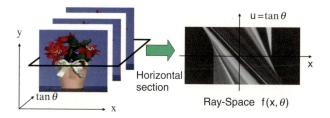

Fig. 5 Example of orthogonal ray-space

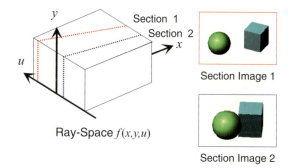

Fig. 6 Generation of view images

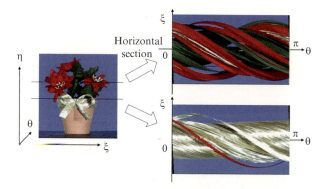

Fig. 7 Example of spherical ray-space

The hierarchical ray-space [10] is defined for scalable expression of 3D scene. Figure 8 shows the concept of the hierarchical ray-space. Figure 9 shows free viewpoint image generation at various distances using the hierarchical ray-space.

There are other representation methods of rays such as light field [17] and concentric mosaic [18]. The ray-space was proposed in 1994 [8]. On the other hand, the light field and the concentric mosaic were proposed in 1996 and 1999, respectively. The light field is the same as the orthogonal ray-space as explained in Fig. 10. The concentric mosaic is the same as the spherical ray-space. It is explained in Fig. 11. The spherical ray-space is obtained by arranging the orthographic images of Fig 11a

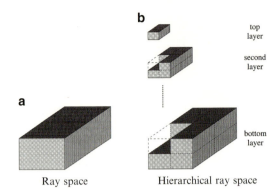

Fig. 8 Concept of hierarchical ray-space

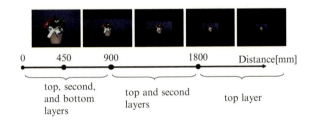

Fig. 9 Free viewpoint image generation at various distances using hierarchical ray-space

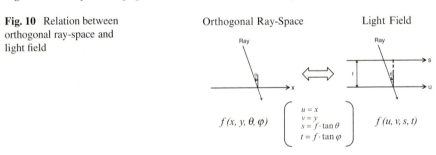

Fig. 10 Relation between orthogonal ray-space and light field

as shown in Fig. 11c It is the same as the concentric mosaic obtained by arranging the slit images of Fig 11b as shown in Fig. 11d. Therefore, the ray-space is a very important original method that includes the light field and the concentric mosaic.

3 FTV System

3.1 Configuration of FTV System

Figure 12 shows a basic configuration of the FTV system. At the sender side, a 3D scene is captured by multiple cameras. The captured images contain the misalignment and luminance differences of the cameras. They must be corrected to construct

Free-Viewpoint Television

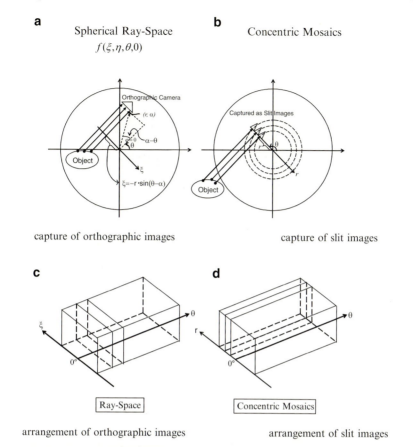

Fig. 11 Relation between spherical ray-space and concentric mosaics

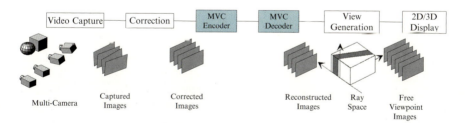

Fig. 12 Basic configuration of FTV system

the ray-space. The corrected images are compressed for transmission and storage by the MVC (Multi-view Video Coding) encoder.

At the receiver side, reconstructed images are obtained by the MVC decoder. The ray-space is constructed by arranging the reconstructed images and interpolating them. Free-viewpoint images are generated by cutting the ray-space vertically and are displayed on a 2D/3D display.

The function of FTV was successfully demonstrated by generating photo-realistic, free-viewpoint images of the moving scene in real time. Each part of the process shown in Fig 12 is explained in greater detail below.

3.2 Capture

We constructed a 1Darc capturing system shown in Fig. 13 for a real-time FTV system covering the complete chain of operation from video capture to display [19, 20]. It consists of 16 cameras, 16 clients and 1 server. Each client has one camera and all clients are connected to the server with Gigabit Ethernet.

A "100-camera system" has been developed to capture larger space by Nagoya University (Intelligent Media Integration COE and Tanimoto Laboratory) [21]. The system consists of one host-server PC and 100 client PCs (called 'nodes') that are equipped with JAI PULNiX TM-1400CL cameras. The interface between camera and PC is Camera-Link. The host PC generates a synchronization signal and distributes it to all of the nodes. This system is capable of capturing not only high-resolution video with 30 fps but also analog signals of up to 96 kHz. The specification of the 100-camera system is listed in Table 1.

The camera setting is flexible as shown in Fig. 14. MPEG test sequences "Rena" and "Akko & Kayo" shown in Fig. 15 were taken by camera arrangements (a) and (c), respectively.

Fig. 13 1D-arc capturing system

Table 1 Specification of 100-camera system

Image resolution	1392(H) × 1040(V)
Frame rate	29.4118 fps
Color	Bayer matrix
Synchronization	Less than 1 μs
Sampling rate of A/D	96 kS s^{-1} maximum
Maximum number of nodes	No limit (128 max for one sync output)

Fig. 14 100-camera system

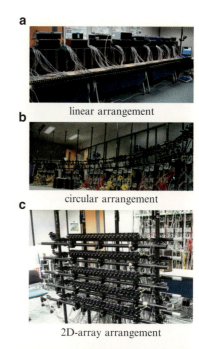

linear arrangement

circular arrangement

2D-array arrangement

Fig. 15 MPEG test sequences

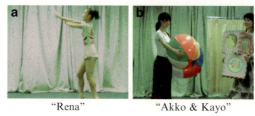

"Rena"   "Akko & Kayo"

3.3 Correction

The geometric correction [22, 23] and color correction [24] of multi-camera images are performed by measuring the correspondence points of images. This measurement is made once the cameras are set.

An example of the geometric correction is shown in Fig. 16 [23]. Here, the geometric distortion of a 2 dimensional camera array is corrected by the affine transform. It is seen that the trajectory of correspondence point becomes square by the geometric correction.

An example of color correction is shown in Fig. 17.

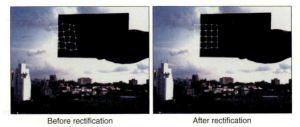

Fig. 16 An example of geometric correction

Fig. 17 An example of color correction

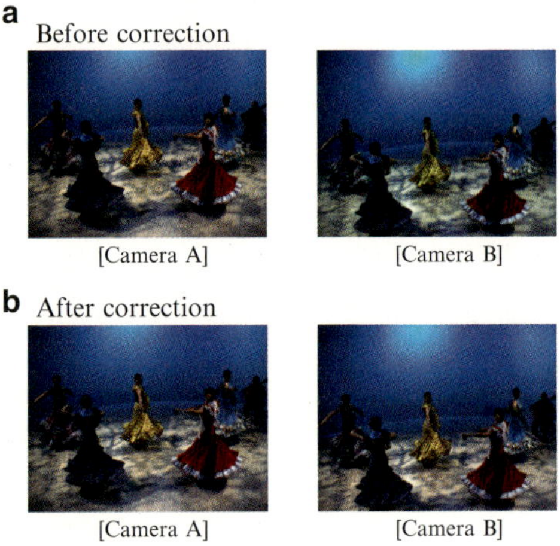

3.4 MVC Encoding and Decoding

An example of time and view variations of multi-view images is shown in Fig. 18. They have high temporal and interview correlations. MVC (Multi-view Video Coding) reduces these correlations [15, 25, 26]. The standardization of multi-camera image compression is progressing with MVC (Multi-view Video Coding) in MPEG. Details are described in Sect. 5.

3.5 View Generation

Ray-space is formed by placing the reconstructed images vertically and interpolating them. Free-viewpoint images are generated by making a cross-section of the ray-space.

Examples of the generated free-viewpoint images are shown in Fig 19. Complicated natural scenes, including sophisticated objects such as small moving fish, bubbles and reflections of light from aquarium glass, are reproduced very well.

Free-Viewpoint Television

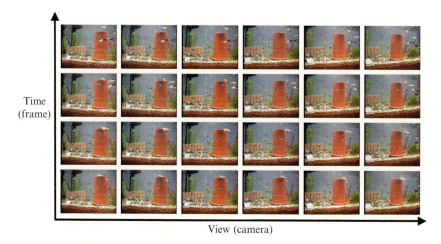

Fig. 18 Time and view variations of multi-view images

Fig. 19 An example of generated FTV images at various times and viewpoints

The quality of the generated view images depends on the view interpolation. The ray-space interpolation is achieved by detecting depth information pixel by pixel from the multi-view video. We proposed several interpolation schemes of the ray-space [27–32].

The free-viewpoint images were generated by a PC cluster in [19, 20]. Now, they can be generated by a single PC, and FTV on a PC can be accomplished in real time [31].

3.6 2D/3D Display

FTV needs a new user interface to display free-viewpoint images. Two types of display, 3D display and 2D/3D display with a viewpoint controller, are used for FTV as shown in Fig. 20.

Viewpoint control by head-tracking is shown here. Many head-tracking systems have been proposed using magnetic sensors, various optical markers, infrared cameras, retroreflective light from retinas, etc. Our head-tracking system uses only a

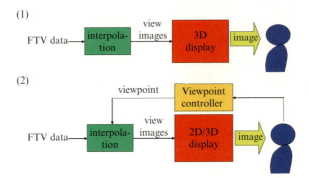

Fig. 20 Display of FTV

Fig. 21 2D display with eye tracking

conventional 2D camera and detects the position of a user's head by image processing. The user doesn't need to attach any markers or sensors.

In the user interface using a 2D display, the location of the user's head is detected with the head-tracking system and the corresponding view image is generated. Then, it is displayed on the 2D display as shown in Fig. 21.

Automultiscopic displays enable a user to see stereoscopic images without special glasses. However, there are two problems: a limited viewing zone and discreteness of motion parallax. Because the width of the viewing zone for each view approximates the interpupillary distance, the view image does not change with the viewer's movement within the zone. On the other hand, when the viewer moves over the zone, the view image changes suddenly.

In the user interface using the automultiscopic display, the function of providing motion parallax is extended by using the head-tracking system. The images fed to the system change according to the movement of the head position to provide small motion parallax, and the view channel for feeding the images is switched for handling large motion. This means that binocular parallax for the eyes is provided by automultiscopic display, while motion parallax is provided by head tracking and changing the image adaptively as shown in Fig. 22.

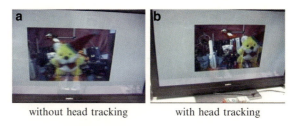

Fig. 22 3D display with and without head tracking

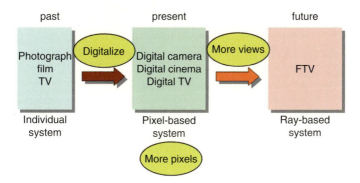

Fig. 23 Evolution of image systems

4 Creation of Ray-Based Image Engineering

4.1 Evolution of Image Systems

Figure 23 shows the evolution of image systems. In the past, image systems such as photography, film and TV were individual systems. At present, they are digitized and can be treated on the same platform as pixel-based systems. These pixel-based systems are developing toward using more pixels. This trend is exemplified by Super High-Definition TV [33]. Although Super HDTV has about 100 times the pixels of SDTV, there is still only one view.

In the future, the demand for more pixels will be saturated, and more views will be needed. This will result in the evolution from a pixel-based system to a ray-based system. We have been developing FTV according to this scenario. Roughly speaking, we can achieve SD-FTV by using the technologies of HDTV or Super HDTV and balancing the number of pixels and views.

4.2 Ray Capture and Display

We are developing ray-reproducing FTV to create ray-based image engineering. Ray-reproducing FTV consists of ray capture, ray processing and ray display.

Fig. 24 Mirror-scan ray capturing system

We developed a ray capturing system [34] shown in Fig. 24. It acquires a dense ray-space without interpolation in real time. In this capturing system, a high-speed camera and a scanning optical system are used instead of multiple cameras. The important feature of this configuration is that the spatial density of a multi-camera setup is converted to a time-density axis, i.e. the frame rate of the camera. This means that we can increase the density of the camera interval equivalently by increasing the frame rate of the camera. The scanning optical system is composed of a double-hyperbolic mirror shell and a galvanometric mirror. The mirror shell produces a real image of an object that is placed at the bottom of the shell. The galvanometric mirror in the real image reflects the image in the camera-axis direction. The reflected image observed from the camera varies according to the angle of the galvanometric mirror. This means that the camera can capture the object from various viewing directions that are determined by the angle of the galvanometric mirror. To capture the time-varying reflected images, we use a high-speed camera with an electronic shutter that is synchronized with the angle of the galvanometric mirror. We capture more than 100 view images within the reciprocation time of the galvanometric mirror. The collection of the view images is then mapped to the ray-space.

However, this system can capture the spherical ray-space with the viewing zone of only 55°.

Then, we have developed a 360° ray capturing system as shown in Fig. 25 [35, 36]. This system uses two parabolic mirrors. Incident rays that are parallel to the axis of a parabolic mirror gather at the focus of the parabolic mirror. Hence, rays that come out of an object placed at the focus of the lower parabolic mirror gather at the focus of the upper parabolic mirror. Then, the real image of the object is generated at the focus of the upper parabolic mirror and a rotating aslope mirror scans rays at the focus of the upper parabolic mirror. Finally, the image from the aslope mirror is captured by a high-speed camera. By using this system, we can capture all-around convergent views of an object as shown in Fig. 26.

Figure 27 shows SeeLINDER [37], a 360°, ray-producing display that allows multiple viewers to see 3D FTV images. Structure of the SeeLinder is shown in Fig. 28. It consists of a cylindrical parallax barrier and one-dimensional light-source arrays. Semiconductor light sources such as LEDs are aligned vertically for the one-dimensional light-source arrays. The cylindrical parallax barrier rotates quickly, and the light-source arrays rotate slowly in the opposite direction. If the aperture width of the parallax barrier is sufficiently small, the light going through the aperture

Free-Viewpoint Television

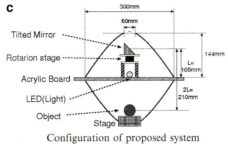

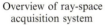

Overview of ray-space acquisition system

Ray-space acquisition system without upper parabolic mirror

Fig. 25 360° mirror-scan ray capturing system

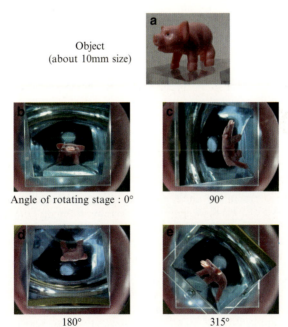

Fig. 26 Object and captured images of 360° ray capturing system

Fig. 27 The SeeLINDER, a 360° ray-reproducing display

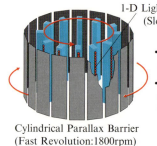

Fig. 28 Structure of the SeeLinder

becomes a thin flux, and its direction is scanned by the movement of the parallax barrier and the light-source arrays. By synchronously changing the intensity of the light sources with the scanning, pixels whose luminance differs for each viewing direction can be displayed. We can see the 3D image naturally, and the images have the strong depth cues of natural binocular disparity. When we move around the display, the image changes corresponding to our viewing position. Therefore, we perceive the objects just as if they were floating in the cylinder.

We are going to connect these two systems directly in real time.

Figure 29 shows the progress made in 3D capturing and display systems. In this figure, the ability of 3D capturing and display is expressed as a factor of the pixel-view product, defined as (number of pixels) × (number of views). In Fig. 29, ① denotes the factor of the 100-camera system mentioned earlier. We have also been developing new types of ray-capturing and display systems. Their factors are indicated by ② and ③; ② is a mirror-scan ray-capturing system [34] and ③ is the 360°, ray-reproducing SeeLINDER display [37].

In Fig. 29, the progress of space-multiplexing displays follows Moore's Law because it is achieved by miniaturization. The factor of the time-multiplexing display is larger than that of the space-multiplexing display. The difference is a result of time-multiplexing technology. The progress of capturing does not follow

Free-Viewpoint Television

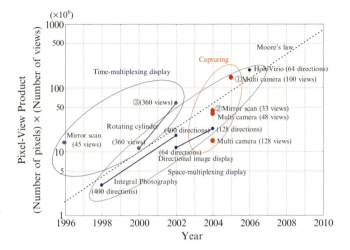

Fig. 29 Progress in increasing pixel-view product for 3D capturing and display

Fig. 30 Typical example of orthogonal ray-space and a horizontal cross-section

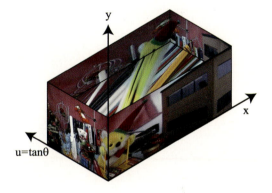

Moore's Law because it depends on camera resolution and the number of cameras used. Furthermore, the pixel-view product has increased very rapidly year after year in both capture and display. This development strongly supports our scenario.

4.3 Ray Processing

Typical example of orthogonal ray-space with a horizontal cross-section is shown in Fig. 30. The horizontal cross-section has a line structure. The slope of the line corresponds to the depth of object.

The ray-space is a platform of ray processing. Various kinds of signal processing can be done in the ray-space. Vertical cross-sections of the ray-space give real view

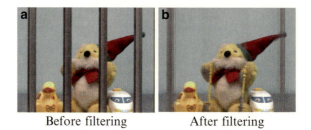

Before filtering   After filtering

Fig. 31 An example of ray-space processing: object elimination by non-linear filtering

Fig. 32 Scene composition by ray-space processing

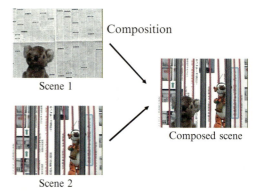

Fig. 33 Ray-space processing for scene composition

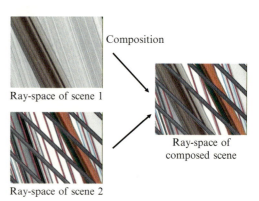

images at the corresponding viewpoints. Manipulation, division and composition of 3D scenes are also performed by ray-space processing.

Figure 31 shows an example of the ray-space processing. Bars in the scene of Fig. 31a are eliminated in Fig. 31b by applying non-linear filtering to the ray-space [38].

Composition of 2 scenes shown in Fig. 32 is performed by ray-space processing as shown in Fig. 33 [39].

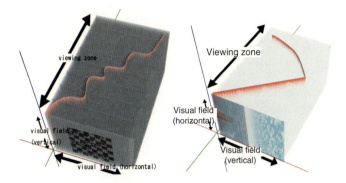

Fig. 34 Cutting ray-space with curved planes for image generation with optical effects

Images with optical effects are generated by cutting the ray-space with a curved plane as shown in Fig. 34. The shape of the curved plane is determined due to an optical effect to be realized. Artistic images shown in Fig. 35 are generated by cutting the ray-space with more general planes [40, 41].

5 International Standardization

We proposed FTV to MPEG in December 2001. Figure 36 shows the history of FTV standardization at MPEG. In the 3DAV (3D Audio Visual) group of MPEG, many 3D topics such as omni-directional video, FTV [42], stereoscopic video and 3DTV with depth disparity information were discussed. The discussion was converged on FTV in January 2004. Then, the standardization of the coding part of FTV started as MVC (Multi-view Video Coding) [15]. The MVC activity moved to the Joint Video Team (JVT) of MPEG and ITU for further standardization processes in July 2006. The standardization of MVC is based on H.264/MPEG4-AVC and was completed in March 2009. MVC was the first phase of FTV. FTV cannot be constructed by coding part alone. We proposed to standardize the entire FTV [43] and MPEG started a new standardization activity of FTV in April 2007 [44].

The function of view generation in Fig. 12 is divided into depth search and interpolation. As shown in Fig. 37, an FTV system can be constructed in various ways, depending on the location of depth search and interpolation. In case A, both depth search and interpolation are performed at the receiver side as in Fig. 12. In case B, depth search is performed at the sender side and interpolation is performed at the receiver side. In case C, both depth search and interpolation are performed at the sender side. Case B is suitable for the download/package and broadcast services since processing at the sender side is heavy and processing at the receiver side is light.

Case B was adopted by the FTV reference model [45] shown in Fig. 38. Possible standardization items are FTV data format, decoder and interpolation. Interpolation

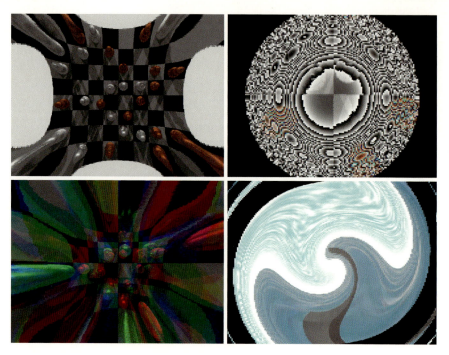

Fig. 35 Examples of artistic images generated by cutting the ray-space with more general planes

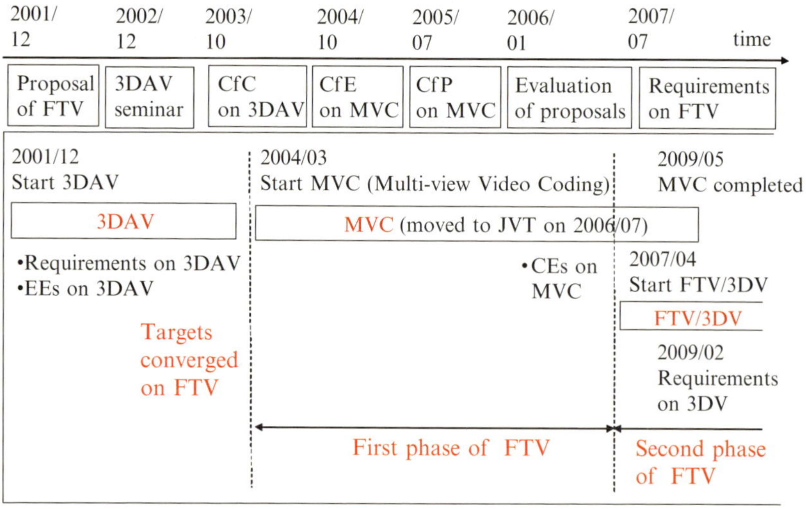

Fig. 36 History of FTV Standardization at MPEG

Free-Viewpoint Television

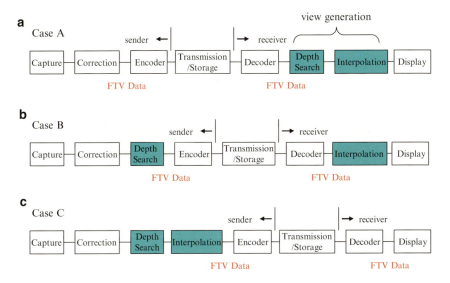

Fig. 37 3 cases of FTV configuration based on the positions of depth search and interpolation

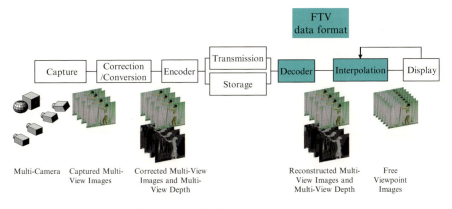

Fig. 38 FTV reference model and standardization items

might be informative. Thus, FTV is a new framework that includes a coded representation for multi-view video and depth information to support the generation of high-quality intermediate views at the receiver. This enables free viewpoint functionality and view generation for 2D/3D displays. Free-viewpoint images can be generated efficiently from multi-view images using multi-view depth information [46, 47].

In January 2008, MPEG-FTV targeted the standardization of 3DV (3D Video). 3DV is the second phase of FTV and a standard that targets serving for a variety of 3D displays. The introduction to 3DV is described in the reference [16].

3DV has two application scenarios [48]. One is enabling stereo devices to cope with varying display types and sizes, and different viewing preferences. This includes the ability to vary the baseline distance for stereo video to adjust the depth perception, which could help to avoid fatigue and other viewing discomforts. Another is support for high-quality auto-stereoscopic displays, such that the new standard enables the generation of many high-quality views from a limited amount of input data. These two scenarios can be realized by decoupling the capture and display using the view synthesis function of FTV.

Multi-view test sequences with depth have been collected and the reference softwares for depth estimation and view synthesis have been determined.

For further information on MPEG-FTV activity, please follow the link and subscribe MPEG-FTV. https://mailman.rwth-aachen.de/mailman/listinfo/mpeg-ftv

6 Conclusion

We have developed FTV that enables users to view a 3D scene by freely changing the viewpoints. FTV is the most advanced visual system and will bring an epochal change in the history of television.

FTV is an ultimate 3DTV that captures and transmits all visual information of a 3D scene, a natural interface between human and environment, and an innovative tool to create new types of content and art.

FTV will find many applications in the fields of broadcast, communication, amusement, entertainment, advertising, exhibition, education, medicine, art, archives, security, surveillance, and so on.

The most essential element of visual systems is ray. FTV is not a conventional pixel-based system but a ray-based system. FTV needs advanced technologies of ray capture, processing and display. We are creating ray-based image engineering through the development of FTV.

The international standardization of FTV has been conducted in MPEG. The introduction of FTV is not far because the interactive view generation of FTV has already been realized in real time. Therefore, we can enjoy FTV on a PC or a portable player if the standards of FTV are available and FTV contents are delivered over internet or with packaged media.

Acknowledgements This research is partially supported by Strategic Information and Communications R&D Promotion Programme (SCOPE) of the Ministry of Internal Affairs and Communications, and National Institute of Information and Communications Technology, Japan (NICT).

References

1. Tanimoto, M.: Free viewpoint television. J. Three Dimensional Images **15**(3), 17–22 (2001) (in Japanese)
2. Tanimoto, M.: Free viewpoint television – FTV. In: Picture Coding Symposium 2004, Special Session 5, December 2004
3. Tanimoto, M.: FTV (free viewpoint television) creating ray-based image engineering. In: Proc. of ICIP2005, pp. II-25–II-28, September 2005

4. Tanimoto, M.: Overview of free viewpoint television. Signal Proces. Image Commun. **21**(6), 454–461 (2006)
5. Tanimoto, M.: Free viewpoint television. In: OSA Topical Meeting on Digital Holography and Three-Dimensional Imaging, DWD2, June 2007 (Invited Paper)
6. Tanimoto, M.: FTV (free viewpoint TV) and creation of ray-based image engineering. ECTI Trans. Electr. Eng., Electron. Commun. **6**(1), 3–14 (2008) (Invited Paper)
7. Tanimoto, M.: FTV (free viewpoint TV) and ray-space technology. IBC2008, The Cutting Edge, Part Two, September 2008
8. Fujii, T.: A basic study on integrated 3-D visual communication. Ph.D dissertation in engineering, The University of Tokyo (1994) (in Japanese)
9. Fujii, T., Kimoto, T., Tanimoto, M.: Ray space coding for 3D visual communication. In: Picture Coding Symposium 1996, pp. 447–451, March 1996
10. Tanimoto, M., Nakanishi, A., Fujii, T., Kimoto, T.: The hierarchical ray-space for scalable 3-D image coding. In: Picture Coding Symposium 2001, pp. 81–84, April 2001
11. Fujii, T., Tanimoto, M.: Free-viewpoint television based on the ray-space representation. In: Proc. SPIE ITCom 2002, pp. 175–189, August 2002
12. Sekitoh, M., Toyota, K., Fujii, T., Kimoto, T., Tanimoto, M.: Virtual bird's-eye view system based on real image. EVA 2000 Gifu, 8, pp. 8-1–8-7, October 2000
13. Sekitoh, M., Fujii, T., Kimoto, T., Tanimoto, M.: Bird's eye view system for ITS. In: IEEE, Intelligent Vehicle Symposium, pp. 119–123, May 2001
14. Tehrani, M.P., Niwa, K., Fukushima, N., Hirano, Y., Fujii, T., Tanimoto, M., Takeda, K., Mase, K., Ishikawa, A., Sakazawa, S., Koike, A.: 3DAV integrated system featuring arbitrary listening-point and viewpoint generation. In: Proc. of IEEE Multimedia Signal Processing, MMSP 2008, PID-213, pp. 855–860 (October 2008). (Best paper award)
15. Introduction to multi-view video coding. ISO/IEC JTC 1/SC 29/WG11, N7328, July 2005. http://www.chiariglione.org/mpeg/technologies/mp-mv/index.htm
16. Introduction to 3D video. ISO/IEC JTC1/SC29/WG11 N9784, May 2008. http://www.chiariglione.org/mpeg/technologies/mp3d/index.htm
17. Levoy, M., Hanrahan, P.: Light field rendering. In: Proc. SIGGRAPH (ACM Trans. Graphics), pp. 31–42, August 1996
18. Shum, H.Y., He, L.W.: Rendering with concentric mosaics. In: Proc. SIGGRAPH (ACM Trans. Graphics), pp. 299–306, August 1999
19. Na Bangchang, P., Fujii, T., Tanimoto, M.: Experimental system of free viewpoint television. In: Proc. IST/SPIE Symposium on Electronic Imaging, vol. 5006–66, pp. 554–563, Jan. 2003
20. Na Bangchang, P., Tehrani, M.P., Fujii, T., Tanimoto, M.: Realtime system of free viewpoint television. J. Inst. Image Inform. Televis. Eng. (ITE) **59**(8), 63–701 (2005)
21. Fujii, T., Mori, K., Takeda, K., Mase, K., Tanimoto, M., Suenaga, Y.: Multipoint measuring system for video and sound: 100-camera and microphone system. In: IEEE 2006 International Conference on Multimedia & Expo (ICME), pp. 437–440, July 2006
22. Matsumoto, K., Yendo, T., Fujii, T., Tanimoto, M.: Multiple-image rectification for FTV. In: Proc. of 3D Image Conference 2006, P-19, pp. 171–174, July 2006
23. Fukushima, N., Yendo, T., Fujii, T., Tanimoto, M.: A novel rectification method for two-dimensional camera array by parallelizing locus of feature points. In: Proc. of IWAIT2008, B5-1, January 2008
24. Yamamoto, K., Yendo, T., Fujii, T., Tanimoto, M.: Colour correction for multiple-camera system by using correspondences. J. Inst. Image Inform. Televis. Eng. **61**(2), 213–222 (2007)
25. He, Y., Ostermann, J., Tanimoto, M., Smolic, A.: Introduction to the special section on multiview video coding. IEEE Trans. Circ. Syst. Video Tech. **17**(11), 1433–1435 (2007)
26. Yamamoto, K., Kitahara, M., Yendo, T., Fujii, T., Tanimoto, M., Shimizu, S., Kamikura, K., Yashima, Y.: Multiview video coding using view interpolation and color correction. IEEE Trans. Circ. Syst. Video Tech. **17**(11), 1436–1449 (2007)
27. Nakanishi, A., Fujii, T., Kimoto, T., Tanimoto, M.: Ray-space data interpolation by adaptive filtering using locus of corresponding points on epipolar plane image. J. Inst. Image Inform. Televis. Eng. (ITE) **56**(8), 1321–1327 (2002)

28. Droese, M., Fujii, T., Tanimoto, M.: Ray-space interpolation based on filtering in disparity domain. In: Proc. 3D Conference 2004, pp. 213–216, Tokyo, Japan, June 2004
29. Droese, M., Fujii, T., Tanimoto, M.: Ray-Space Interpolation Constraining Smooth Disparities Based On Loopy Belief Propagation. In: Proc. of IWSSIP 2004, pp. 247–250, Poznan, Poland, Sept. 2004
30. Fukushima, N., Yendo, T., Fujii, T., Tanimoto, M.: Real-time arbitrary view interpolation and rendering system using ray-space. In: Proc. SPIE Three-Dimensional TV, Video, and Display IV, vol. 6016, pp. 250–261, Nov. 2005
31. Fukushima, N., Yendo, T., Fujii, T., Tanimoto, M.: An effective partial interpolation method for ray-space. In: Proc. of 3D Image Conference 2006, pp. 85–88, July 2006
32. Fukushima, N., Yendo, T., Fujii, T., Tanimoto, M.: Free viewpoint image generation using multi-pass dynamic programming. In: Proc. SPIE Stereoscopic Displays and Virtual Reality Systems XIV, vol. 6490, pp. 460–470, Feb. 2007
33. Sugawara, M., Kanazawa, M., Mitani, K., Shimamoto, H., Yamashita, T., Okano, F.: Ultrahigh-definition video system with 4000 scanning lines. SMPTE Mot. Imag. J. **112**, 339–346 (2003)
34. Fujii, T., Tanimoto, M.: Real-time ray-space acquisition system. SPIE Electron. Imaging **5291**, 179–187 (2004)
35. Manoh, K., Yendo, T., Fujii, T., Tanimoto, M.: Ray-space acquisition system of all-around convergent views using a rotation mirror. In: Proc. of SPIE, vol. 6778, pp. 67780C-1–8, September 2007
36. Fujii, T., Yendo, T., Tanimoto, M.: Ray-space transmission system with real-time acquisition and display. In: Proc. of IEEE Lasers and Electro-optics Society Annual Meeting 2007, pp. 78–79, October 2007
37. Endo, T., Kajiki, Y., Honda, T., Sato, M.: Cylindrical 3-D video display observable from all directions. In: Proc. of Pacific Graphics 2000, pp. 300–306, October 2000
38. Takano, R., Yendo, T., Fujii, T., Tanimoto, M.: Scene separation in ray-space. In: Proc. of IMPS2005, pp. 31–32, November 2005
39. Takano, R., Yendo, T., Fujii, T., Tanimoto, M.: Scene separation and synthesis processing in ray-space. In: Proc. of IWAIT2007, P6–23, pp. 878–883, January 2007
40. Chimura, N., Yendo, T., Fujii, T., Tanimoto, M.: New visual arts by processing ray-space. In: Proc. of Electronic Imaging & the Visual Arts (EVA) 2007 Florence, pp. 170–175, March 2007
41. Chimura, N., Yendo, T., Fujii, T., Tanimoto, M.: Image generation with special effects by deforming ray-space. In: Proc. of NICOGRAPH, S1–4, November 2007
42. Tanimoto, M., Fujii, T.: FTV—free viewpoint television. ISO/IEC JTC1/SC29/WG11 M8595, July 2002
43. Tanimoto, M., Fujii, T., Kimata, H., Sakazawa, S.: Proposal on requirements for FTV. ISO/IEC JTC1/SC29/WG11, M14417, April 2007
44. Applications and requirements on FTV. ISO/IEC JTC1/SC29/WG11, N9466, October 2007
45. AHG on FTV (free viewpoint television). ISO/IEC JTC1/SC29/WG11 N8947, April 2007
46. Tanimoto, M., Fujii, T., Suzuki, K.: Experiment of view synthesis using multi-view depth. ISO/IEC JTC1/SC29/WG11, M14889, October 2007
47. Mori, Y., Fukushima, N., Yendo, T., Fujii, T., Tanimoto, M.: View generation with 3D warping using depth information for FTV. Signal Process. Image Commun. **24**(1–2) 65–72 (2009)
48. Applications and requirements on 3D video coding. ISO/IEC JTC 1/SC 29/WG11, N11061, October 2009

Free-Viewpoint Video for TV Sport Production

Adrian Hilton, Jean-Yves Guillemaut, Joe Kilner, Oliver Grau, and Graham Thomas

Abstract Free-viewpoint video in sports TV production presents a challenging problem involving the conflicting requirements of broadcast picture quality with video-rate generation of novel views, together with practical problems in developing robust systems for cost effective deployment at live events. To date most multiple view video systems have been developed for studio applications with a fixed capture volume, controlled illumination and backgrounds. Live outdoor events such as sports present a number of additional challenges for both acquisition and processing. Multiple view capture systems in sports such as football must cover the action taking place over an entire pitch with video acquisition at sufficient resolution for analysis and production of desired virtual camera views. In this chapter we identify the requirements for broadcast production of free-viewpoint video in sports and review the state-of-the-art. We present the *iview* free-viewpoint video system which enables production of novel camera views of sports action for use in match commentary, for example the referees viewpoint. Automatic online calibration, segmentation and reconstruction is performed to allow rendering of novel viewpoints from the moving match cameras. Results are reported of production trials in football (soccer) and rugby which demonstrate free-viewpoint video with a visual quality comparable to the broadcast footage.

1 Introduction

Multiple view video systems are increasingly used in sports broadcasts for production of novel *free-viewpoint* camera views and annotation of the game play. This chapter identifies the requirements for free-viewpoint video in sports TV broadcast

---

A. Hilton (✉), J.-Y. Guillemaut, and J. Kilner
University of Surrey, Surrey, UK
e-mail: a.hilton@surrey.ac.uk, j.guillemaut@surrey.ac.uk, j.kilner@surrey.ac.uk

O. Grau and G. Thomas
BBC R&D, London, UK
e-mail: oliver.grau@bbc.co.uk, graham.thomas@bbc.co.uk

R. Ronfard and G. Taubin (eds.), *Image and Geometry Processing for 3-D Cinematography*, Geometry and Computing 5, DOI 10.1007/978-3-642-12392-4_4,
© Springer-Verlag Berlin Heidelberg 2010

production, reviews previous work in relation to the requirements and presents a system designed to meet these requirements. Free-viewpoint video in sports TV production is a challenging problem involving the conflicting requirements of broadcast picture quality with video-rate generation. Further practical challenges exist in developing systems which can be cost effectively deployed at live events. To date most multiple view video systems have been developed for studio applications with a fixed capture volume, controlled illumination and backgrounds. Live outdoor events such as sports present a number of additional challenges for both acquisition and processing. Multiple view capture systems in sports such as football must cover the action taking place over an entire pitch with video acquisition at sufficient resolution for analysis and production of desired virtual camera views. The *iview* system presented in this chapter is based on use of the live broadcast cameras as the primary source of multiple view video. In a conventional broadcast for events such as premier league football these cameras are manually operated to follow the game play zooming in on events as they occur. Advances are presented in real-time through the lens camera calibration to estimate both the camera pose, focus and lens distortion from the pitch marks. Free-viewpoint video is then produced starting with a volumetric reconstruction followed by a view-dependent refinement using information from multiple views. Results are presented from production trials at international soccer and rugby sports events which show high-quality virtual camera views. This article is intended to be of benefit to the research community in defining broadcast industry requirements and identifying open problems for future research.

There is a demand from the broadcast industry for more flexible production technologies at live events such as sports. Currently cameras are operated manually with physical restrictions on viewpoint and sampling rate. The ability to place *virtual* cameras at any location around or on the pitch is highly attractive to broadcasters greatly increasing flexibility in production and enabling novel delivery formats such as mobile TV. Examples of the type of physically impossible camera views which could be desirable are the goal keepers view, a player tracking camera or even a ball camera. Virtual Replay[1] is an example of an existing technology using synthetic graphics and manual match annotation derived from broadcast footage of premier league soccer which allows viewers to select a virtual camera view. Manual annotation is labour intensive and the resulting virtual replay using generic player models and approximate movements is far from realistic. Ultimately the challenge is to achieve the flexibility of synthetic graphics with the visual-quality of broadcast TV in an automated approach.

This article presents the *iview*[2] system which is being developed to address the requirements of broadcast production. The system primarily utilises footage from match cameras used for live TV broadcast. Match cameras provide wide-baseline views and are manually controlled to cover the play. Typically major sporting events will have 12–18 manually operated high-definition cameras in the stadium.

---

[1] Virtual Replay, http://www.bbc.co.uk/virtualreplay.

[2] iview, http://www.bbc.co.uk/rd/iview.

However, only a fraction of these will be viewing specific events of interest for production of free-viewpoint renders to enhance the commentary, other cameras are used for coverage of the pitch, crowd, coaches and close-up shots of players. Robust algorithms are required for both recovery of camera calibration from the broadcast footage and wide-baseline correspondence between views for reconstruction or view interpolation. At major sports events direct access to match cameras is limited as they are often operated by a third-party. Therefore *iview* uses a method for real-time camera calibration from the match footage [45]. Player segmentation is performed using chroma-key and difference matting techniques independently for each camera view [15]. Automatic calibration and player segmentation for moving broadcast cameras results in errors of the order of 1–3 pixels which is often comparable to the size of players arms and legs in the broadcast footage. Robust reconstruction and rendering of novel viewpoints is achieved in the *iview* system by an initial conservative visual-hull reconstruction followed by a view-dependent refinement. View-dependent refinement simultaneously refines the player reconstruction and segmentation exploiting visual cues from multiple camera views. This achieves free-viewpoint rendering with pixel accurate alignment of neighbouring views to render novel views with a visual quality approaching that of the source video. Results of the current *iview* system in production trials are presented. Limitations of the existing approaches are identified with respect to the requirements of broadcast production. This leads to conclusions on future research advances required to achieve free-viewpoint technology which is exploited by the broadcast industry for use in production.

2 Background

Over the past decade there has been significant interest in the interpolation of novel viewpoints from multiple camera views. Recently research in this field for novel viewpoint interpolation from multiple view image sequences has been referred to as *video-based rendering, free-viewpoint video* or *3DTV*. In this section we review the principal methodologies for synthesis of novel views with a focus on the application to view-interpolation in sports. We first review the state-of-the-art in studio based systems where multiple cameras are used in an environment with controlled illumination and/or backgrounds. Studio systems have been the primary focus of both research and film production to capture a performance in a limited volume and allow playback of novel view sequences as free-viewpoint video. The extension of multiple camera systems to sports TV production requires capture of much larger volumes with relatively uncontrolled illumination and backgrounds. Here we review the current state-of-the-art in sports production of live events such as football and identify the limitations of existing approaches in the context of TV production.

2.1 Methodologies for Free-Viewpoint Video

Three principal methodologies have been investigated to rendering novel views of scenes captured from two or more camera viewpoints: interpolation; reconstruction; and tracking.

Interpolation: View interpolation directly estimates the scene appearance from novel viewpoints without explicitly reconstructing the 3D scene structure as an intermediate step [5, 38]. Interpolation between camera views requires the 2D-to-2D image correspondence. Given the correspondence image morphing and blending techniques can be used to interpolate intermediate views [38]. Commonly multiple view epipolar and tri-focal constraints on visual geometry are used in the estimation of correspondence to reduce complexity and ambiguities which implicitly exploit the 3D scene structure. Both sparse feature based and dense stereo techniques have been used to estimate correspondence for view interpolation. View interpolation avoids the requirement for explicit 3D reconstruction but is in general limited to rendering viewpoints between the camera views. This has the advantage of circumventing inaccuracies in explicit reconstruction due to errors in camera calibration. The quality of rendered views is dependent on the accuracy of correspondence to align multiple view observations. Extrapolation of novel views has also been investigated based on the colour consistency of observations from multiple views without explicit reconstruction [24]. A comprehensive survey of image-based rendering techniques for novel view synthesis is given in [40]. Recent research [44] has demonstrated perceptually convincing rendering of novel viewpoints for dynamic scenes by purely image-based interpolation between real camera views.

Reconstruction: Reconstruction of the 3D scene structure from multiple view images is commonly used as a basis for rendering novel views. The intermediate reconstruction of scene structure provides a geometric proxy which can be used to combine observations of scene appearance from multiple views in order to render images from novel viewpoints. The visual geometry of multiple views is now well understood [21]. Given multiple views of a dynamic scene such as a moving person a number of approaches have been used for reconstruction: visual-hull; photo-hull; stereo; and global shape optimisation. Visual-hull reconstruction intersect silhouette cones from multiple views [10, 29–31, 34] to reconstruct the maximal volume occupied by the scene objects. This requires prior segmentation of the foreground scene objects, such as a person, from the background. The photo-hull [39] is the maximal photo consistent volume between multiple views. Given the visual-hull as the maximum occupied volume the photo-hull is a sub-volume. An advantage of the photo-hull is that it does not require prior segmentation of the foreground. The photo-hull relies on the availability of a photo-consistency measure to distinguish different surface regions, in real scenes this may result in poor results. To overcome limitations of the visual and photo hull approaches a number of probabilistic approaches to volumetric reconstruction from multiple views have recently been introduced [3, 11]. A survey of volumetric reconstruction approaches with further details can be found in [41]. Both the visual and photo hull are maximal approximations of the scene structure and therefore commonly result in significant visual artifacts in rendering novel views due to inaccurate geometry and resulting mis-alignment of the observed images. Stereo correspondence has been used for surface reconstruction from image pairs, a survey of approaches is given in [36]. Stereo reconstruction from multiple views has been used to reconstruct dynamic scenes of moving people [25, 48]. Dense correspondence from stereo ensures reconstruction of a surface which align the multiple view images reducing artifacts in rendering of novel views. However, stereo correspondence requires local variation in appear-ance across the scene surface and is ambiguous in regions of uniform appearance. To overcome this limitation research has investigated the combination of volumet-ric and stereo reconstruction in a global optimisation framework to ensure robust reconstruction in areas of uniform appearance and accurate alignment of images from multiple views [32, 42]. A comparison of approaches for reconstruction of static scenes from multiple views is presented in [37].

Tracking: An alternative approach to rendering novel viewpoints of a known scene such as a moving person is model-based 3D tracking as proposed for free-viewpoint video [4]. Model-based 3D tracking of human motion from multiple view video has received considerable attention in the field of computer vision [33]. A generic 3D humanoid model approximating the surface shape and underlying skeletal structure is aligned with the image observations to estimate the pose at each frame. Typically an analysis-by-synthesis framework is used to optimise the pose estimate at each frame which minimises the re-projection error of the 3D humanoid model with the multiple view image observations. State-of-the-art multi-ple view 3D tracking of human motion [4,8] achieves reconstruction of gross human pose but does not accurately reconstruct detailed movement such as hand rotations. Given the estimated pose sequence the model can be rendered from an arbitrary viewpoint. Shape reconstruction and observed appearance of the subject over time can be used to render realistic free-viewpoint video of the subjects performance. Model-based 3D tracking has the advantage over direct reconstruction of providing a compact structured representation. However, the shape reconstruction is limited by the constraints of the prior model. Recent approaches [1, 47] have exploited mod-els constructed from examples of the performer to more accurately represent body shape and incorporate loose clothing and long-hair. Current 3D human tracking systems from video are limited to individual people in a controlled studio environ-ment with high-quality imagery. If these limitations are overcome this approach has the potential to provide an alternative for free-viewpoint rendering in more general scenes such as sports with multiple players and inter-player occlusion.

Given multiple camera views of a scene containing multiple moving people 3D scene reconstruction is commonly used to provide an explicit intermediate representation for rendering novel views. In subsequent sections we review the state-of-the-art in both studio and outdoor reconstruction with respect to the requirements of free-viewpoint video.

2.2 Studio Production of Free-Viewpoint Video

Over the past decade there has been extensive research in multiple camera systems for reconstruction and representations of dynamic scenes. Following the pioneering work of Kanade et al. [26] introducing Virtualized Reality^{TM} there has been extensive research on acquisition of performances to allow replay with interactive control of a virtual camera viewpoint or *free-viewpoint video*. This system used 51 cameras over a 5-m hemisphere to capture an actors performance. Reconstruction is performed by fusion of stereo surface reconstruction from multiple pairs of views. Novel viewpoints are then rendered by texture mapping the reconstructed surface. Other multiple camera studio systems with small numbers of cameras (6–12) have used the visual-hull from multiple view silhouettes [30, 35] and photo-hull [46]. Real-time free-viewpoint video with interactive viewpoint control has also been demonstrated [12, 30]. An advantage of these volumetric approaches over stereo correspondence is the use of widely spaced views reducing the number of cameras required. However, due to visual ambiguity with a limited number of views both visual and photo-hull approaches are susceptible to phantom volumes and extraneous protrusions from the true scene surface. All of these approaches typically result in a reduction of visual quality relative to the camera image sequences in rendering novel views. This is due to the approximate scene geometry which results in misalignment of images between views.

A number of subsequent approaches have exploited temporal consistency in surface reconstruction resulting in improved visual quality by removing spurious artifacts. Volumetric scene flow [46] was introduced to enforce photo-metric consistency over time in surface reconstruction. Spatio-temporal volumetric surface reconstruction from multiple view silhouettes using level-sets to integrate information over time has also been investigated [13]. Incorporation of temporal information improves the visual quality by ensuring smooth variation in surface shape over time.

Recent advances have achieved offline production of free-viewpoint video with a visual quality comparable to the captured video. Zitnick et al. [48] presented high-quality video-based rendering using integrated stereo reconstruction and matting with a 1D array of eight cameras over a 30° arc. Results demonstrate video-quality rendering comparable to the captured video for novel views along the 30° arc between cameras. High-quality rendering for all-round 360° views has also been demonstrated for reconstruction from widely spaced views (eight cameras with 30–45° between views) using global surface optimisation techniques which integrate silhouette constraints with wide-baseline stereo [32, 42, 43]. This approach refines an initial visual-hull reconstruction to obtain a surface which gives accurate alignment between widely spaced views. As an alternative to refinement of geometry reconstruction direct texture correspondence and interpolation using optical flow techniques has been used to correct misalignment due to errors in geometry and achieve high-quality rendering on relatively low-resolution models [9]. This approach could be advantageous in the presence of camera calibration error as it avoids the requirement for a single surface reconstruction which is consistent between all views.

In film production rigs of high-resolution digital cameras have been used to produce high-quality virtual camera sweeps at a single time instant, as popularised by the *bullet-time* shots in the Matrix. These systems typically use very narrow baseline configurations (approximately 3° spacing) of hundreds of cameras. Semi-automatic tools are used in post-production for interpolation between camera views to render a continuous free-viewpoint camera move. This technology is now in wide-spread use for film and broadcast advertising with acquisition of both studio and outdoor scenes, such as dolphins jumping, to produce film quality sequences (http://www.timeslicefilms.com, http://www.digitalair.com). Virtual views are limited to interpolating along pre-planned paths along which the cameras are positioned. This restricts the general use of current technologies used in film production for free-viewpoint video.

2.3 Free-Viewpoint Video in Sports and Outdoor Scenes

Transfer of studio technology to sports events requires acquisition over a much larger area with uncontrolled conditions such as illumination and background. Studio systems typically have all cameras focused on a fixed capture volume together with constant illumination and background appearance. To date the majority of studio systems have used fixed cameras with similar focal lengths. Achieving broadcast quality free-viewpoint video of live events such as sports taking place over large areas presents a significant additional challenge.

Initial attempts have been made to transfer studio-based reconstruction methodologies to acquisition and reconstruction of outdoor events. The Virtualized Reality™ technology [26] was used in the EyeVision[3] system to produce virtual camera sweeps as action replays for Super Bowl XXXV in 2001. In this system the pan, tilt and zoom of more than 30 cameras spaced around the stadium were slaved to a single manually controlled camera to capture the same event from different viewpoints. Switching between cameras was used in the Super Bowl television broadcast to produce sweep shots with visible jumps between viewpoints.

More recently a number of groups have investigated volumetric [16] and image-based interpolation techniques [6, 23, 28] for rendering novel views in sports. Grau et al. [16] used a texture mapped visual-hull reconstruction obtained from multiple view silhouettes to render novel views of the players. Their approach utilises a set of 15 static cameras position around one end of a football pitch. The reconstruction allows flexible production of free-viewpoint rendering from any virtual camera viewpoint. Rendered views are lower visual quality than the captured video due to inaccuracies in the visual hull geometry causing misalignment between views which results in blur and double exposure. This work forms the basis for the free-viewpoint rendering system described in further detail in Sect. 4.

---

[3] EyeVision, http://www.ri.cmu.edu/events/sb35/tksuperbowl.html.

Interpolation of novel views between the real cameras without explicit 3D reconstruction has also been investigate in the context of sports [6, 23, 28]. Inamoto and Saito [23] allow free-viewpoint video synthesis in football by segmenting the observed camera images into three layers: dynamic foreground (players); pitch; and background (stadium). Segmentation of dynamic foreground regions corresponding to players is performed based on motion and colour. Image-based novel view synthesis is performed by morphing of the foreground layer between adjacent pairs of views. Morphing is achieved by interpolation along the corresponding intervals of the epipolar line for the foreground layer. Results from four static camera views of the penalty area in football demonstrate interpolation of intermediate views. The approach does not take into account inter-player occlusion which may result in visual artifacts. A layered representation for the spatio-temporal correspondence and occlusion of objects for pairs of views is proposed in [6] and applied to football view interpolation. A MAP formulation is presented using the EM algorithm to estimate the object layer segmentation and parameters for correspondence and occlusion. This approach implicitly represents the 3D scene structure using correspondence and does not require prior camera calibration to interpolate intermediate views. Results are presented for interpolation of views between a pair of fixed cameras covering the goal area in football. The layered representation allows view interpolation of intermediate views in the presence of significant inter-player occlusion with a visual quality comparable to the source video. Interpolation based approaches to novel view rendering in sports are limited to the generation of intermediate views between the match cameras. Potential advantages include avoiding the requirement for camera calibration or explicit 3D reconstruction and direct rendering of novel views from the camera images. Extrapolation of novel viewpoints away from the captured views is limited due to the lack of explicit geometry and may result in unrealistic image distortions.

All current systems for free-viewpoint video in sports use special rigs of auxiliary cameras to capture footage over a wide-area and result in a visual quality below that acceptable for TV broadcast. In addition due to the use of static cameras players are captured at relatively low resolution. One solution to this is the remote controlled camera slaved to a single operator used in the EyeVision system allowing close-ups of the critical action from multiple views. The use of auxiliary cameras in addition to the match cameras adds a significant cost to rigging which may prohibit deployment of free-viewpoint video at all but key sports events such as cup finals or the SuperBowl. The added-value in broadcast production depends on the trade-off between potential use and costs such as rigging additional cameras.

An attractive alternative would be to use the manually controlled match cameras used for live broadcast which primarily focus on the play action of interest. Use of match cameras would also avoid the prohibitive costs associated with rigging of additional cameras. Subsequent sections of this article specify the multiple camera capture requirements for sports and identify practical limitations in using either match cameras or special camera rigs. LiberoVision[4] recently introduced a commercial system for interpolation between pairs of match camera views in football broadcast. This system has the advantage of only using the existing broadcast cameras. Current technology is limited to viewpoint interpolation at a single static frame and does not allow free-viewpoint rendering of the game play from novel views such as the referee viewpoint. The system is currently in use for half-time and post-match commentary of football. The Piero[5] system developed by BBC R&D allows annotation of the broadcast video footage together with limit change in viewpoint using player billboards together with camera calibration to extrapolate views around a single camera. Red Bee Media commercialised the Piero system and have introduced free-viewpoints of static plays using manually posed player models to illustrate the action. Current commercial technologies are limited to view-interpolation at a single frame and do not allow novel camera views of the action.

The *iview* free-viewpoint video system presented in this chapter aims to allow rendering of live action from informative viewpoints which add to the broadcast coverage of a sporting event. The system allows rendering of viewpoints on the pitch such as the referees or goal keepers view of events using the broadcast match cameras together with additional auxiliary cameras to increase coverage if available. *iview* has introduced automatic methods for online calibration, segmentation, reconstruction and rendering of free-viewpoint video for sports broadcast production.

3 Specification of Requirements for TV Sports Production

There are three critical issues for use of free-viewpoint video in sports TV broadcasts: visual quality; timing; and cost. In this section we identify the requirements and constraints for use of free-viewpoint video in sports TV production.

3.1 Visual Quality for Broadcast Production

The hardest technical constraint for free-viewpoint video of novel views in TV sports production is visual quality. Broadcast video quality equivalent to the live footage from the broadcast cameras is ideally required to be acceptable to the viewing public. Visual artifacts such as flicker and jitter are unacceptable. In addition, the visual resolution of the broadcast footage should be preserved without spatial blur or ghosting artifacts. The occurrence of visual artifacts due to reconstruction

---

[4] LiberoVision, http://www.liberovision.com.

[5] Piero, http://www.bbc.co.uk/rd/projects/virtual/piero/.

errors such as false volumes due to multiple player occlusion is also unacceptable. In broadcast production the visual quality is a primary consideration which must be satisfied for a new technology to be widely used.

High-definition (HD) cameras are now widely used for acquisition at live sports events together with increasing use of HD transmission to the viewer. Free-viewpoint rendering needs to achieve HD quality for rendering of full-screen shots. For sports commentary free-viewpoint video rendering of novel viewpoints may also be used as in-picture inserts at a lower resolution to illustrate different views of the play requiring a lower resolution. However, the visual-quality at the displayed resolution should be equivalent to that of conventional TV broadcast footage.

3.2 Production Requirements on Timing

The time taken to produce free-viewpoint video is critical to the potential uses in broadcast. From a production standpoint free-viewpoint video would ideally be available at video broadcast rate (<40 ms per frame) with 100% reliability on visual quality allowing the sports producer to select free-viewpoint video streams as additional cameras for broadcast. In practice due to both algorithm reliability and computational delay there are three critical time points where free-viewpoint video could be exploited in production:

Action Replay: Within seconds of an event happening (e.g. a player is fouled, and a novel view is offered in place of conventional *instant* replays.

Action Review: Within a few minutes, such that a novel view can be presented during half time or immediately that a match finishes.

Match Analysis: After many minutes, such that a novel view sequence made available for use as part of a post-match analysis programme (which may be later that day or week).

Aiming for the *replay* timeframe is an extreme challenge. Calibration, definition of viewpoint and rendering must all be done within a small number of seconds. Additionally, the production crew must make split second decisions about which replays to present before the match action continues. Consequently, rendered free-viewpoint sequences must be 100% reliable to be offered as additional shots. In the case of review or analysis, time pressure is much more relaxed. Particularly for analysis, it may even be possible to prepare novel view sequences remotely, having had video sequences transmitted from the outside broadcast venue. Reliability levels required for non time-critical post-match analysis are significantly lower as it is possible to have an operator check sequence quality. The added value to sports broadcast production of free-viewpoint video decreases rapidly with the time-taken for shot production. Consequently the level of expenditure available for production systems will also decrease depending on the production time.

3.3 Acquisition Requirements

Production of sports events such as football for live broadcast typically use 12–18 match cameras at key locations around the stadium. This number may be increased at major sports events such as the SuperBowl or a cup final. In the 2006 FIFA World Cup[6] 26 HD cameras were used for coverage at each stadium as illustrated in Fig. 1. The main broadcast cameras are typically located one side and on the ends of the stadium to avoid disorienting the viewer with switches to reverse views. Additional high-speed, overhead and reverse side camera views are typically used for action replays in commentary. All cameras are manually controlled by individual operators to cover both the action on the pitch and the crowd. This leads to the problem that even with 15–20 broadcast cameras only a small number will be covering the same area during normal play. An exception to this is events such as penalties where more camera angles may be available on a particular player. In addition the framing varies between cameras from zoomed in shots of individual players with little background visible for calibration to wider shots of groups of players. Figure 2 shows a typical

Fig. 1 Typical stadium broadcast camera layout for a major sporting event. Out of 26 cameras 1–4, 10, 11, 19, 20 provide potentially useful views, 5–8, 18, 21 are high-speed cameras and the remainder are at pitch level providing insufficient coverage for calibration or reconstruction

---

[6] http://www.fifa.com/worldcup/archive/germany2006/news/newsid=13449.html.

Fig. 2 Broadcast match camera views for a penalty event during a cup-final. Of the 15 match cameras only the views shown were of the penalty

set of shots from match cameras for a penalty event during a cup-final match where there were 15 match cameras in total only 8 of which captured the event.

Ideally the broadcast match camera views would be used for free-viewpoint video production. However, as shown in Fig. 2 only a small proportion of these will have overlapping fields of view even for a specific event in the game where normal play is suspended. In addition as the match cameras are individually operated it is necessary to calibrate the camera orientation and zoom from the broadcast footage. Calibration requires fixed features such as pitch markings to be visible in the shot which may not be the case for zoomed in shots of individual players or cameras at pitch level. Therefore, in practice it may be desirable to use auxiliary cameras in addition to the match cameras to ensure sufficient coverage for high-quality free-viewpoint video production. Placement of cameras may be restricted to the pre-defined camera positions due to access restrictions. The requirement for additional cameras introduces a significant cost for rigging and operation. This cost must be justified in terms of the additional value of the free-viewpoint video shots.

In addition to the camera placement simultaneous access to the broadcast camera footage from multiple camera feeds is required. Commonly only a limited number of camera views of interest for a particular time instant are stored. These are typically stored in a compressed format for subsequent editing in post match commentary. Acquisition and storage of multiple camera views is required for subsequent free-viewpoint video rendering.

3.4 Production Requirements for Use of Free-Viewpoint Video

An important consideration in the development of free-viewpoint technology for sports is the end-use in production. To justify the additional cost and complexity of free-viewpoint video technology the content produced must add value to the programme. The added value to broadcasters and viewers must go beyond the short-term novelty of a special-effect such as the virtual camera fly-through shots produced from the EyeVision system at the SuperBowl. Added value requires the production of free-viewpoint camera shots which add to the experience of viewing the game and the associated match commentary. Examples of shots which might add value in a football (soccer) match include the referees view during a specific event, the goal-keepers view or a players viewpoint. Virtual cameras able to generate free-viewpoint video of such views during or shortly after a particular event are more likely to add-value and be exploited in the programme production.

4 A Free-Viewpoint System for Sports TV Production

This section presents a system being developed for free-viewpoint video in TV production of sports events. The system is being developed to utilise footage from both the manually operated match cameras and fixed auxiliary cameras if available to ensure full stadium coverage. Automatic camera calibration from the pitch markings has been developed to allow combination of footage from multiple camera views including the moving and zooming match cameras. In this section we review both the acquisition system and algorithms developed to facilitate broadcast quality free-viewpoint video production. The current prototype system has been used for production trials capturing test footage of live events for off-line processing to evaluate the performance with respect to the broadcast production requirements, as specified in Sect. 3. Results of evaluation for both international soccer and rugby are presented.

4.1 System Overview

An overview of the free-viewpoint TV production system is shown in Fig. 3. Capture is performed using time synchronised acquisition from both auxiliary and match cameras. Synchronisation using genlock is a standard process in conventional broadcast acquisition. Uncompressed camera footage is stored directly to disk for offline

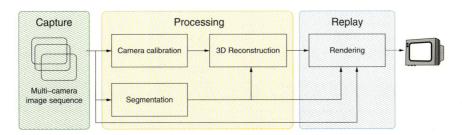

Fig. 3 Overview of the iview free-viewpoint video system

processing. Automatic calibration of all cameras is performed from the pitch lines of the captured footage. This avoids the need for prior camera calibration and allows the use of footage from match cameras. The calibration is capable of real-time operation for use during live match footage. Calibration estimates the extrinsic and intrinsic parameters of each camera including lens distortion. Matting of foreground (players) from the background (pitch) is performed using chroma and difference key matting. This allows the approximate segmentation of the foreground players for subsequent processing to produce free-viewpoint video. Aperture correction is also applied to each video sequence to correct for the camera edge enhancement used in standard broadcast footage.

The calibration of moving match cameras achieves an rms error of 1–2 pixels. Likewise matting in relatively uncontrolled outdoor conditions with changing illumination achieves a segmentation within 1–2 pixels of the true foreground with the addition of background clutter for the crowd, hoardings and on-pitch advertising. The accumulation of errors from calibration and matting can cause gross errors in the reconstruction of the scene such as loss of limbs as their image size may also be of the order of a few pixels. Therefore, robust algorithms have been developed for scene reconstruction and free-viewpoint rendering in the presence of matting and calibration errors.

Initial reconstruction from multiple views is performed using a conservative visual-hull algorithm, taking into account the maximum error to ensure that the foreground is within the reconstructed volume. This allows robust reconstruction of a coarse approximation of the scene structure. However, scene rendering of novel views using the conservative visual hull surface results in visual artifacts due to errors in the geometry resulting in poor alignment of the foreground between views. Given the conservative visual hull as an initial reconstruction techniques have been developed for view-dependent refinement of both the surface reconstruction and foreground matting. Initialisation of the view-dependent reconstruction from a conservative approximation of the scene structure ensures that the resulting refined segmentation accurately separates the foreground and background in regions of visual ambiguity by integrating information from multiple views. View-dependent surface refinement enables robust reconstruction in the presence of camera calibration error for the rendering of high-quality novel viewpoints. The replay module renders the captured scene in realtime using the computed 3D model and the original camera images deploying view-dependent texture mapping [7]. The replay module is designed to work at interactive rates allowing free-viewpoint camera path planning for editorial shot production.

Currently the capture, calibration, matting and visual-hull together with interactive rendering of novel views can be performed in real-time. Reconstruction refinement is currently performed as an off-line process taking several minutes per frame. The capture, reconstruction and rendering components are integrated in a pipeline for use in live production.

4.2 Video-Rate Calibration of Live Broadcast Footage

One way in which camera calibration data can be derived is by performing an initial off-line calibration of the position of the camera mounting using a theodolite or range-finder, and mounting sensors on the camera and the lens to measure the pan, tilt, and zoom. However, this is costly and sometimes very difficult, for example if it is not possible to gain access to the camera mounts, or if cameras are mounted on non-rigid structures such as a crane. It is often the case in international sports that the cameras are controlled by a host broadcaster and only access to the match camera feeds are available.

A more attractive way of deriving calibration data is by analysis of the camera image sequence. The lines on a sports pitch are usually in known positions, and these can be used to compute the camera pose. In some sports, such as football, the layout of some pitch markings (such as those around the goal) are fully specified, but the overall dimensions vary between grounds. Pitches are also often raised in the middle by 30–50 cm for drainage. For soccer the Football Association specifies that the pitch length must be in the range 90–120 m, and the width 45–90 m; for international matches, less variation is allowed (length 100–110 m and width 64–75 m). It is thus necessary to obtain a measurement of the actual pitch.

For free-viewpoint video in sports TV production we have developed a real-time (50–60 Hz) camera pose estimation system for the live broadcast cameras [45]. The online calibration estimates the camera position, orientation, focal length and lens distortion from the match footage using pitch markings. Camera calibration is computed to minimise the reprojection error of observed pitch lines. Calibration is based on line tracking using a multi-hypothesis approach based on edge points closest to the predicted line position. This provides robustness to the appearance of other nearby edge points. The method includes an automatic initialisation process implemented in such a way that the process can be carried out in about one second. We also take advantage of the fact that TV cameras at outside broadcasts are often mounted on fixed pan/tilt heads, so that their position remains roughly constant over time. Further details can be found in [45], the stages of the calibration are as follows:

Initial Estimation of Camera Position: As broadcast cameras are commonly mounted in a fixed location on a pan and tilt head an initial estimate of the camera position is obtained from multiple images with a wide range of camera orientations. The camera position, orientation and field-of-view is estimated with the position constrained to a common value for all images. This initial calibration from images over a wide range of orientations significantly reduces the ambiguity between distance of the camera from reference features and the focal length. Evaluation of this approach found that an accuracy of 0.3 m for the initial camera position estimate was obtained with 10–20 images across the pitch. This estimate computed offline is used to initialise the camera tracking and can be automatically refined during tracking from the online calibration.

Initialisation: A Hough transform is used to quickly establish how well the image matches the set of lines that would be expected to be visible from a given pose.

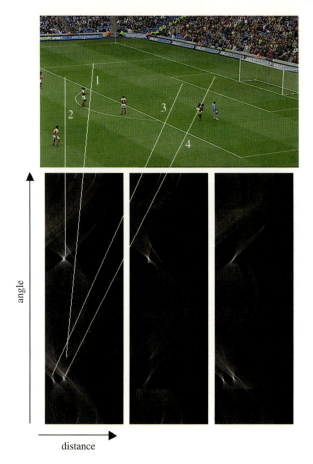

Fig. 4 Calibration initialisation using Hough space: (*top*) original image, (*bottom*) Hough space for full image (*left*), top or left half of image (*centre*) and bottom or right half (*right*)

An exhaustive search process is used to establish the pose which gives the best line matches to the observed image. For each pre-determined camera position, we search over the full range of plausible values of pan, tilt, and field-of-view, calculating the match value by summing the values in the bins in the Hough transform that correspond to the line positions that would be expected. Figure 4 presents examples of the Hough space corresponding to the single frame shown. The initialisation has been found to reliably give an initial estimate of the camera calibration on sports footage in less than 1 s from an unknown orientation and focal-length [45]. Figure 5 shows some example images successfully used for initialisation of the calibration.

Tracking: The tracking process uses the pose estimate from the previous image, and searches a window of the image centred on each predicted line position for points likely to correspond to pitch lines. A straight line is fitted through each

Fig. 5 Online calibration from pitch markings for two match cameras showing a synthetic grid superimposed on the pitch

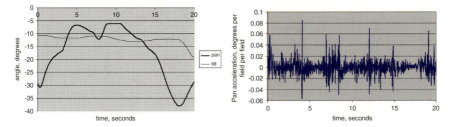

Fig. 6 Calibration parameters for a 20 s sequence (*left*) Pan and tilt angles (*right*) second derivative of pan

set of points, and an iterative process is used to minimise the distance in the image between the ends of the observed line and the corresponding line in the model. Figure 6 shows the estimated camera angles for a 20 s sequence. Evaluation of the real-time tracking on sports footage [45] shows that the noise error in the estimated pan angle is approximately 0.02° which corresponds to one image line.

4.3 Foreground Segmentation

For the segmentation of players colour-based methods, like chroma-keying against the green of football and rugby pitches have been considered. However, the colour of grass varies significantly on pitches. This is due to inhomogeneous illumination in the uncontrolled environment and anisotropic effects in the grass caused by the process of mowing in alternating directions. Under these conditions chroma-key gives a segmentation that is too noisy to achieve a high-quality visual scene reconstruction. Therefore two improved methods have been implemented and tested: A global colour-based k-nearest-neighbour classifier and a motion compensated difference key. After segmentation we compute the foreground/background colours for 'mixed pixels' with a method similar to that described by Hillman in [22].

K-nearest neighbour classifier: This classifier is controlled by a simple GUI: The user clicks on positions in an image that represent background. The RGB colour values of that pixel are stored as a prototype $P_i = I$ into a list. All pixels in the image that are within a radius in RGB space r_1 of the colour prototype are then marked as background. The user continues to select background pixels until the resulting segmentation is satisfactory. The segmentation $S_{k-nearest}$ is computed by finding the nearest colour prototype P_{best} from the list with the minimum RGB colour distance d of the pixel RGB values I:

$$d = argmin_I \{d_{rgb}(P_{best}, I)\} \quad (1)$$

The segmentation is then given by:

$$S_{k-nearest} = \begin{cases} 0, & d < r_1 \\ 1, & \text{otherwise} \end{cases} \quad (2)$$

In order to get continuous values a soft key can be obtained using a second radius r_2. See [17] for details.

Motion compensated difference keying: Difference keying is often used as a simple segmentation technique. It is based on the difference in colour space between a pixel I of the image and the corresponding pixel I_{bg} in the background plate (Fig. 7). The background plate can be created by either taking a picture of the scene without any foreground objects or if this is not possible the background plate can be generated by applying a temporal median filter over a sequence to remove moving foreground objects. The difference between I and I_{bg} can be computed in any colour space. We used the difference in RGB space here:

$$\delta = d_{rgb}(I, I_{bg}) \quad (3)$$

The segmentation S_{diff} is computed as a binarisation with threshold σ:

$$S_{diff} = \begin{cases} 0, & \delta < \sigma \\ 1, & \text{otherwise} \end{cases} \quad (4)$$

Difference keying assumes correspondences between image pixel I and I_{bg} in the background plate and therefore requires static cameras. However, under known

Fig. 7 Difference keying. The camera image (*middle*) is compared against a spherical panorama of the scene (*left*) giving the key (*right*)

Fig. 8 Image of a sport scene from a broadcast camera (*left*) and detail (*right*)

nodal (pan, tilt) movement of the camera a background plate can be constructed by piecewise projection of the camera images into a spherical map. This transformation is derived from the camera parameters, as computed in the camera calibration (Sect. 4.2). A clear plate of foreground objects is created by applying a temporal median filter on the contributing patches. The colour distance δ defined in (4) is computed by projecting each pixel I into the spherical map using the camera parameters of the image.

4.3.1 Aperture Correction in Broadcast Cameras

Broadcast cameras have a control known as aperture correction or sometimes *detail*. The aperture correction is used to *sharpen* an image to emphasise high-frequency image components and is therefore a high-boost filter. Figure 8 shows an example of a broadcast image. The image was taken during a rugby match with a Sony HDC-1500 high definition camera.

In sport productions it is quite common to add a lot of *detail*. The effect can be seen in the detail picture (Fig. 8 bottom) as over- and undershoots of the image signal (visible as black haloes of white shirts against the background). In broadcast this feature is intended to give a better perceived contrast of objects. For segmentation the effect is a challenge since significant colour changes take place at the edge of objects of up to 3–5 pixels and can therefore be a significant problem.

The effect of the aperture correction can be compensated by a symmetric low-pass filter applied to the luminance channel of the broadcast image. We propose to compute the segmentation on the compensated image to improve the robustness and quality of the key.

Figure 9 shows results of the difference keying applied to the original broadcast image (shown in Fig. 9a,b) and after compensation of the aperture correction (Fig. 9c,d). In contrast to the global colour-based methods the pitch lines are suppressed except in areas with shadows. As expected the segmentation in the compensated image is more precisely aligned to object edges. This is clearly visible in the right image of Fig. 9. A more detailed analysis and description of the implemented compensation filter can be found in [14].

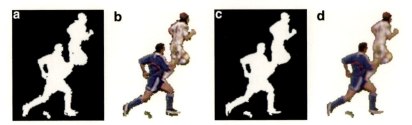

Fig. 9 Detail of difference key computed on broadcast image (**a**, **b**) and after compensation of aperture correction (**c**, **d**)

4.4 Free-Viewpoint Video Production from Match Cameras

Free-viewpoint video production for outdoor sports scenes captured over large areas must be robust to errors in the online camera calibration and natural scene matting. Typically for moving match cameras and chroma-key or difference matting errors are of the order 1–3 pixels. In typical camera footage the foreground players are relatively small 10–20 pixel width with arms and legs of the order 3–6 pixels. Direct multiple view reconstruction from erroneous foreground mattes can result in gross errors such as missing arms and legs. Due to the wide-baseline between cameras direct stereo matching between adjacent views is also problematic. In this section we present algorithms developed for robust rendering of novel views from multiple cameras in the presence of matting and calibration errors. The approach comprises two stages: conservative visual-hull reconstruction to recover a coarse scene approximation which encloses the true scene surfaces; and view-dependent optimisation to simultaneously refine the surface reconstruction and foreground segmentation to estimate a scene approximation which aligns images across multiple adjacent wide-baseline views and accurately segments the foreground player boundary. This approach achieves high-quality rendering of novel views in the presence of calibration and matting errors.

4.4.1 Conservative Visual-Hull

The conservative visual-hull (CVH) [27] is a volumetric approximation of the scene from multiple view image silhouettes up to a maximum error in the camera calibration and matting. The CVH is a global multiple view reconstruction which encloses the true scene surface. In the presence of camera calibration errors there is no single global scene reconstruction which corresponds to the observations from all views. A conservative visual hull with an n-pixel tolerance is obtained by dilating the image silhouettes by n pixels prior to silhouette intersection. Typically in this work $n = 3$ gives an upper-bound on the combined calibration and image segmentation errors. This ensures that the true scene surface projects to inside the dilated silhouettes for all views given errors in calibration and matting.

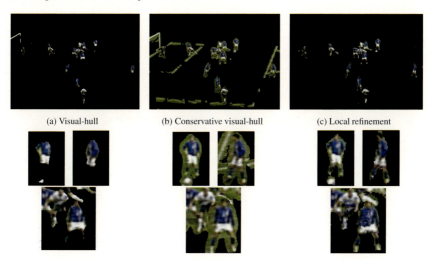

Fig. 10 Novel view rendering for scene reconstructions in the presence of camera calibration and matting errors

Figure 10 presents examples of novel view rendering of a football match for the visual-hull and conservative visual-hull. The visual-hull demonstrates the problem of global reconstruction from multiple views in the presence of calibration and matting errors. A number of players have missing arms or legs. The CVH in Fig. 10b reconstructs a surface which enclosed the true foreground scene objects including narrow limbs. However, significant visual artifacts occur with the CVH rendering as the surface is over extended at the boundaries and does not accurately align adjacent camera views for rendering. In our system the CVH provides an initial global scene reconstruction which is then locally refined for accurate image alignment and boundary extraction.

4.4.2 Local View-Dependent Visual-Hull and Segmentation Refinement

The CVH provides a robust initial estimate of a surface which encloses the true scene surfaces. This surface is then locally refined with respect to a specific camera viewpoint to obtain a surface approximation which aligns the adjacent images and accurately segments the foreground boundaries. The CVH surface is an initial approximation which provides constraints to enable wide-baseline stereo matching between adjacent views. Stereo correspondence is constrained to lie inside the CVH reducing the likelihood of false matches. This approach was previously introduced for wide-baseline reconstruction in multiple camera studios [42] and has recently been extended for refinement in outdoor sports scenes [20, 27]. The critical advance required for high-quality rendering in sports production is the simultaneous refinement of both the initial 2D image segmentation and initial CVH surface reconstruction. Refinement in a view-dependent framework is robust to errors in the global camera calibration where there is no global reconstruction which is consistent with all camera views. Local view-dependent refinement combines information from multiple views together with priors on background appearance to achieve robust segmentation and improvements in the surface reconstruction. Simultaneous matting and reconstruction from multiple views of sports scenes was introduced in [20]. This approach has been extended to incorporate prior information on background and foreground appearance and improved optimisation of multiple view image cues.

The problem of simultaneous segmentation and reconstruction from multiple views is formulated in a Bayesian framework [20] and extended to incorporate multiple visual cues [18]. The maximum likelihood solution leads to the minimisation of an energy function:

$$
\begin{aligned}
E(l,d) = \ & \lambda_{colour} E_{colour}(l) + \lambda_{match} E_{match}(l,d) \\
& + \lambda_{contrast} E_{contrast}(l) + \lambda_{smooth} E_{smooth}(l,d)
\end{aligned}
\tag{5}
$$

where $E_{colour}(l)$ and $E_{match}(l,d)$ are likelihood terms for image foreground and background layer assignments based on colour models and for depth assignments based on stereo matching scores. $E_{contrast}(l)$ and $E_{smooth}(l,d)$ are contrast and smoothness priors on the labelling. The parameters λ_{colour}, λ_{match}, $\lambda_{contrast}$ and λ_{smooth} control the contribution of each term.

The colour energy is defined in terms of the probability $P(\mathbf{I_p}|l = l_i)$ that a pixel $\mathbf{p}$ belongs to a foreground or background layer l_i as:

$$
E_{colour}(l) = \sum_{\mathbf{p} \in \mathcal{P}} -\log P(\mathbf{I_p}|l),
\tag{6}
$$

A reference background image is constructed a priori from the image sequence by mosaicing known background pixels and computing the local per pixel mean and variance (similar to the approach in Sect. 4.3). The probability of a pixel belonging to a foreground or background layer is evaluated according to a combination of local per-pixel and a global Gaussian mixture colour models.

The contrast energy term encourages changes in layer label to occur in regions of high image contrast and is based on the difference in image values for adjacent pixels p and q as:

$$
E_{contrast}(l) = \sum_{(\mathbf{p},\mathbf{q}) \in \mathcal{N}} \frac{\pi}{4} e_{contrast}(l, \mathbf{p}, \mathbf{q}),
\tag{7}
$$

where

$$
e_{contrast}(l, \mathbf{p}, \mathbf{q}) = \begin{cases} 0 & \textit{if } l_{\mathbf{p}} = l_{\mathbf{q}}, \\ \exp\left(-\beta C(\mathbf{I_p}, \mathbf{I_q})\right) & \textit{otherwise}. \end{cases}
\tag{8}
$$

Free-Viewpoint Video for TV Sport Production 99

$\mathcal{N}$ denotes the pixel neighborhood, and $C(\cdot, \cdot)$ is a squared colour distance between adjacent pixels.

The matching energy combines sparse and dense correspondence between camera views:

$$E_{match}(l, d) = \sum_{\mathbf{p} \in \mathcal{P}} e_{dense}(l, d, \mathbf{p}) + \sum_{\mathbf{p} \in \mathcal{P}} e_{sparse}(l, d, \mathbf{p}) \tag{9}$$

where $e_{dense}(l, d, \mathbf{p})$ is the photo-consistency measure with respect to adjacent camera views at a depth d along the optical ray passing through pixel $\mathbf{p}$. A robust photo-consistency score is evaluated between the reference and adjacent camera. The sparse correspondence energy $e_{sparse}(l, d, \mathbf{p})$ is introduced to constrain the depth where feature correspondences have been detected between views using a Hessian-affine feature detector with SIFT descriptors which is robust to changes in viewpoint and illumination respectively.

The smoothness term constrains the local continuity of the reconstructed surface as:

$$E_{smooth}(l, d) = \sum_{(\mathbf{p}, \mathbf{q}) \in \mathcal{N}} w_{\mathbf{p}, \mathbf{q}} e_{smooth}(l, d, \mathbf{p}, \mathbf{q}), \tag{10}$$

where

$$e_{smooth}(l, d, \mathbf{p}, \mathbf{q}) = \tag{11}$$
$$\begin{cases} \min(|d_{\mathbf{p}} - d_{\mathbf{q}}|, d_{max}) & \text{if } l_{\mathbf{p}} = l_{\mathbf{q}} \text{ and } d_{\mathbf{p}}, d_{\mathbf{q}} \neq \mathcal{U}, \\ 0 & \text{if } l_{\mathbf{p}} = l_{\mathbf{q}} \text{ and } d_{\mathbf{p}}, d_{\mathbf{q}} = \mathcal{U}, \\ d_{max} & \text{otherwise.} \end{cases}$$

where $\mathcal{U}$ is the unknown label. Discontinuities between layers are assigned a constant smoothness penalty equal to d_{max}, while within a layer the penalty is defined as a truncated linear distance. This defines a discontinuity preserving function which does not over-penalise large discontinuities within a layer. Iso-surfaces of the conservative visual hull are sampled which results in a reconstructed surface which is biased towards the visual hull shape. This is useful in the case where other reconstruction cues are weak which commonly occurs in regions of uniform appearance such as the players shirts. Optimisation of the energy defined by (5) is performed using the α-expansion graph-cut [2] to obtain an efficient approximation to the global solution.

4.5 Free-Viewpoint Rendering Results

Production trials of the iview system have been conducted for football and rugby. In both cases free-viewpoint video sequences were generated for events of editorial value identified by sports producers. Sports events were captured using both the

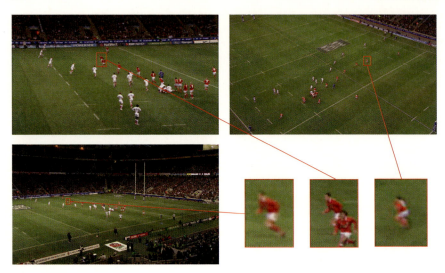

Fig. 11 Two moving broadcast camera views (*top*) and a locked-off camera view (*bottom-left*) from a rugby match

moving broadcast cameras and a small number (4–6) of additional fixed cameras to give coverage of the entire pitch area. Typically in a high-profile football or rugby match with 12–18 cameras only 6–8 cameras are viewing the events of interest for free-viewpoint production. All cameras in the production trials were captured in uncompressed HD-SDI 4:2:2 format for subsequent processing. Figure 11 shows camera views from an international rugby match for two match cameras and one static camera with a typical wide-baseline and difference in framing between views. The close-ups of one player show the difference in resolution of the players between camera views. This illustrates the variation in viewpoint and scale for a set of multiple view cameras. Players are at a relatively small scale in the static cameras due to the requirement to cover the complete pitch. In the broadcast camera views players are at a larger scale but the scale varies between views and there is a high-degree of motion blur due to camera movement.

A comparison of results for segmentation algorithms for soccer and rugby is presented in Fig. 12. Chroma-key and difference key are 2D image segmentation techniques which show errors in segmentation as both additional foreground clutter and areas of the foreground (arms/legs) which are incorrectly classified as background. In contrast the multiple view segmentation and reconstruction refinement algorithm presented in Sect. 4.4.2 produces a clean foreground segmentation with correct classification of the foreground objects even in ambiguous situations where the foreground and background have similar colour ie lines or muddy parts of the players legs. This is due to the combination of information from multiple views to overcome single view visual ambiguities. The refined segmentation

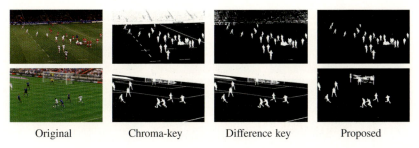

Original   Chroma-key   Difference key   Proposed

Fig. 12 Example of segmentation results on rugby (*top*) and soccer (*bottom*) data (see attached video for full sequence)

allows improvements in the reconstruction quality and subsequent free-viewpoint rendering.

Free-viewpoint rendering results for rugby and soccer trials are shown in Figs. 13 and 14. In both cases the virtual camera sequences were produced for production trials to generate specific camera views which give added value to the match commentary. In the case of the rugby sequence shots show specific game plays and for soccer the shot was specified to view the offside line in a contentious incident. Free-viewpoint camera moves can either take place at a single frame or whilst the action is taking place according to the production requirements. All sequences were generated with automatic calibration, 2D segmentation, reconstruction and refinement. View-dependent rendering is performed using the view-dependent geometry to render images from the adjacent views with dynamic feathering over cameras rendered from different views [19]. The stadium backgrounds are manually modelled using either images from the captured sequences as in Fig. 13a or a synthetic appearance Fig. 13c. Rugby and soccer present different challenges for free-viewpoint production, rugby is particularly challenging as the players are distributed across the field and groups of players form rucks and malls where individual players come into contact and cannot be isolated. The approach developed in the iview project does not make any prior assumptions on player shape allowing high quality free viewpoint rendering of both isolated and tightly packed groups of players.

Free-viewpoint rendering with the proposed approach achieves an image quality comparable to that of the input image sequences as demonstrated in the closeup of Fig. 13b. Degradation in image quality will occur if there are no real cameras which see a part of the scene or there are insufficient views for reconstruction. The proposed approach is robust to the wide-baseline moving camera views at different resolutions which occur in broadcast coverage. The iview free-viewpoint rendering system takes advantage of the manually operated broadcast cameras which generally frame the play to give higher player resolution than the static cameras. The current iview system can operate from the match cameras only but this limits coverage and virtual camera viewpoints to sections of the play where there are sufficient views. Addition of a small number of auxiliary cameras adds to the production cost but ensures complete coverage of the game play and increased range of views for

(a) Rugby virtual camera sequences at 20 frame intervals

(b) Virtual camera closeup

(c) Rugby virtual camera sequences with a virtual stadium model

Fig. 13 Free-viewpoint video rendering of rugby to show pitch level views for commentary

free-viewpoint production. The correct trade-off between coverage and cost will be determined by the production requirements for a specific sport or event. Production trials have demonstrated free-viewpoint shots which add value to the commentary and are of a quality suitable for broadcast.

Fig. 14 Free-viewpoint video rendering of soccer to show an offside incident

5 Conclusion

Free-viewpoint video in live sports broadcast production presents significant challenges to achieve a visual quality comparable to captured video with minimal delay from the manually controlled moving and zooming match cameras. In addition, capture of stadium sports such as soccer and rugby requires acquisition over a large area with relatively uncontrolled conditions. Advances over previous studio based free-viewpoint video are necessary to achieve robust reconstruction for wide-baseline camera views in the presence of calibration and segmentation errors for manually operated moving camera which capture the scene at different resolutions. The iview free-viewpoint video system presented in this chapter incorporates automatic calibration, segmentation and reconstruction from the broadcast cameras. Online calibration uses the pitch lines to estimate extrinsic and intrinsic parameters. Scene reconstruction is performed in two stages: a conservative visual-hull reconstruction provides a robust initial reconstruction in the presence of calibration and matting errors; and view-dependent refinement of reconstruction and segmentation has been introduced which combines cues from multiple views. This approach is robust to online calibration and 2D image segmentation errors enabling reconstruction from wide-baseline moving and zooming broadcast cameras. Production trials of the iview system have been conducted on soccer and rugby to generate novel camera sequences which add to the match coverage. The iview system allows automatic reconstruction and free-viewpoint rendering from the match cameras. Results demonstrate free-viewpoint rendering with a visual quality comparable to the captured video.

A number of open-problems remain to achieve widespread deployment in broadcast production: calibration and use of close-up and pitch level camera views where pitch lines are not visible for calibration; rendering quality of close-up free-viewpoint shots which are limited by the available camera resolution; validated accuracy of free-viewpoint rendering for match decisions (offside); temporal coherence of free-viewpoint rendering and representation for moving scenes; robust free-viewpoint rendering of desired views from match cameras only; and interfaces for rapid free-viewpoint shot production.

Acknowledgements This work was supported by TSB Technology Programme Project TP/3/DSM/6/l/15515 and EPSRC EP/D033926/1 "iview: Free-viewpoint video for entertainment content production".

References

1. de Aguiar, E., Stoll, C., Theobalt, C., Ahmed, N., Seidel, H.-P., Thrun, S.: Performance capture from sparse multi-view video. In: Proc. ACM SIGGRAPH, **27**(3) (2008)
2. Boykov, Y., Veksle, O., Zabih, R.: Fast approximate energy minimization via graph cuts. IEEE Trans. Pattern Anal. Mach. Intell. **23**(11), 1222–1239 (2001)
3. Broadhurst, A., Drummond, T.W., Cipolla, R.: A probabilistic framework for space carving. In: International Conf. on Computer Vision, pp. 388–393 (2001)
4. Carranza, J., Theobalt, C., Magnor, M., Seidel, H.-P.: Free-viewpoint video of human actors. In: Proc. ACM SIGGRAPH, pp. 565–577 (2003)
5. Chen, S.E., Williams, L.: View interpolation for image synthesis. In: Proc. ACM SIGGRAPH (1993)
6. Connor, K., Reid, I.: A multiple view layered representation for dynamic novel view synthesis. In: British Machine Vision Conference (2003)
7. Debevec, P., Yu, Y., Borshukov, G.: Efficient view-dependent image-based rendering with projective texture-mapping. In: 9th Eurographics Rendering Workshop, pp. 105–116 (1998)
8. Deutscher, J., Davidson, A., Reid, I.: Automatic partitioning of high-dimensional search spaces associated with articulated body motion capture. In: Conference on Computer Vision and Pattern Recognition, pp. 669–676 (2001)
9. Eisemann, M., Decker, B., Magnor, M.A., Bekaert, P., Aguiar, E., Ahmed, N., Theobalt, C., Sellent, A.: Floating textures. Comput. Graph. Forum **27**(2), 409–418 (2006)
10. Franco, F.S., Boyer, E.: Exact polyhedral visual hulls. In: British Machine Vision Conference, pp. 329–338 (2003)
11. Franco, F.S., Boyer, E.: Fusion of multi-view silhouette cues using a space occupacy grid. In: International Conf. on Computer Vision, pp. 1747–1753 (2005)
12. Franco, F.S., Menier, C., Boyer, E., Raffin, B.: A distributed approach for real-time 3D modeling. In: CVPR Workshop on Real-Time 3D Sensors and their Applications (2004)
13. Goldluecke, B., Magnor, M.: Space–time isosurface evolution for temporally coherent 3D reconstruction. In: Conference on Computer Vision and Pattern Recognition, pp. S–E, Washington, DC, USA, July (2004). IEEE Computer Society.

14. Grau, O., Easterbrook, J.: Effects of camera aperture correction on keying of broadcast video. In: Proc. of the 5rd European Conference on Visual Media Production (CVMP) (2008)
15. Grau, O., Hilton, A., Kilner, J., Miller, G., Sargeant, T., Starck, J.: A free-viewpoint video system for visualisation of sports scenes. SMPTE Motion Imag. J. **116**(5–6), 213–219 (2007)
16. Grau, O., Prior-Jones, M., Thomas, G.: 3D modelling and rendering of studio and sport scenes for TV applications. In: Proc. of WIAMIS (2005)
17. Grau, O., Thomas, G.A., Hilton, A., Kilner, J., Starck, J.: A robust free-viewpoint video system for sport scenes. In: Proceeding of 3DTV conference 2007, Kos, Greece (2007)
18. Guillemaut, J.-Y., Kilner, J., Hilton, A.: Robust graph-cut scene segmentation and reconstruction for free-viewpoint video of complex dynamic scenes. In: IEEE Int.Conf. on Computer Vision, ICCV (2009)
19. Guillemaut, J.-Y., Kilner, J., Starck, J., Hilton, A.: Dynamic feathering: minimising blending artefacts in view dependent rendering. In: IET European Conference on Visual Media Production, pp. 1–8 (2007)
20. Guillemaut, J.Y., Hilton, A., Starck, J., Kilner, J.J., Grau, O.: A Baysian framework for simultaneous reconstruction and matting. In: IEEE Int.Conf. on 3D Imaging and Modeling (2007)
21. Hartley, R., Zisserman, A.: Multiple View Geometry in Computer Vision. Cambridge University Press, Cambridge (2000)
22. Hillman, P., Hannah, J., Renshaw, D.: Foreground/background segmentation of motion picture images and image sequences. IEE Trans. Vis. Image Signal Process. **152**(4), 387–397 (2005)
23. Inamoto, N., Saito, H.: Virtual viewpoint replay for a soccer match by view interpolation from multiple cameras. IEEE Trans. Multimedia **9**(6), 1155–1166 (2007)
24. Irani, M., Hassner, T., Anandan, P.: What does the scene look like from a scene point? In: European Conference on Computer Vision (2002)
25. Kanade, T., Rander, P.: Virtualized reality: Constructing virtual worlds from real scenes. IEEE MultiMedia **4**(2), 34–47 (1997)
26. Kanade, T., Rander, P.W., Narayanan, P.J.: Virtualized reality: constructing virtual worlds from real scenes. IEEE MultiMedia **4**(1), 34–47 (1997)
27. Kilner, J.J., Starck, J., Hilton, A., Guillemaut, J.Y., Grau, O.: Dual mode deformable models for free-viewpoint video of outdoor sports events. In: IEEE Int. Conf. on 3D Imaging and Modeling (2007)
28. Kimura, K., Saito, H.: Player viewpoint video synthesis using multiple cameras. In: IEE European Conference on Visual Media Production, pp. 112–121 (2005)
29. Laurentini, A.: The visual hull concept for silhouette based image understanding. IEEE Trans. Pattern Anal. Mach. Intell. **16**(2), 150–162 (1994)
30. Matusik, W., Buehler, C., Raskar, R., Gortler, S.J., McMillan, L.: Image-based visual hulls. In: Proceedings of ACM SIGGRAPH, pp. 369–374 (2000)
31. Miller, G., Hilton, A., Starck, J.: Interactive free-viewpoint video. In: IEE European Conf. on Visual Media Production, pp. 50–59 (2005)
32. Miller, G., Starck, J.R., Hilton, A.: Projective surface refinement for free-viewpoint video. In: IET European Conference on Visual Media Production, pp. 153–162 (2006)
33. Moeslund, T., Hilton, A., Kruger, V.: A survey of advances in vision-based human motion capture and analysis. Comput. Vis. Image Underst. **104**(2–3), 90–127 (2006)
34. Moezzi, S., Katkere, A., Kuramura, D.Y., Jain, R.: Reality modeling and visualization from multiple video sequences. IEEE Comput. Graph. Appl. **16**(6), 58–63 (1996)
35. Moezzi, S., Tai, L.C., Gerard, P.: Virtual view generation for 3D digital video. IEEE MultiMedia **4**(1), 18–25 (1997)
36. Scarstein, D., Szeliski, R.: A taxonomy and evaluation of dense two-frame stereo correspondence algorithms. Int. J. Comput. Vis. **47**(1), 7–42 (2002)
37. Seitz, S.M., Curless, B., Diebel, J., Scharstein, D., Szeliski, R.: A comparison and evaluation of multi-view stereo reconstruction algorithms. In: Conference on Computer Vision and Pattern Recognition, pp. 519–528 (2006)
38. Seitz, S.M., Dyer, C.R.: View morphing: synthesizing 3D metamorphosis using image transforms. In: Proc. ACM SIGGRAPH, pp. 21–30 (1996)

39. Seitz, S.M., Dyer, C.R.: Photorealistic scene reconstruction by voxel coloring. Int. J. Comput. Vis. **35**(2), 1–23 (1997)
40. Shum, H.-Y., Kang, S.B., Chan, S.-C.: Survey of image-based representations and compression techniques. IEEE Trans. Circuits Syst. Video Technol. **13**(11) (2003)
41. Slabaugh, G., Culbertson, W.B., Malzbender, T., Stevens, M., Schafer, R.: A survey of methods for volumetric scene reconstruction from photographs. In: Volume Graphics (2001)
42. Starck, J., Hilton, A.: Virtual view synthesis of people from multiple view video. Graph. Models **67**(6), 600–620 (2005)
43. Starck, J., Hilton, A.: Surface capture for performance-based animation. IEEE Comput. Graph. Appl. **27**(3), 21–31 (2007)
44. Stich, T., Linz, C., Wallraven, C., Cunningham, D., Magnor, M.: Perception-motivated interpolation of image sequences. In: Proc. ACM APGV, pp. 97–106 (2008)
45. Thomas, G.A.: Real-time camera pose estimation for augmenting sports scenes. In: European Conference on Visual Media Production, pp. 10–19 (2006)
46. Vedula, S., Baker, S., Rander, P., Collins, R., Kanade, T.: Three-dimensional scene flow. IEEE Trans. Pattern Anal. Mach. Intell. **27**(3), 475–480 (2005)
47. Vlasic, D., Baran, I., Matusik, W., Popović, J.: Articulated mesh animation from multi-view silhouettes. In: Proc. ACM SIGGRAPH, pp. 1–9 (2008)
48. Zitnick, C.L., Kang, S.B., Uyttendaele, M., Winder, S., Szeliski, R.: High-quality video view interpolation using a layered representation. In: Proc. ACM SIGGRAPH, pp. 600–608 (2004)

Challenges for Multi-View Video Capture

Bennett Wilburn

Abstract We discuss challenges in large scale multi-view video capture and use broadcast soccer matches as a motivating application. How far away are we from watching a soccer match at home and changing the virtual viewpoint smoothly from full field views to closeups of individual players, from any angle? We start with a discussion of the total number of pixels required to capture this event for two different kinds of systems: a fixed array of high-resolution cameras, and an active array of low-resolution cameras. Next we look at strategies for increasing the effective frame rate of such systems to allow high-speed capture. Most multi-camera systems are synchronized so all cameras trigger simultaneously, but staggering the triggers samples more efficiently in time, with no additional cost in data bandwidth. We then move from sampling issues to rendering realism and look at one open challenge that is critical for video-based rendering of people: the capture and rendering of moving human hair. Finally, we consider the real-time processing and broadcast of this media for a live soccer match.

1 Overview

This chapter will focus on open challenges in multi-view video capture and rendering. This is not an attempt to survey past work in the field, and as such we will not make an attempt to exhaustively cite prior art. Instead, we will focus on unsolved problems, discuss some prior results when appropriate, and when necessary add a mix of back of the envelope calculations and suppositions. To motivate this discussion, we consider a "holy grail" application: remotely viewing a professional soccer match, with viewer control over the virtual viewpoint.

How far away are we from watching a soccer match at home while changing the virtual viewpoint smoothly, from full-field views to closeups of individual players,

---

B. Wilburn
Microsoft Research Asia, Beijing, China
e-mail: bennett.wilburn@microsoft.com

R. Ronfard and G. Taubin (eds.), *Image and Geometry Processing for 3-D Cinematography*, Geometry and Computing 5, DOI 10.1007/978-3-642-12392-4_5,
© Springer-Verlag Berlin Heidelberg 2010

from any angle? This sort of virtual viewpoint control can be almost arbitrarily difficult; imagine attempting to view a match from the viewpoint of one of the players. In the following section, we'll explore the spatial resolution requirements for a slightly less ambitious goal: multiview video capture of an entire soccer field with enough spatial resolution to match our usual broadcast viewing experience. In Sect. 3, we turn our attention from spatial to temporal resolution, and discuss how to increase the effective frame rate of a multiple camera array. We show a capture strategy that increases the effective frame rate of a multiple camera system without increasing the data bandwidth. After that, we consider open challenges in realism for multi-view interpolation, specifically the capture and rendering of moving human hair. As simple as it may sound, hair is a major challenge for 3-D video of real people, and only very basic steps have been made towards capturing and rendering real, moving hair in a convincing fashion. Finally, we briefly explore the processing and transmission requirements for a real-time broadcast.

2 Spatial Resolution

The view of televised soccer matches can vary from a perspective covering the entire field to a closeup view of a single player. Of course, home viewers are limited to the viewpoints provided by the broadcaster. How could we capture the entire field so viewers could choose their own view? Is this possible or affordable given today's technologies? For the purposes of the discussion, we will limit the viewpoints to typical television viewpoints: from the stands, with fields of view ranging from the entire field to individual players. This precludes creating a view from the ground level, or zooming in for a high-resolution image of only a player's face. This section estimates the pixel resolution requirements for this task and investigates two different approaches to meeting those requirements.

2.1 Resolution Requirements

To start, consider capturing the entire soccer field from multiple camera positions. Figure 1 shows a typical scenario. Suppose a system with a ring of video cameras around the stadium, with each camera capturing the entire field every frame. We would like to virtually zoom in on individual players, so the portion of the entire image occupied by a player should have roughly the same resolution as our display device. This cropped image of the player (shown on the right in the figure) corresponds to a certain area on the field, indicated by the dashed lines.

To approximate the pixel resolution required to capture this match, we consider that cropped area as an image tile, and ask how many such tiles are required to fill the entire field. We will do this calculation for a camera at midfield, at a thirty degree vertical angle from the center of the field. For the perspective projection of typical cameras, this tile-based estimate is only roughly correct, but under some

Challenges for Multi-View Video Capture

Fig. 1 Image resolution requirements for full-field, multi-view video of a soccer match. We will consider a ring of cameras around the stadium, with each camera capturing the entire field every frame. The resolution should be high enough that a cropped region filled by a player (shown on the right) has roughly the same pixel resolution as the viewer's display

circumstances it is nearly exact. Some "view cameras", for example, allow the film plane to tilt relative to the lens plane. The Scheimpflug rule states that it is possible to focus on a plane not parallel to the lens plane if the lens plane, focal plane and film plane all intersect in one line. In this case, not only could the soccer field be the plane of focus, but the image of parallel lines on the field would also remain parallel, and our tile-based approximation would be nearly exact.

The dimensions of an international soccer field are roughly 110 m × 80 m. From the camera's perspective, a 2 m tall player covers a depth of 4 m on the field. Assume our display devices are high-definition (HD), which we will take to be 1,280 × 720 pixels. At this aspect ratio, the horizontal extent of our image tile on the field would be $4 * 1{,}280/720 \approx 7$ m. Filling a 110 m × 80 m field with 7 m × 4 m tiles requires a 19 × 16 grid of tiles, for a total of 304 tiles. Although the real number might be larger (to capture the area around the corner flags, the player benches, and so on), we will say we need roughly 300 HD image tiles, per camera, to cover the entire field.

2.2 Ring of Cameras

What are the implications of trying to film a soccer match with a ring of fixed cameras, each with 300 times high-definition resolution? Wide-baseline camera arrays for image-based rendering vary in size from small numbers of cameras [14, 15, 23] to tens [9] or even 100 [26]. Eight views is an accepted minimum number of cameras for many algorithms, but with only four cameras on any side of a player, occlusions could be troublesome. More cameras increase the hardware costs and bandwidth requirements, but reduce the demands on the accuracy of the view interpolation methods. The "Eye Vision" multiple camera system captured Super Bowl XXXV with a ring of over 30 cameras [6]. Even with that many cameras, the transition from one view to an adjacent one produces a noticeable jump. We will err on the side of visual fidelity and consider a 32-camera system. Each camera in this proposed system streams video with 300 times HD resolution. (300 tiles) $*$ (1,280 $*$ 720 pixels/tile) $=$ 276,480,000 pixels, or roughly 300 megapixels (MP) per frame. If the cameras run at 30 frames per second (fps), record two bytes of data per pixel, and use 30:1 video compression, then each one will generate 300 MP $*$ 30 Hz $*$ 2B/30 $=$ 600 MB/s of video data, or 19 GB/s for the 32 camera system. An entire 95 min game would require 110 TB of storage media. Storage is increasingly inexpensive, and today 1 TB hard drives can be bought for as little as $100 USD, so we could store an entire game using only $11,000 worth of hard drives. Even assuming less aggressive image compression, or extra memory capacity for error protection or backups, storing this multi-view is remarkably inexpensive.

Unfortunately, actually capturing the video is not nearly as affordable. Video cameras with comparable resolution and frame rate have been or are being developed, primarily for military surveillance applications [1, 2], but they are not commercially available today. We can do some rough calculations to understand why. Broadcast quality video cameras must have good noise performance (even at night!), so such a camera would probably need pixel widths of at least five microns. A 300 MP image sensor with five micron pixels would have an area of 70 cm^2. Building such a large image sensor is impractical, which is why existing systems proposed for military use reach high resolutions using camera arrays [1], or hybrid systems with multiple lenses focusing onto arrays of image sensors [2]. Such video cameras would still need the equivalent of 300 HD video encoders running in parallel to transmit the data in real-time, or 10,000 encoders for the full 32-viewpoint camera system.

At this point, one must ask why we are capturing all of this data. At any given time, most of the field is occupied only by grass.[1] In Sect. 2.3, we explore a much more efficient capture system: using many active cameras to follow all of the interesting subjects in a soccer game.

2.3 Active Cameras

Rather than record the entire field, from every camera, every frame, let us instead consider using active cameras – motorized cameras which automatically pan, tilt

---

[1] My father, an American uninterested in soccer, claims that watching soccer is nearly the same as watching grass grow. He may be correct in a literal sense, but I am still trying to convince him he's watching the wrong part of the game.

and zoom – to capture all of the interesting content in a soccer field. What would using an active camera system for a professional match entail? We will say the interesting content consists of 27 subjects: 22 players, one referee, two linesman, the soccer ball, and a wide-angle view of the entire field for context. Assume that for each subject we have a separate set of 32 HD video cameras around the stadium, and that all the cameras in a set automatically pan, tilt, and zoom to follow their subject around the field. For simplicity, we will neglect possible extra cameras to capture fixed features like the goal cages, corner flags, and so on. This brings us to a total of 864 HD video cameras.

864 active cameras might seem vastly more complicated than 32 fixed cameras, but the active system has major efficiency advantages. With fixed cameras, we would capture the equivalent of 300 HD images from each of 32 cameras spaced around the stadium. The active system captures only 27 HD images from each of 32 views. This is over an order of magnitude reduction in the total amount of captured pixels, data bandwidth, and processing power required at each view. Now, the captured video from each match could be stored using only $1 K of storage media. A tenfold decrease in data bandwidth brings the system video bandwidth down to roughly 2 GB/s of compressed data, almost within reach of today's consumer bus protocols.

The advantage of the active array comes with a cost. Physically arranging the system is non-trivial. One might imagine two-dimensional, 27-camera arrays spaced roughly ten meters apart around the stadium. Active cameras must be dynamically calibrated as they move, increasing the processing requirements for image-based rendering at each frame. Active cameras must also dynamically and un-erringly follow their subjects around the field, although one could imagine adding manual assistance. The mechanics and optics for active cameras will certainly add cost to the system, but these devices benefit from economies of scale, while 300 MP video cameras will not.

With a ring of 32 high-resolution, fixed video cameras, we could duplicate some of the advantages of the active system by throwing away the static pixels (i.e. most of the grass) in each frame. However, the simple act of capturing all of the pixels in the first place, only to discard them later, places a huge unnecessary cost on the system. Active cameras are a much more efficient solution. The chapter in this book by Takashi Matsuyama offers more information on the challenges and current technologies for active camera systems.

3 High-Speed Multi-View Video

We have looked at the economics and efficiency of two different systems for capturing multi-view video of a soccer match. For that exercise, we assumed a system that would capture imagery of players at HD resolution, from 32 different viewpoints, at 30 fps. Now let's consider another aspect of sports television: high-speed video. Many sports broadcasts are enhanced with imagery from high-speed cameras. How can we add this functionality to the multi-view video systems envisioned in the previous section? The most straightforward design would simply replace the cameras in the previous system with faster ones. Of course, high-speed cameras add cost. Data rates also scale linearly with the frame rate, increasing the computational and storage requirements for the system. In this section, we will explore spatiotemporal sampling aspects of multi-view video capture, and show how to increase the effective temporal sampling rate of the system without capturing more images.

3.1 Spatiotemporal Sampling for Camera Arrays

Each pixel in a set of cameras samples what is known as the "plenoptic function"; the pixel value is a measurement of the light passing through a certain point in a certain direction. For the hypothetical camera array in this discussion, our rough sampling requirements have been set as HD-resolution video from a fixed number of views around our subjects. Chai et al. derived much more precise limits on camera spacing for Light Field Rendering [12] given the characteristics of the scene, cameras, and rendered images [7]. In this chapter, we will look at camera array sampling issues, but with an added dimension: time. With few exceptions, video camera arrays in research and industry are designed so the cameras all trigger simultaneously. This has many advantages: if we switch between adjacent views, the scene appears fixed, and if we try to analyze the structure of the scene from images taken at the same time, we can reason about a static scene without considering motion. As we will see, however, from a temporal sampling perspective, simultaneous triggering is very inefficient [26].

To analyze space-time sampling strategies for moving scenes, we must first decide how to measure sampling performance. One way is to look at image motion, and in particular to try to capture images in such a way that the image motion between any one image and its neighbors (i.e. images captured at nearby positions and times) is minimized. This is reasonable because image-based rendering methods and vision algorithms like optical flow and multi-view stereo generally perform best when the motion between input images is small. Motion between images from a multiple video camera array is due to two sources: parallax motion as the viewpoint changes from one camera to another, and object motion as the scene moves over time.

For the rest of this discussion, we will consider a planar camera array, a common arrangement for light field rendering and multi-view video capture. We assume that images from all cameras are aligned to a common fronto-parallel reference plane, shown in Fig. 2. This is equivalent to projecting each camera's image onto the reference plane, then imaging it from a reference camera, a transformation of image coordinates that can be represented with a homography. Figure 2 shows how motion in the images on the reference plane is related to the scene geometry and velocities. We assume the scene has near and far depth limits with signed distances ΔZ_{near} and ΔZ_{far} from the reference plane, and the reference plane is optimally placed at a depth Z_0 as described by Chai et al. [7]. For a camera spacing of ΔX, the parallax

Fig. 2 Image motion

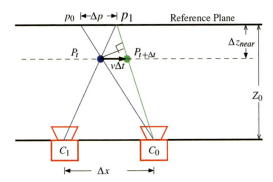

Δp in the aligned images for a point P at a distance ΔZ_p from the reference plane is $\Delta p = \Delta x * \Delta Z_p = (\Delta Z_p + Z_0)$. If we define the *relative depth d* of the point to be $\Delta Z_p = (\Delta Z_p + Z_0)$, this simplifies to $\Delta p = \Delta X * d$.

The worst-case parallax occurs at the near and far depth planes. The worst case temporal motion will occur if P is moving at the maximum velocity in the scene, v, on the near-depth plane, such that the vector $P_t P_{t+1}$ is orthogonal to the projection ray from C_0 at time $t + 1$. If we assume a narrow field of view for our lenses, we can approximate this with a vector parallel to the focal plane, shown as $v\Delta t$. If P has velocity v, the maximum temporal motion of its image in C_0 is $v\Delta t Z_0 = (Z_0 + \Delta Z_{near})$. Equating this motion to the maximum parallax for P in a neighboring camera yields

$$\Delta t = \Delta X \Delta Z_{near} v Z_0 \qquad (1)$$

This is the time step for which maximum image motion between views at the same camera equals the maximum parallax between neighboring views. If we represent a view by two spatial (x, y) coordinates and one time coordinate t, measuring time in increments of the time step Δt and space in units of camera spacings provides a normalized set of axes to relate space-time views. Because motion due to parallax and temporal motion are not orthogonal, the true distance measure is the Euclidean spatial distance plus the temporal distance. Minimizing this distance measure between views minimizes the maximum image motion.

This metric gives us a method to optimize our distribution of samples in space and time, as well as a way to evaluate a given sampling strategy. Figure 3 plots the (x, t) coordinates of captured views for a linear camera array with different values of Δx and Δt. Figure 3(a) compares two sampling strategies for a scene with rapid object motion. The red dots show views captured using a linear array with all cameras triggering simultaneously. Because the object velocities are fast, the distance between horizontal lines of red dots is large; objects are moving a lot between subsequent video frames. In this case, synchronized time samples is one of the worst sampling patterns, because it creates dense rows of samples with large areas blank. As the blue dots show, simply staggering the samples in time results in a much more even sampling pattern using the same number of image samples.

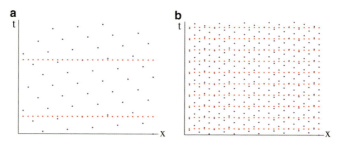

Fig. 3 Space-time sampling for a linear video camera array. The plots show space and time coordinates for images captured by a linear array of video cameras. The x coordinate represents camera position along the array, and the t coordinate is the time the frame was captured by that camera. Both figures show a uniform time sampling in red and an optimal distribution of samples in blue. (**a**) For scenes with small camera spacings or high velocities, uniform sampling creates dense rows of samples and leaves most of the space-time area unsampled. (**b**) For scenes with large camera spacings or very slow motion, time shifting of the cameras makes little difference

In scenes with little motion (Fig. 3(b)), the temporal pattern makes little difference, since the main image motion is from parallax. Since the object motion is often not known in advance, we want a sampling that works for a wide variety of motion vectors. For fast moving scenes, the best timing for the cameras is one where the available time resolution increases with increasing parallax distance from the main sample. For N cameras, an optimized sample pattern starts each camera at $Q * i$ mod N, where i is the index of the camera, and Q is chosen to be roughly 1/3 and also relatively prime with N. Over the entire baseline, every one of the frame-time/N possible starting times is used. This is the pattern shown in blue in Fig. 3(a). Using this offset timing pattern does not hurt in the low object velocities, since the changes in time make little difference in the images that are formed.

Without fundamentally changing the hardware or increasing the data bandwidth, we can improve the spatiotemporal sampling efficiency of the system. The total number of images captured by the array is exactly the same, and the camera frame rates are unchanged. We have simply added fixed delays to triggers of different cameras. In some senses, this increase in temporal resolution is "free".

3.2 Spatiotemporal View Interpolation

Staggered camera triggers may produce a more even sampling of images in space and time, but the reader may be concerned that multi-view video captured in this way is more difficult to analyze. Better spatiotemporal sampling would improve even the simplest view interpolation algorithms like blending, but the sampling densities required for ghost-free images using blending are prohibitive. Here, we present one example algorithm for spatiotemporal view interpolation that exploits improved sampling from staggered triggers: view interpolation using a multibaseline, spatiotemporal optical flow variant [26]. We will again consider a planar camera array, with each image described by an (x, y, t) coordinate. This method will be approximately correct for arrays of other shapes, for example inward-pointing cameras on a sphere, if the shape can be locally approximated as a sphere.

Optical flow typically computes flow between two images by iteratively warping the first image to the second based on the flow field computed for pixels in the second image. Because we have several input views, we choose to solve for a flow field at the (x, y, t) location of our desired virtual view. Similar approaches had been used previously for optical flow from a single video [10, 28]. Computing flow at a frame halfway between two images in a video sequence is more accurate and avoids the hole-filling problems of forward-warping when creating new views. For view interpolation, we compute a flow field for the desired view in our normalized (x, y, t) view space. The results in this chapter were produced by modifying the robust optical flow estimator of Black and Anandan [4], using code available on the author's web site. We iteratively warp the nearest (according to the space-time sampling distance described above) four captured images toward the virtual view and minimize the weighted sum of pairwise robust data and smoothness error terms.

The standard intensity constancy equation for optical flow is

$$I(i, j, t) = I(i + u\delta t, j + v\delta t, t + \delta t).$$

Here, (i, j, t) represent the pixel image coordinates and time, and u and v are the horizontal and vertical image velocities of a point in the image. To be precise, (u, v) is the projection of the point's velocity onto the image plane. Typically optical flow is computed between subsequent images at times δt apart in a single video. Because we are analyzing motion between images captured at different times and from different camera positions, we also use the relative depth d of each pixel to account for parallax motion between views. As before, the parallax motion of a point between images from different camera positions $d * (\delta x, \delta y)$, where $(\delta x, \delta y)$ is the vector from one camera position to the other in the camera array plane. This leads to the following intensity constancy equation representing constancy between the virtual view and a nearby captured image at some offset $(\delta x, \delta y, \delta t)$ in the space of source images:

$$I_{virtual}(i, j, x, y, t) = I_{source}(i + u\delta t + d\delta x, j + v\delta t + d\delta y, t + \delta t)$$

The flow components are separated into parallax motion, determined by a point's relative depth d and the spatial distance between views, and temporal motion, the product of the time between views and the projection (u, v) of the temporal motion onto the virtual view's image plane.

For each virtual view, we choose input views for the flow algorithm by computing a three-dimensional Delaunay triangulation of the camera sampling points and selecting the views from the tetrahedron which encloses the desired (x, y, t) view. These images are progressively warped toward the common virtual view at each iteration of the algorithm. We cannot test the intensity constancy equation for each warped image against a virtual view, so we instead minimize the error between the four warped images themselves, using the sum of the pairwise robust intensity constancy error estimators. This produces a single flow map, which can be used to warp the four source images to the virtual view. One could attempt to reason about occlusions, but for the images in this chapter we simply blend the flowed images using barycentric weights defined by the virtual view coordinates and the tetrahedron formed by the input views.

Figures 4, 5, and 6 compare the results of synthesizing new views with different methods. Figure 4 is a simple cross-dissolve between two subsequent images from the same camera. This is not a satisfactory interpolation method because of the object motion between frames. Figure 5 shows a new view created using a camera array with staggered triggers. The new view was synthesized by blending the four views captured from different positions nearest in space and time to the virtual view. The ghosting artifacts are significantly reduced, but still objectionable. Using the spatiotemporal optical flow algorithm, we can remove the ghosting artifacts, and seen in Fig. 6.

The optical flow based, spatiotemporal view interpolation method just described is an example of how better spatiotemporal sampling allows us to create new views using a relatively simple, image-based algorithm. The method, however, relies on dense camera spacing, so it will not work for the widely separated cameras proposed earlier for soccer matches. The results suggest, however, that we could use more sophisticated algorithms to take advantage of staggered camera timings for widely spaced cameras. Such techniques would offer a powerful way to extend the performance of any multiple video camera system.

Fig. 4 Blending two sequential images from the same camera is a poor interpolation strategy when the object motion between frames is large

Challenges for Multi-View Video Capture

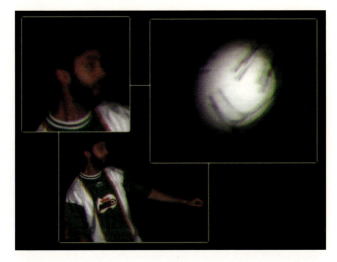

Fig. 5 Using staggered camera triggers and nearest neighbor interpolation produces an interpolated image significantly better than a simple cross-dissolve (*blending*)

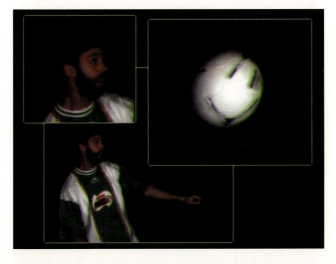

Fig. 6 Using staggered camera triggers and the multibaseline spatiotemporal view interpolation method, we can synthesize a new view without ghosting artifacts

4 Realism

Sections 2 and 3 discuss different aspects of data collection for multi-view video systems, and describe one method for interpolating new virtual views. In this section, we look more closely at generating photorealistic videos from a user-chosen, smoothly moving virtual viewpoint. Given enough cameras, one could create the illusion of smooth camera motion by simply hopping from one camera's video to its neighbor's, or blending between the two views' images. This basic image-based rendering approach, however, requires a high density of input views for visually pleasing results [7]. Instead, using computer vision and graphics techniques, we can model scenes captured from fewer views, then synthesize images that appear to have been taken from new viewpoints. A review of such image-based modeling and rendering techniques easily fills books, not to mention a mere chapter section. Two such references for interested readers are the books by Shum et al. [19] and Magnor [13]).

Although image-based modeling and rendering techniques have grown increasingly powerful, some objects and materials are still difficult to capture and render. Rather that catalogue all of these items, this subsection deals with just one such material that is ubiquitous in our lives and media, and essential for convincing images of real people: hair. With rare exceptions, the multi-view video community avoids the challenges of rendering real dynamic hair by constraining their subjects to have simple, easily-modeled hairstyles. For example, the main characters in *The Matrix*, the movie which popularized view interpolation as a special effect, all had short or slicked-back hairstyles. The hairstyle choice may have had nothing to do with the special effects, but it certainly made the view interpolation simpler. Many of our most-watched celebrities, however, are known for their long hairstyles. More often than not female entertainers and athletes have long hair. For men, long hair has often been used to express a rebellious attitude. The 1980s hairstyle of tennis player Andre Agassi is one example; from the world of soccer, the iconic blonde afro of soccer player Valderrama is another.

Hair poses a challenge for image-based rendering techniques because there are so many of them on typical human heads [2], their structure is much too fine to capture at typical camera image resolutions, and their optical properties are very complex. View interpolation using purely image-based methods is difficult because changes in the relative position between the viewer, hair and lights can cause dramatic changes in the hair appearance, and also because the numerous occlusions and disocclusions in different views of some hairstyles is not amenable to simple blending or image warping. Standard multi-view stereo methods [21] are in general not appropriate for hair because the hair width is such a tiny fraction of the typical image pixel dimension.

Because more general-purpose image-based modeling methods do not apply well to hair, researchers have developed systems specifically for hair. Readers are referred to the survey by Ward et al. [24] for a review of the current state of the art in graphics and vision for hair modeling. Several methods are capable of modeling static hairstyles from real images. Some use constraints provided by varying illumination [8, 16], while others use multiple view geometry [11, 25]. Wei et al. [22] capture many images of a static hairstyle from different viewpoints and measure 2D hair orientations in the images to determine 3-D hair orientations. Their algorithm attempts to grow hairs, starting from the scalp, in such a way that the hair

---

[2] or none, in the case of this author

orientations match the input views. Yamaguchi et al. [27] extended this method to multi-view video of long moving hair, but produced only basic results for simple motions and hairstyles.

What is the state of the art for photorealistic view interpolation of real, dynamic hair? One indication is the latest SIGGRAPH publications on acquisition of real hairstyles. Paris et al. presented a system called Hair Photobooth [17] in 2008. They use a 16-camera array, three DLP projectors, and 150 LED lights on a dome to capture very convincing models of the structure and appearance of real hairstyles. The results are excellent, but the system requires that the object hold still, and the hardware is not appropriate for an unconstrained event like a live soccer match. Moreover, they show results only for short hairstyles. This work produces an unprecedented level of realism for its intended use (static, relatively short and simple hairstyles), but also shows how far we still are from a general capture and rendering method for live performances.

5 Real-Time Processing and Broadcast

Assume we have a system that can capture a soccer match from multiple views, and that we can analyze the content of the videos well enough to render realistic images or individual players or the entire field, from smoothly interpolated viewpoints around the stadium. What other challenges remain for providing this experience to the home user? Sports media is notoriously time-sensitive; fans want to see matches live, and the value of the media is worth substantially less the day after the match. The multi-view video must therefore be processed in real-time, with latency on the order of minutes, and certainly under an hour. Given real-time processing, we face another challenge: how do we distribute this media to the home viewer? In this section, we consider the challenges of real-time processing and broadcast.

5.1 Processing

The throughput of a live, broadcast multi-view camera system must be real-time, say 30 fps. The permissible latency depends on how much delay the average soccer fan is willing to tolerate in a live broadcast. Let us say that a 5 min delay is acceptable. To get a sense for how close the state of the art is to satisfying this constraint on 5 min latency, and real-time throughput, we will look at one representative technology: multi-view stereo. Generally speaking, some form of multi-view stereo, depth estimation, or motion estimation is required for high quality view interpolation.

Multi-view stereo is a well-researched field, and rather than summarize it here we refer the reader to the survey by Seitz et al. [21]. The Middlebury Multi-view Stereo web site [20] which accompanies that survey includes an evaluation that is still used today as a benchmark for comparing multi-view stereo algorithms. Fortunately, this site reports execution times as well as accuracy measures, so we can use it to compare the execution times of state of the art algorithms with the constraints for a real-time broadcast.

We have been discussing a system with 32 views of the field (either for the whole field, or for each individual player). As of this writing, the Middlebury web site uses two datasets: "Temple", with 312, 47, or 16 views; and "Dino", with 363, 48, or 16 views. For the 16-view datasets, there are 23 methods compared for the Temple data, and 22 for the Dino. The execution times have all been approximately normalized to 3 GHz performance on their processors. A quick scan shows that most algorithms take between 30 min and a few hours to analyze the data. This is prohibitively long – 30 min is five orders of magnitude too slow for real-time! For the 48-image datasets, essentially every method gets more accurate results, but the run times are even longer.

Luckily, as the quality of multi-view stereo methods improved, researchers have begun to focus on execution time, too. This is important for multi-view stereo for static scenes, and critical for multi-view video. Currently, two of the methods on the Middlebury evaluation show both high accuracy and high speed. Bradley et al. [5] produce the most accurate results for the 16-view Temple dataset in the least time (only 3 min, 33 s), and Pon et al. [18] produce very good results for the Dino set in only 3 min. These datasets are VGA resolution (640×480 pixel). For $1,280 \times 720$ pixel resolution video, one would expect results three times slower, or roughly 9 min. This is good news; the latency for multi-view stereo, using existing techniques (but ignoring problematic areas like hair), is roughly the right order of magnitude.

Although the latency for these fast multi-view stereo methods is acceptable, the throughput for a single computer is not. The two methods cited above run at roughly 16,000 times real-time performance for one set of 32 HD videos. For our hypothetical stadium with active cameras tracking the players, ball, referee and linesman, the throughput demands will be 26 times higher. Using existing technology, the system throughput is 400,000 times too slow.

Closing a gap of five orders of magnitude to reach real-time performance will require advances in algorithms and hardware, but is not inconceivable. Processors with a hundred cores or more will be common in a few years, and for sporting events a server farm of 100 machines or more would be economically feasible. This might yield a throughput improvement of 1,000 times or more simply through brute force parallelism. Another avenue for accelerating multi-view video processing is exploiting the temporal coherence of the content. Each frame of multi-view video is very similar to the previous one, and thus the structure of one frame is a very good approximation for the next one. It seems possible, but non-trivial, to reach real-time throughput within a few years using a combination of algorithmic and hardware advances.

5.2 Broadcast

Analyzing multi-view video in real-time is challenging, but transmitting that data to the home viewer might be even more challenging. Transmission offers the least flexibility because it must use existing national or global infrastructure. We might be able to instrument a single stadium with as many cameras and computers as we like, but it is unlikely that we will be able to radically change existing digital cable or satellite television broadcast systems.

To get an idea of whether existing broadcast systems could support a soccer match, let us make an assumption: for the active camera system, regardless of the number of views of each subject, each multi-view video stream can be compressed to the equivalent of two high-definition video streams. This may be too optimistic, and readers are welcome to substitute more conservative estimates. For 27 total objects (players, referees, the ball and a wide-angle view of the field), this means the bandwidth equivalent of 54 HD video streams for a single match. Given this approximation, let us examine whether or not we could broadcast a match using existing systems.

As of this printing, Sky TV [22], the most-watched satellite television distributor in the United Kingdom, offers roughly 540 channels, of which 33 are high-definition. Sky states that HD channels have up to four times the resolution of regular channels, so we can estimate the total bandwidth of the Sky TV broadcast as $500/4 + 33 \approx 150$ channels. Sky would have to dedicate one third of its total video bandwidth to broadcast a single soccer match. The situation in the USA is similar: DirectTV [3] offers roughly 150 HD channels and 150 regular definition channels. Again, a single soccer match would occupy a significant fraction of the total satellite bandwidth.

Although it seems technically feasible to broadcast the holy grail soccer match, it may be economically impractical for the near future. Distributors will most likely be unwilling to trade fifty revenue-generating HD channels for a single soccer match. Also, one match may be feasible, but four soccer matches played at the same time would fill the system bandwidth. Broadcasting the holy grail soccer match will require a major evolution in television distribution. Then again, maybe this sort of immersive media will prove to be the killer application that motivates deployment of higher bandwidth media delivery systems using either cable TV, the internet, or higher bandwidth satellite television.

6 Discussion

It appears that we have much work to do before we can broadcast a live soccer match in such a way that home viewers can view the match from different perspectives in the stands, and smoothly zoom in to and pan around players of their choice. Some aspects of this brief survey, however, are encouraging. Using today's technologies, albeit with a large financial investment, we could certainly capture the match in high-resolution. Active camera arrays will make this relatively efficient, and if desired, staggered camera triggers will increase the effective frame rate without adding more video data. Storing the compressed data for this match is not only possible, but easily affordable today. It seems technically possible, although financially impractical, to deliver this media using existing television distribution systems.

The critical challenge to overcome is understanding and modeling the scene well enough to smoothly interpolate new views, and doing this analysis in real-time. The best methods for recovering 3-D scene structure are still several orders of magnitude too slow for a real-time system. Still, one might imagine that we can accelerate current methods sufficiently using parallel computing hardware. Even accelerating our best existing methods to real-time rates, however, would be insufficient. There are still scenes and content that we cannot handle properly. This chapter mentioned one of the most challenging materials, hair, but there are certainly others (faces, fingers, and so on). Before we can consider a real-time broadcast, we must be able to automatically create convincing virtual views of the most challenging subjects, regardless of the execution time.

Multi-view video broadcast is a very broad topic, and one book chapter is certainly too short to do it justice. The chapter only briefly mentions critical components like compression and rendering, and includes no discussion of more exotic approaches to multi-video view interpolation. We have also avoided the question of what kind of experience viewers would really want to see. Interpolating views along a ring in the stadium may not be compelling enough to warrant major infrastructure investments. Watching a soccer match from a virtual perspective on the field almost certainly would be. It will be interesting to watch the evolution of multi-view video technologies and see at what point these technologies become compelling, practical, and affordable enough to enter our living rooms as mainstream products.

Acknowledgements The spatiotemporal sampling and view interpolation work described in this chapter, and the images shown, are the results of collaboration with members of the Stanford Graphics Lab. In particular, the author would like to acknowledge Neel Joshi, Vaibhav Vaish, and Professors Mark Horowitz and Marc Levoy.

References

1. Argusight. http://argusight.com/ (2009)
2. Autonomous real-time ground ubiquitous surveillance - imaging system (argus-is) (2009)
3. Directtv. http://www.directtv.com/ (2009)
4. Black, M., Anandan, P.: A framework for the robust estimation of optical flow. In: IEEE International Conference on Computer Vision (ICCV), pp. 231–236 (1993)
5. Bradley, D., Boubekeur, T., Heidrich, W.: Accurate multi-view reconstruction using robust binocular stereo and surface matching. In: Proceedings of the IEEE Conference on Computer Vision and Pattern Recognition (CVPR) (2008)
6. Carnegie mellon goes to the super bowl. http://www.ri.cmu.edu/events/sb35/tksuperbowl.html

7. Chai, J.X., Tong, X., Chan, S.C., Shum, H.Y.: Plenoptic sampling. In: ACM SIGGRAPH 2000, Proceedings of the 27th Annual Conference on Computer Grapics, pp. 307–318. Orleans, LA, USA (2000)
8. Grabli, S., Sillion, F., Marschner, S., Lengyel, J.: Image-based hair capture by inverse lighting. In: Proceedings of the Graphics Interface, pp. 51–58 (2002)
9. Kanade, T., Rander, P., Narayanan, P.: Virtualized reality: constructing virtual worlds from real scenes. IEEE Multimed. 4(1), 34–47 (1997)
10. Kang, S.B., Uyttendaele, M., Winder, S., Szeliski, R.: High Dynamic Range Video, pp. 319–325 (2003)
11. Kong, W., Takahashi, H., Nakajima, M.: Generation of 3D hair model from multiple pictures. In: Proceedings of the Multimedia Modeling, pp. 183–196 (1997)
12. Levoy, M., Hanrahan, P.: Light field rendering. In: Proceedings of the ACM SIGGRAPH'96 (1996)
13. Magnor, M.: Video-Based Rendering. AK Peters (2005)
14. Matsuyama, T., Wu, X., Takai, T., Nobuhara, S.: Real-time 3D shape reconstruction, dynamic 3D mesh deformation, and high fidelity visualization for 3D video. Int. J. Comput. Vis. Image Underst. 96(3), 393–434 (2004)
15. Matusik, W., Buehler, C., Raskar, R., McMillan, L., Gortler, S.: Image-based visual hulls. In: SIGGRAPH 2000 (2000)
16. Paris, S., Briceno, H.M., Sillion, F.X.: Capture of hair geometry from multiple images. ACM Trans. Graph. 23(3), 712–719 (2004)
17. Paris, S., Chang, W., Jarosz, W., Kozhushnyan, O., Matusik, W., Zwicker, M., Durand, F.: Hair photobooth: Geometric and photometric acquisition of real hairstyles. ACM Trans. Graph. 27(3) (2008)
18. Pons, J.P., Keriven, R., Faugeras, O.: Modelling dynamic scenes by registering multi-view image sequences. In: Proceedings of the IEEE Conference on Computer Vision and Pattern Recognition (CVPR) (2005)
19. Shum, H., Chan, S., Kang, S.B.: Image-Based Rendering. Springer (2007)
20. Seitz, S., Curless, B., Diebel, J., Scharstein, D., Szeliski, R.: Multi-view stereo web page. http://vision.middlebury.edu/mview/
21. Seitz, S., Curless, B., Diebel, J., Scharstein, D., Szeliski, R.: A comparison and taxonomy of multi-view stereo reconstruction algorithms. In: IEEE Conference on Computer Vision and Pattern Recognition, vol. 1, pp. 519–526 (2006)
22. Sky tv. http://www.sky.com/ (2009)
23. Starck, J., Hilton, A.: Surface capture for performance based animation. IEEE Comput. Graph. Appl.
24. Ward, K., Bertails, F., Kim, T., Marschner, S., Cani, M., Lin, M.: A survey on hair modeling: styling, simulation, and rendering. IEEE Trans. Vis. Comput. Graph. 13(2), 213–234 (2007)
25. Wei, Y., Ofek, E., Quan, L., Shum, H.: Modeling hair from multiple views. ACM Trans. Graph. 24(3), 816–820 (2005)
26. Wilburn, B., Joshi, N., Vaish, V., Talvala, E., Antunez, E., Barth, A., Adams, A., Horowitz, M., Levoy, M.: High performance imaging using large camera arrays. ACM Trans. Graph. 24(3), 765–776 (2005)
27. Yamaguchi, T., Wilburn, B., Ofek, E.: Video-based modeling of dynamic hair. In: Proceedings of the 3rd Pacific Rim Symposium on Advances in Image and Video Technology (PSIVT), Lecture Notes In Computer Science (LNCS), vol. 5414, pp. 585–596. Springer (2009)
28. Zhao, W.Y., Sawhney, H.S.: Is super-resolution with optical flow feasible? In: Heyden, A., et al. (eds.) Eurorean Conference on Computer Vision (ECCV), LNCS, vol. 2350, pp. 599–613. Springer (2002)

Part II
Recent Developments in Geometry

Performance Capture from Multi-View Video

Christian Theobalt, Edilson de Aguiar, Carsten Stoll, Hans-Peter Seidel, and Sebastian Thrun

Abstract Nowadays, increasing performance of computing hardware makes it feasible to simulate ever more realistic humans even in real-time applications for the end-user. To fully capitalize on these computational resources, all aspects of the human, including textural appearance and lighting, and, most importantly, dynamic shape and motion have to be simulated at high fidelity in order to convey the impression of a realistic human being. In consequence, the increase in computing power is flanked by increasing requirements to the skills of the animators. In this chapter, we describe several recently developed performance capture techniques that enable animators to measure detailed animations from real world subjects recorded on multi-view video. In contrast to classical motion capture, performance capture approaches don't only measure motion parameters without the use of optical markers, but also measure detailed spatio-temporally coherent dynamic geometry and surface texture of a performing subject. This chapter gives an overview of recent state-of-the-art performance capture approaches from the literature. The core of the chapter describes a new mesh-based performance capture algorithm that uses a combination of deformable surface and volume models for high-quality reconstruction of people in general apparel, i.e. also wide dresses and skirts. The chapter concludes with a discussion of the different approaches, pointers to additional literature and a brief outline of open research questions for the future.

1 Introduction

Today, photo-realistically rendered virtual humans are becoming ever more important elements of feature films. They can perform almost any type of action or stunt at no risk of fatality, as long as an animator is capable of creating the desired effect.

C. Theobalt (✉), E. de Aguiar, C. Stoll, and H.-P. Seidel
MPI Informatik, Saarbruecken, Germany
e-mail: theobalt@mpii.de, theobalt@cs.stanford.edu, edeaguia@mpi-inf.mpg.de, stoll@mpi-inf.mpg.de, hpseidel@mpi-inf.mpg.de

C. Theobalt, S. Thrun
Stanford University, Stanford, CA, USA
e-mail: thrun@stanford.edu

R. Ronfard and G. Taubin (eds.), *Image and Geometry Processing for 3-D Cinematography*, Geometry and Computing 5, DOI 10.1007/978-3-642-12392-4_6,
© Springer-Verlag Berlin Heidelberg 2010

In recent years, ever more powerful computing hardware and rendering algorithms have made it feasible to display detailed realistic humans not only in big-budget feature films, but even in real-time applications available to the end-user at home. For instance, it is foreseeable that in the near future computer game engines will be able to display characters with detailed texture and dynamic geometry, such as correctly deforming cloth. Another application that will gain increasing importance is 3-D video, a new form of media where either the user or the broadcasting company can instantaneously change the viewpoint on a displayed scene. In both cases, it will be important to be able to capture detailed time-varying 3-D models of humans.

Unfortunately, currently available acquisition technology frequently falls short of capturing such rich 3-D scene descriptions that would be directly applicable in the application scenarios mentioned above. Motion capture systems have been around for many years, but they are merely able to measure skeletal motion under controlled conditions. Currently, they are unable to capture shape, motion and appearance of actors in general everyday apparel. Image-based rendering techniques have been proposed to create novel view points of scenes by computationally combining views taken from a few input video streams. However, as we will see later in this chapter, many of these approaches fail to fulfill the visual quality requirements that most professional productions have.

This chapter therefore describes a new category of algorithms, performance capture methods, which are able to fulfill these requirements. Performance capture methods retrieve highly-detailed dynamic shape and motion of moving subjects from (usually) only a handful of unmodified video recordings, i.e. actively placed visual markers are not required. In contrast to previous methods from the literature they are able to handle people in general everyday apparel, such as a skirt or a dress. Also, they are able to capture spatio-temporally coherent geometry, a characteristic that sets them apart from many previous methods from the literature, in particular image-based rendering approaches. Spatio-temporal coherence is an important feature since only if correspondences between reconstructed poses over time are known it is easy to post-process, store and modify the captured data.

In the following chapter, we will first review general related work from the fields of motion capture and image-based rendering. Thereafter, we will discuss four representative, but conceptually different performance capture methods. The first method retrieves detailed time-varying geometry of pieces of apparel from multi-view video using a combination of stereo and cross-parameterization. Along a similar line of thinking, the second approach described employs a combination of visual hulls, multi-view stereo and spatio-temporal cross-parameterization to reconstruct complete performances of humans. The third method described differs from these two approaches in that it employs a template model and skeleton-based pose-fitting to visual hull sequences to measure full human performances. The core of the chapter is a new performance capture approach that takes an unconventional, yet very effective alternative route. Instead of relying on a classical skeleton-based representation of humans, it exploits deforming meshes to faithfully capture the dynamic appearance of actors in arbitrary general apparel. The paper concludes with a discussion and some pointers to additional reading.

2 Paving the Way for Performance Capture: Motion Capture, Image-Based Rendering and 3-D Video Approaches

Modern Performance Capture algorithms can capitalize on a body of related methods which focused on solving sub-problems of the overall performance capture problem. In the following we give a brief overview of important categories of such techniques.

Marker-based optical motion capture systems are the workhorses in many game and movie production companies for measuring motion of real performers [29]. Despite their high accuracy, their very restrictive capturing conditions (that often require the subjects to wear skin-tight body suits and reflective markings) make them incapable of capturing shape and texture simultaneously with motion. Park et al. [35] try to overcome part of this limitation by using several hundred markers to extract a model of human skin deformation. While their animation results are very convincing, manual mark-up and data cleanup times can be tremendous in such a setting and generalization to normally dressed subjects is difficult. In contrast, marker-free performance capture algorithm require a lot less setup time and enable *simultaneous* capture of shape, motion and texture of people wearing everyday apparel.

Marker-less motion capture approaches are designed to overcome some restrictions of marker-based techniques and enable performance recording without optical scene modification [31,39]. Although they are more flexible than intrusive methods, it remains difficult for them to achieve the same level of accuracy and the same application range. Furthermore, since most approaches employ kinematic body models, it is hard for them to capture motion, let alone detailed shape, of people in loose everyday apparel. Some methods, such as [42] and [4] try to capture more detailed body deformations in addition to skeletal joint parameters by adapting the models closer to the observed silhouettes, or by using captured range scan data [2]. But both algorithms require the subjects to wear tight clothes. Only few approaches, such as the work by [40], aim at capturing humans wearing more general attire, e.g. by jointly relying on kinematic body and cloth models. Unfortunately, these methods typically require hand-crafting of shape and dynamics for each individual piece of apparel. Also, they focus on joint parameter estimation under occlusion rather than accurate geometry capture, and therefore the shape quality of the captured performers is typically very crude.

Other related work explicitly reconstructs highly-accurate geometry of moving cloth from video [43,56]. However, these methods require visual interference with the scene in the form of specially tailored color patterns on each piece of garment which renders simultaneous shape and texture acquisition infeasible.

A slightly more application-driven concept related to performance capture is put forward by *3-D video* methods which aim at rendering the appearance of reconstructed real-world scenes from new synthetic camera views never seen by any real camera. Early shape-from-silhouette methods reconstruct rather coarse approximate 3-D video geometry by intersecting multi-view silhouette cones [19, 28].

Despite their computational efficiency, the moderate quality of the textured coarse scene reconstructions often falls short of production standards in the movie and game industry. To boost 3-D video quality, researchers experimented with image-based methods [52], multi-view stereo [61], multi-view stereo with active illumination [55], or model-based free-viewpoint video capture [10]. In contrast to performance capture approaches, the first three methods do not deliver spatio-temporally coherent geometry or full 360 degree shape models, which are both essential prerequisites for animation post-processing. At the same time, previous kinematic model-based 3-D video methods were unable to capture performers in general clothing.

Data-driven 3-D video methods synthesize novel perspectives by a pixel-wise blending of densely sampled input viewpoints [57]. While even renderings under new lighting can be produced at high fidelity [15], the complex acquisition apparatus requiring hundreds of densely spaced cameras makes practical applications often difficult. Further on, the lack of geometry makes subsequent editing a major challenge.

3 Performance Capture Approaches

Performance capture approaches differ from the methods described in the previous section in a few key aspects. First, they aim at reconstruction of highly detailed dynamic scene geometry. By this we mean that the quality of the reconstructed shape should be of such high fidelity that it can even be used without original texture, e.g. for rendering under new artificial lighting and surface material. In consequence, even subtle aspects of shape, such as folds in attire, have to be measured at a sufficient level of detail.

Second, performance capture approaches reconstruct spatio-temporally coherent shape sequences. Here, coherence means that the correspondences between surface points over time are known. This is an important feature since it allows for simpler post-processing, editing and representation of the captured performances. Establishing these correspondences is one of the hardest problems in visual scene reconstruction. As we will see later, different strategies have been explored to achieve coherence. One class of methods uses spatio-temporal cross-parameterization techniques. Another class of approaches starts off with a detailed shape model of the performer, e.g. from a laser scan, that is then deformed to match the input multi-view video data.

Finally, performance capture approaches require no optical modification of the captured scene, e.g. in the form of intentionally placed visual markings, and they impose little restrictions on the type of apparel that a person can wear. As we will show, the majority of algorithms can even handle people in wide and wavy apparel, such as skirts or dresses. This puts them apart from the vast majority of marker-based and marker-less motion capture approaches that have been proposed up to now. In the following sections, we review a few representative examples of performance capture algorithms, and go into a slight bit more detail about a mesh-deformation-based approach that we have developed as part of our research.

3.1 Garment Capture

Capturing the motion of garment is a sub-problem of performance capture by the previously given definition. However, due to their complex deformation behavior, pieces of apparel are among the most difficult elements of dynamic scenes to be reconstructed from video. A recently presented algorithmic recipe to approach the problem shares many similarities to full performance capture, and it is therefore instructive to include it into our overview.

Most previously proposed methods for garment capture require active scene modification, e.g. in the form of color patterns printed on the captured attire (see also Sect. 2). Therefore, despite good results, they fall short to fulfill one of the main characteristics of what we call performance capture approaches in this chapter. In contrast, the recent approach by Bradley et al. captures spatio-temporally coherent geometry of moving pieces of apparel from multi-view video without any marker pattern [8]. In their method, a person wearing the piece of apparel to be reconstructed moves in front of a multi-view video camera setup. The method starts by reconstructing a 3-D mesh of the piece of garment at each time step of video by means of a multi-view stereo approach, that captures the detailed geometry of the fabric, including folds and creases, at each time step of video. Naturally, the meshes found at each time step may contain holes due to occlusions, and there is no spatio-temporal coherence in the mesh connectivity over time. To obtain a spatio-temporally coherent 3-D model representation and to fill in holes, Bradley et al. suggest a spatio-temporal cross parametrization approach that remaps the geometry from each time step to a template 3-D model. Figure 1 shows the acquisition setup, a test subject wearing a t-shirt to be reconstructed, and a 3-D mesh model of a reconstructed shirt illustrating nicely that both the overall shape as well as dynamic folds can be faithfully reconstructed.

Fig. 1 Garment capture from multi-view video using the method of Bradley et al. [8]. (**a**, **b**) Input camera setup with test person. (**c**) Reconstructed 3-D mesh model of the t-shirt at the same time step. (Images courtesy of Derek Bradley, University of British Columbia, Vancouver)

3.2 Surface Capture

One of the first full performance capture approaches in the literature, by this we mean a method to capture entire humans, is the work by Starck and Hilton [48]. Input to their algorithm are eight HD video streams from a fully-calibrated camera setup. Via chroma-keying, the silhouette of the person in each frame is extracted.

In a first pass, their algorithm reconstructs an individual 3-D geometry model for each time step of multi-view video. To this end, a combination of visual hull and stereo reconstruction is used. The visual hull defines an outer boundary for the shape. By using a combination of sparse multi-view line feature matching and a graph-cut based stereo reconstruction, the very coarse visual hulls can be refined and concavities in the surfaces recovered, See Fig. 2 for an example. A surface texture for each captured pose can be created by projectively blending the input video frames on the 3-D surface.

Also here, one of the biggest challenges is to establish spatio-temporal correspondences. Similar to Bradley et al., Sect. 3.1, Starck and Hilton also use a spatio-temporal re-parameterization approach to remesh the individual triangle

Fig. 2 Surface Capture method by Starck and Hilton [48]: 3-D models are reconstructed from multi-view video by means of a combination of shape-from-silhouette and stereo constraints. Spatio-temporal coherence in the meshes (at least for sub-sequences) is established during post-processing by means of spatio-temporal re-parametrization. (Images courtesy of Jonathan Starck, University of Surrey)

Performance Capture from Multi-View Video 133

meshes from each time step to a temporally consistent triangulation [46]. In essence, they cut the surface open to achieve a genus zero surface that can then be parameterized over a sphere. On the sphere an adaptive subdivision and remeshing is performed such that eventually a mesh with the same graph structure is used to represent at least subsequences of an entire multi-view data set. Spatio-temporal reparameterization is a non-trivial problem and it is not guaranteed that under all circumstances the quality of the correspondences will be sufficient. Therefore, other researchers resorted to some form or prior model that is matched to each frame of video. This way, spatio-temporal correspondences are implicitly established.

A method similar to the one by Starck and Hilton has been proposed by Nobuhara et al. [33]. They also reconstruct shape-from-silhouette volumes and employ a deformation-based correspondence finding approach to establish spatio-temporal coherence. Their results show that they are able to successfully handle some cases of topology change.

3.3 Simultaneous Surface and Skeleton Capture

One approach that uses such a prior model is the work by Vlasic et al. [53]. Their approach also uses synchronized multi-view video sequences of human performers as input. The main conceptual difference to the previous two approaches lies in the fact that it employs a form of template model whose motion is tracked. This model comprises a surface triangle mesh and an underlying kinematic skeleton that is coupled to the surface via linear-blend skinning. The surface mesh is either reconstructed by means of a shape-from-silhouette approach, or obtained from a full-body laser scan of the person.

The algorithm commences by reconstructing a shape-from-silhouette 3-D model for each time step of video. The actual performance capture pipeline comprises two stages. In the first stage, only the skeleton part of the model is fitted into each visual hull, in order to capture the general body pose of the actor. The tracker minimizes an energy functional that drives the skeleton close to the medial axis of each visual hull, enforces temporal coherence, and ensures that the extremities of the skeleton are correctly positioned into the respective parts of the visual hulls. An additional term in the energy function takes into account user-defined position constraints that are required in difficult postures where automatic pose determination is likely to fail. To improve tracking accuracy, the authors suggest to use both a forward and a backward tracking pass.

The second stage of the pipeline deforms the surface of the template model such that the silhouettes in all camera views correctly match the outline of the reprojected model. This surface adaptation comprises of several sub-steps itself. First, the template surface is deformed into a new pose via skinning only. In general, this will not bring the mesh into agreement with the silhouettes, since, for instance, nonrigid deformations are only coarsely approximated. Further, skinning deformation artifacts may have deteriorated the surface. The authors therefore suggest an iterative deformation scheme which starts off the skinning pose of the mesh, but purposefully reduces its geometric complexity and iteratively deforms the reduced complexity meshes to match the silhouette boundaries. While iterating, high-frequency geometric surface detail of the template mesh is gradually re-introduced. To serve this purpose, a variant of Laplacian surface deformation [7] is used which allows for such gradual control of surface detail.

The final output is a sequence of skeleton poses together with the template mesh deformed in such a way that the multi-view silhouettes are matched. Overall, the visual quality of the results is very high. The algorithm has been shown to also handle sequences with more complex clothing, such as woman wearing a skirt.

The method has several advantages over the previous two approaches. It captures a rigged skeleton-based character which directly matches the animation pipeline used in most applications. It is also comparably fast, requiring only several tens of seconds of computation time per frame. This is a significant performance benefit over algorithms involving multi-view stereo. One of the disadvantages is that the tracking process of most sequences will require supervision by the user, since in difficult poses manual correction may be required. Further on, even though the quality of the recovered geometry is very convincing in general, it can naturally only capture the deformation of the surface as it is visible in the silhouette boundaries. True waving of cloth (including true folds and creases), as it is mostly observed in the interior of silhouettes, is not actually captured but in a sense pretended by the employed surface adaptation approach. In other words, high-frequency shape detail stays fixed and deforms with the underlying base surface. Here, multi-view stereo approaches are able to capture more true shape detail. Another potential problem is that the skeleton model, although it facilitates tracking, also imposes a prior on motion which is incorrect for wide pieces of apparel whose motion is not explained by skinning. Some of these problems were attacked by another template-based performance capture approach which is detailed in the following sections.

4 Mesh-Based Performance Capture

Template-based performance capture approaches bear several advantages over algorithms doing without strong a priori model assumptions. The template imposes a prior on geometry and motion which can be exploited to make scene reconstruction more robust and correspondence finding easier. The price to be paid is often measured in loss of flexibility since only scenes for which a template is easy to obtain can be reconstructed. Nonetheless, template-based approaches prove very successful for reconstructing performances of humans, as it was shown in the previous section where a kinemtic body model was used. However, a kinematic skeleton with surface skinning is obviously not the right prior model for representing wavy cloth. Although the previous method has shown that on a coarse scale cloth tracking is feasible with a kinematic prior and surface deformation, cloth tracking artifacts are likely to occur.

The method in this chapter intentionally abandons the skeleton-component of the template and uses a deformable surface model as scene representation [11]. This idea has been motivated by the fact that recently many new animation design [7], animation editing [58], deformation transfer [51] and animation capture methods [5] have been proposed that employ shape deformation approaches with great success. The explicit abandonment of kinematic parameterizations makes performance capture a much harder problem, but bears the striking advantage that it enables more reliable capturing of both rigidly and non-rigidly deforming surfaces with the same underlying technology.

First approaches that implemented this idea in the context of full body performance capture were suggested by de Aguiar et al. [12, 13]. Both approaches reconstruct a deformable human template model from a laser scan of the subject to be tracked. The mathematical deformation approach used in either case is a variant of Laplacian surface editing. Performances are retrieved by extracting features from the multi-view image streams and using their 3-D trajectories as deformation handles to change the model pose. Although these approaches can track performances of people wearing complex apparel at high reliability, the algorithms are subject to a few important limitations. Surface-based deformation represents a relatively "weak" prior on 3-D motion that may lead to erroneous deformation if the measured features are starkly noise contaminated, or if there are large regions with no deformation handle at all. This frequently happens when the motion is very fast and thus the image displacement of features is big. Therefore, rapid movements are hard to track with both these approaches. Further on, both methods share the limitation of the skeleton-based approach from the previous section that true high-frequency shape detail cannot be reconstructed.

This chapter describes a new deformation-based performance capture method that exceeds the abilities of the aforementioned algorithms in several ways. First, a new analysis-through-synthesis tracking framework enables capturing of motion that shows a very high complexity and speed. Secondly, we propose a volumetric deformation technique that greatly increases robustness of pose recovery. Finally, in contrast to previous related methods, our algorithm explicitly recovers small-scale dynamic surface detail by applying model-guided multi-view stereo.

Related to our approach are also recent animation reconstruction methods that jointly perform model generation and deformation capture from scanner data [54]. However, their problem setting is different and computationally very challenging which makes it hard for them to generate the visual quality that we achieve by resorting to an explicit prior model. The approaches proposed in [50] and [44] are able to deform mesh-models into active scanner data or visual hulls, respectively. Unfortunately, neither of these methods has shown to match our method's robustness, or the quality and detail of shape and motion data which our approach produces from video only.

4.1 Overview

Input data to our method are a full-body laser scan of the subject in its current apparel and a multi-view video stream of the subject recorded with eight synchronized geometrically and photometrically calibrated video cameras. We perform a color-based background subtraction to all video footage to yield silhouette images of the captured performers.

We convert the raw 3-D scan into a high-quality surface mesh $\mathbf{T}_{tri}$ using a robust surface reconstruction algorithm, which yields a water-tight high quality mesh. We also create a coarser tetrahedral version of the surface scan $\mathbf{T}_{tet}$ by applying a quadric error decimation and a subsequent constrained Delaunay tetrahedralization (see Fig. 3 (r)). Typically $\mathbf{T}_{tri}$ contains between 30,000 and 40,000 triangles, and the corresponding tet-version between 5,000 and 6,000 tetrahedrons. We register both models to the first pose of the actor in the input footage by means of a procedure based on iterative closest points (ICP). Since we asked the actor to strike in the first frame of video a pose similar to the one that she/he was scanned in, pose initialization is greatly simplified, as the model is already close to the target pose.

Since our capture method explicitly abandons a skeletal motion parametrization and resorts to a deformable model as scene representation, we are facing a much harder tracking problem. On the other hand we gain an intriguing advantage: we are now able to track non-rigidly deforming surfaces (like wide clothing) in the same way as rigidly deforming models and do not require prior assumptions about material distributions or the segmentation of a model.

We capture performances in a multi-resolution way to increase reliability. In a first step we employ an analysis-through-synthesis method to estimate the global pose of an actor at each frame on the basis of the tetrahedral input model, Sect. 4.3.

Fig. 3 A surface scan $\mathbf{T}_{tri}$ of an actress (l) and the corresponding tetrahedral mesh $\mathbf{T}_{tet}$ in an exploded view (r)

Afterwards we capture the high-frequency aspects of the performances, Sect. 4.3.4. This is achieved by transferring the pose to the high-detail surface and refining the mesh to fit closely to the input video. The output is a dense representation of the performance in both space and time. One important ingredient to achieve this is a fast and reliable shape deformation framework which we will detail in the following section.

4.2 A Deformation Toolbox

We use two variants of Laplacian shape editing in our performance capture technique. For low-frequency tracking, we use an iterative volumetric Laplacian deformation algorithm which is based on our tetrahedral mesh $\mathbf{T}_{tet}$. For recovery of high-frequency surface details, we transfer the captured pose of $\mathbf{T}_{tet}$ to the high-resolution surface scan. Being already roughly in the correct pose, we can resort to a simpler variant of surface-based Laplacian deformation to infer shape detail from silhouette and stereo constraints.

4.2.1 Volumetric Deformation

We want to deform the tetrahedral mesh $\mathbf{T}_{tet}$ as naturally as possible under the influence of a set of position constraints. To this end, we iterate a linear Laplacian deformation step and a subsequent update step, which compensates the (mainly rotational) errors introduced by the nature of the linear deformation. This procedure minimizes the amount of non-rigid deformation each tetrahedron undergoes, and thus exhibits qualities of an elastic deformation. Our technique implicitly preserves certain shape properties, such as cross-sectional areas, after deformation. This greatly increases tracking robustness since non-plausible model poses (e.g. due to local flattening) are far less likely.

Our algorithm is related to [45] and it is based on the following steps:

- Solve the linear tetrahedral Laplacian system given the current constraints
- Extract the transformation of each tetrahedral element and split it into rotational and non-rotational components
- Update the right hand side of the linear system using the extracted rotations
- Iterate the procedure

This procedure minimizes the amount of non-rigid deformation E_D remaining in each tetrahedron with each iteration. While our subsequent tracking steps would work with any physically plausible deformation or simulation method, our technique has the advantages of being extremely fast, of being very easy to implement, and of producing plausible results even if material properties are unknown. Further details on the deformation technique can be found in [49].

4.2.2 Deformation Transfer

To transfer a pose from the tetrahedral mesh $\mathbf{T}_{tet}$ to the high-resolution mesh $\mathbf{T}_{tri}$, we express the position of each vertex of $\mathbf{T}_{tri}$ as a linear combination of the vertices of the tetrahedral mesh. The coefficients for this are calculated in the rest pose and can be used afterwards to update the pose of the triangle mesh.

The coefficients are calculated as a weighted sum of the barycentric coordinates of nearby tetrahedra for each vertex. Using more than a single set of barycentric coordinates ensures that we get smooth deformations over the whole mesh. The weights for each tetrahedron are based on the respective distance from the initial vertex using a radial basis function.

4.2.3 Surface-Based Deformation

Our surface-based deformation relies on a simple least-squares Laplacian system as it has been widely used in recent years (see [7] for an overview). The linear system is calculated using cotangent weights and is used to deform the surface under the influence of a set of weighted position constraints. This simple surface based Laplacian deformation allows for a much wider and more detailed range of deformations than the tetrahedral deformation presented above.

4.3 Capturing the Global Model Pose

The first step in global model pose capture recovers for each time step of video a global pose of the tetrahedral input model that matches the pose of the real actor. In summary, our framework first computes deformation constraints from each pair of subsequent multi-view input video frames at times t and $t+1$, and then it applies the volumetric shape deformation procedure to modify the pose of $\mathbf{T}_{tet}$ at time t until it aligns with the input recorded data at time $t+1$.

Our pose recovery process is divided into three steps and it begins with the extraction of 3-D vertex displacements from reliable image features which brings our model close to its final pose even if scene motion is rapid or complex. Subsequently, two additional steps are performed that exploit silhouette data to fully recover the global pose. The first step refines the shape of the outer model contours until they match the multi-view input silhouette boundaries and the second one optimizes 3-D displacements of key vertex handles until optimal multi-view silhouette overlap is reached. The additional steps are important since 3-D features on the model surface are dependent on scene structure, e.g. texture, and can, in general, be non-uniform or sparse.

We gain further tracking robustness by subdividing the surface of $\mathbf{T}_{tet}$ into approximately 100–200 regions of similar size during pre-processing [59]. Rather than inferring displacements for each vertex, each individual step is applied to a representative vertex handle for each region, as explained in more details in the following sections.

4.3.1 Pose Initialization from Image Features

Given two sets of multi-view video frames from subsequent time steps, we first extract image features. SIFT features are chosen since they are largely invariant under illumination and out-of-plane rotation and enable reliable correspondence finding even if the scene motion is fast or complex [27].

In order to transform the feature data into deformation constraints, we first associate image features from time t with vertices in the model. After creating the spatial feature associations across camera views, we establish temporal correspondences between the features from time t and $t + 1$, Fig. 4a. Outliers are reduced by using a robust spectral matching [25] technique.

The positions of the 3-D deformation constraints are found by calculating the pseudo-intersection point of the reprojected rays passing through the image feature locations at $t + 1$. The 3-D constraints are applied to deform $\mathbf{T}_{tet}$ using a stepwise procedure which, in practice, is unlikely to converge to implausible model configurations. We resort to the set of regions on the surface of the tet-mesh and find for each one the best handle from all candidate handles that lie in that region. If no handle is found for a region, we constrain the center of that region to its original 3-D position to prevent unconstrained surface areas from arbitrary drifting.

For each region handle, new intermediate target positions are calculated such that the corresponding vertices in $\mathbf{T}_{tet}(t)$ move in directions as similar as possible to their original normal directions. This step-wise deformation is repeated until the multi-view silhouette overlap error $SIL(\mathbf{T}_{tet}, t + 1)$ (computed as pixel-wise XOR) cannot be improved further. At the end of this step, a feature-based pose estimate $\mathbf{T}_{tet}^{F}(t + 1)$ has been obtained.

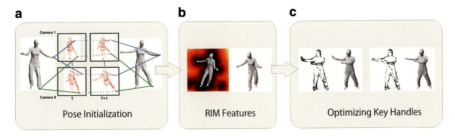

Fig. 4 (a) 3-D correspondences from corresponding SIFT features are used to deform the model into a first pose estimate for $t + 1$. (b) Color-coded distance field and rim vertices with respect to one camera view marked in red on the 3-D model. (c) Model and silhouette overlap after the rim step. Slight pose inaccuracies in the leg and the arms are removed and the model strikes a correct pose after key vertex optimization

4.3.2 Refining the Pose Using Silhouette Rims

In image regions with sparse or low-frequency textures, the pose of $\mathbf{T}_{tet}^F(t+1)$ may not be entirely correct as only few SIFT features could potentially be found. We therefore resort to another constraint that is independent of image texture and has the potential to correct for such misalignments.

We derive additional deformation constraints for a subset of vertices on $\mathbf{T}_{tet}^F(t+1)$ that lie on the silhouette contour. By displacing the constraints along their normals until alignment with the respective silhouette boundaries in 2D is reached, we are able to improve the pose accuracy for the model at $t+1$. The result is a new model configuration $\mathbf{T}_{tet}^R(t+1)$ in which the projections of the outer model contours more closely match the input silhouette boundaries, Fig. 4b.

4.3.3 Optimizing Key Handle Positions

In the majority of cases, the pose of the model in $\mathbf{T}_{tet}^R(t+1)$ is already close to a good match. However, in particular if the scene motion was fast or the initial pose estimate from the first step was not entirely correct, residual pose errors remain. We therefore perform an additional optimization step that corrects such residual errors by globally optimizing the positions of a subset of deformation handles until good silhouette overlap is reached, Fig. 4c.

We only optimize the position of typically 15–25 key vertices, previously selected by the user in a pre-processing step, until the tetrahedral deformation produces optimal silhouette overlap. Tracking robustness is increased by designing our energy function such that surface distances between key handles are preserved, and pose configurations with low distortion energy E_D (see Sect. 4.2.1) are preferred.

Tracking robustness is increased by preserving the distances between key handles, and by generating pose configurations with low distortion energies.

The output of this step is a new configuration of the tetrahedral model $\mathbf{T}_{tet}^O(t+1)$ that captures the overall stance of the model and serves as a starting point for the subsequent surface detail capture.

The above sequence of steps (Sect. 4.3.1–4.3.3) is performed for each pair of subsequent time instants. Typically the second step (silhouette rims) is performed once more after the last silhouette optimization step which, in difficult poses, leads to a better model alignment. Surface detail capture commences after the global poses for all frames were found.

4.3.4 Capturing Surface Detail

After recovering the global pose for each frame we transfer the poses of the tetrahedral mesh $\mathbf{T}_{tet}$ to the triangle mesh $\mathbf{T}_{tri}$ using the algorithm from Sect. 4.2.2. This sequence of high resolution triangle meshes will now be further refined in order to capture small-scale surface detail. We again match the reprojected model to the

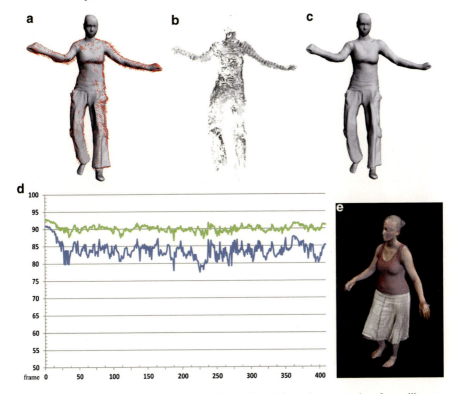

Fig. 5 Capturing small-scale surface detail: (**a**) First, deformation constraints from silhouette contours, shown as red arrows, are estimated. (**b**) Additional deformation handles are extracted from a 3-D point cloud that was computed via model-guided multi-view stereo. (**c**) Together, both sets of constraints deform the surface scan to a highly accurate pose. – **Evaluation**: (**d**) per-frame silhouette overlap in per cent after global pose estimation (*blue*) and after surface detail reconstruction (*green*). (**e**) Blended overlay between an input image and the reconstructed model showing the almost perfect alignment of our result

silhouette rims to better fit the input data and recover deformation detail in the interior of the silhouette with help of a multi-view stereo reconstruction algorithm. The details of the employed stereo reconstruction approach can be found in [11]. We extract position constraints from both of these cues and deform the triangle mesh using our surface Laplacian scheme from Sect. 4.2.3 to match the constraints as closely as possible. A typical set of found position constraints and the result of surface refinement are illustrated in Fig. 5a–c). After a temporal smoothing pass this yields our final output, a dense representation of the performance in both space and time matching the input video as closely as possible.

4.4 Results

The multi-view video data used in our tests comprise of 12 sequences that show four different actors and that feature between 200 and 600 frames each. To show the large application range of our algorithm, the performers wore a wide range of different apparel, ranging from tight to loose, and made of fabrics with prominent texture as well as plain colors only. Also, the recovered set of motions ranges from simple walks, over different dance styles, to fast capoeira sequences. The images in Figs. 6 and 7, as well as the results in the video that can be obtained from

Fig. 6 (**a**) Poses from a fast capoeira performance. (**b**) Jazz dance posture with reliably captured inter-twisted arm motion (Input camera viewpoint and virtual camera viewpoint differ minimally)

Fig. 7 Side-by-side comparison of input and reconstruction of a dancing girl wearing a skirt (input and virtual viewpoints differ minimally). Body pose and detailed geometry of the waving skirt, including lifelike folds and creases visible in the input, have been recovered

http://www.mpi-inf.mpg.de/resources/perfcap/ show that our algorithm faithfully reconstructs this wide spectrum of scenes.

Figure 6a shows two captured poses of a very rapid capoeira sequence in which the actor performs a series of turn kicks. Despite the fact that in our 24 fps recordings the actor rotates by more than 25 degrees in-between some subsequent frames, both shape and motion are reconstructed at high fidelity. The resulting animation even shows deformation details such as the waving of the trouser legs (see video). Furthermore, even with the plain white clothing that the actor wears in the input and which exhibits only few traceable SIFT features, our method performs reliably as it can capitalize on rims and silhouettes as additional sources of information.

The video also shows the captured capoeira sequence with a static checkerboard texture. This result demonstrates that temporal aliasing, such as tangential surface drift of vertex positions, is negligible, and that the overall quality of the meshes remains highly stable.

In Fig. 6b we show two poses from a captured jazz dance performance. As the comparison to the input in image and video shows, we are able to capture this fast and fluent motion. In addition, we can also reconstruct the many poses with complicated self-occlusions, such as the inter-twisted arm-motion in front of the torso.

Figure 7 shows that our algorithm is able to capture the full time-varying shape of a dancing girl wearing a skirt. Even though the skirt is of largely uniform color, our results capture the natural waving and lifelike dynamics of the fabric. In all frames, the overall body posture, and also the folds of the skirt were recovered nicely without the user specifying a segmentation of the model beforehand. We would also like to note that in all skirt sequences the benefits of the stereo step in recovering concavities are most apparent. In the other test scenes, the effects are less pronounced and we therefore deactivated the stereo step (Sect. 4.3.4) there to reduce computation time.

Apart from the scenes shown in the result images, the video contains three more capoeira sequences, two more dance sequences and two more walking sequences.

4.4.1 Validation and Discussion

Table 1 gives detailed average timings for each individual step in our algorithm obtained after code optimization of the version from [11]. These timings were obtained with a single-threaded code running on a Quad Core Intel Xeon Processor E5410 workstation with 2.33 GHz. We still see plenty of room for implementation improvement, and anticipate that parallelization can lead to significant further run time reduction.

To formally validate the accuracy of our method, we have compared the silhouette overlap of our tracked output models with the segmented input frames. We use this criterion since, to our knowledge, there is no gold-standard alternative capturing approach that would provide us with accurate time-varying 3D data. The re-projections of our final results typically overlap with over 85% of the input

Table 1 Average run times per frame for individual steps

Step	Time
SIFT step (Sect. 4.3.1)	$\sim$5 s
Global rim step (Sect. 4.3.2)	$\sim$4 s
Key handle optimization (Sect. 4.3.3)	$\sim$40 s
Capturing Surface Detail (Sect. 4.3.4)	$\sim$34 s

silhouette pixels, already after global pose capture only (blue curve in Fig. 5d). Surface detail capture further improves this overlap to more than 90% as shown by the green curve. Please note that this measure is slightly negatively biased by errors in foreground segmentation in some frames that appear as erroneous silhouette pixels. Visual inspection reveals almost perfect overlap, Fig. 5e.

All 12 input sequences were reconstructed fully-automatically after only minimal initial user input. As part of pre-processing, the user marks the head and foot regions of each model to exclude them from surface detail capture. Even slightest silhouette errors in these regions (in particular due to shadows on the floor and black hair color) would otherwise cause unnatural deformations. Furthermore, for each model the user once marks at most 25 deformation handles needed for the key handle optimization step, Sect. 4.3.3.

In individual frames of two out of three capoeira turn kick sequences (11 out of around 1,000 frames), as well as in one frame of each of the skirt sequences (2 frames from 850 frames), the output of global pose recovery showed slight misalignments in one of the limbs. Please note that, despite these isolated pose errors, the method always recovers immediately and tracks the whole sequence without drifting – this means the algorithm can run without supervision and the results can be checked afterwards. All observed pose misalignments were exclusively due to oversized silhouette areas because of either motion blur or strong shadows on the floor. Both of this could have been prevented by better adjustment of lighting and shutter speed, and more advanced segmentation schemes. In either case of global pose misalignment, at most two deformation handle positions had to be slightly adjusted by the user. In none of the over 3,500 input frames we processed it was necessary to manually correct the output of surface detail capture (Sect. 4.3.4).

Our method is subject to a few further limitations. The current silhouette rim matching may produce erroneous deformations in case the topological structure of the input silhouette is too different from the reprojected model silhouette. However, in none of our test scenes did this turn out to be an issue. In the future, we plan to investigate more sophisticated image registration approaches to solve this problem entirely. Currently, we are recording in a controlled studio environment to obtain good segmentations, but are confident that a more advanced background segmentation will enable us to handle outdoor scenes.

Moreover, there is a resolution limit to our deformation capture scheme. Some of the high-frequency detail in our final result, such as fine wrinkles in clothing or details of the face, has been part of the laser-scan in the first place. The deformation on this level of detail is not actually captured, but this fine detail is "baked in"

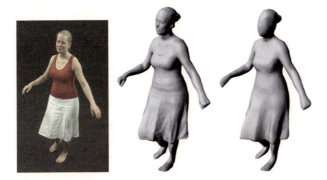

Fig. 8 Input frame (l) and reconstructions using a detailed (m) and a coarse model (r). Although the fine details on the skirt are due to the input laser scan (m), even with a coarse template, our method captures the folds and the overall lifelike motion of the cloth (r)

to the deforming surface. To illustrate the level of detail that we are actually able to reconstruct, we generated a result with a coarse scan only that lacks any fine surface detail. Figure 8 shows an input frame (l), as well as the reconstructions using the detailed scan (m) and the coarse scan (r). While, as noted before, finest detail in Fig. 8(m) is due to the high-resolution laser scan, even with a coarse scan, our method still captures the important lifelike motion and deformation of all surfaces at sufficient detail, Fig. 8(r), in particular cloth motion not visible in the silhouettes alone.

Also, since we rely on a laser scan with fixed topology, our system can currently not track sequences with arbitrarily changing apparent topology (e.g. the movement of hair or deep folds with self-collisions).

Our volume-based deformation technique essentially mimics elastic deformation, thus the geometry generated by the low-frequency tracking may in some cases have a rubbery look. For instance, an arm may not only bend at the elbow, but rather bend along its entire length. Surface detail capture eliminates such artifacts in general, and a more sophisticated yet slower finite element deformation could reduce this problem already at the global pose capture stage.

Despite these limitations, our skeleton-less method can robustly capture a large range of performances at very high detail.

5 Conclusion and Further Reading

Performance capture algorithms enable reconstruction of detailed spatio-temporally coherent scene geometry of scenes from video without having to rely on optical markers. This puts them apart from many previous approaches in the literature and opens up the perspective for many new applications. In this chapter, we presented several recent methods for video-based performance capture, and exemplified the different strategies used to represent geometry and to establish spatio-temporal coherence. The core of the chapter describes a mesh-deformation-based approach and analyzes its benefits and drawbacks in comparison to the other approaches.

The first methods reviewed did not use an a priori template model, but used either stereo, or a combination of stereo and visual hulls to reconstruct a base model to be used as scene representation. From a high-level perspective, the advantage of this strategy is that it makes a method more flexible, and many different scenes can be captured, even if a laser-scan is not available. The conceptual disadvantage is that robustness is much harder to achieve and spatio-temporal coherence is much harder to establish. The methods discussed use some clever yet often computationally expensive cross-parameterization algorithms to solve the latter problem. 3D correspondence finding is itself one of the most challenging problems in dynamic scene reconstruction. The following methods propose a few different strategies to approach this problem that we have not discussed in this chapter [1, 3, 47]. This is not a complete list of references but merely meant to give the reader a starting point.

The second class of algorithms discussed uses a stricter form of a priori shape model to capture performances of humans in general apparel. The first method from this category which we discussed employs a kinematic template model with a loosely deformable surface to retrieve human shape and motion from multi-view video. The kinematic prior greatly helps to make tracking fast and robust, and the silhouette-based surface deformation makes retrieval of cloth motion at a coarse scale feasible. Despite its benefits for tracking the human body itself, however, a skeleton (with surface skinning) generally introduces a wrong bias when tracking cloth regions of a model. The fourth approach discussed in this chapter tries to overcome some of these limitations by explicitly abandoning a skeleton model and using deformable shapes as scene representation. Additionally, the deformation-based approach described also uses a multi-view stereo method to recover true time-varying surface movement also in areas away from silhouette boundaries. This way, more time-varying surface detail than in the purely skeleton-based method can be recovered, lending the final results a more lifelike look. This is particularly visible in the tracking results of the dancer in a skirt shown previously in this chapter where true fold motion, at least at medium resolution, is apparent. The price to be paid, however, is a longer run-time compared to the skeleton-based method and the fact that a kinematic skeleton is not directly available. In our research we were able to show, however, that kinematic skeletons can automatically be learned from moving deforming surfaces which reduces the latter mentioned disadvantage [14].

Overall, this chapter as shown that performance capture techniques open up a new chapter in dynamic scene reconstruction and allow for retrieval of real world performances at such a high level of detail that new levels of quality can be expected in future video game, virtual environment and 3D video productions.

References

1. Ahmed, N., Theobalt, C., Rössl, C., Thrun, S., Seidel, H.P.: Dense correspondence finding for parametrization-free animation reconstruction from video. In: IEEE Conference on Computer Vision and Pattern Recognition (CVPR 2008), pp. 1–8 (2008)
2. Allen, B., Curless, B., Popović, Z.: Articulated body deformation from range scan data. ACM Trans. Graph. **21**(3), 612–619 (2002)
3. Anguelov, D., Koller, D., Srinivasan, P., Thrun, S., Pang, H.C., Davis, J.: The correlated correspondence algorithm for unsupervised registration of nonrigid surfaces. In: Advances in Neural Information Processing Systems (NIPS 2004) (2004)
4. Balan, A.O., Sigal, L., Black, M.J., Davis, J.E., Haussecker, H.W.: Detailed human shape and pose from images. In: Proc. CVPR (2007)
5. Bickel, B., Botsch, M., Angst, R., Matusik, W., Otaduy, M., Pfister, H., Gross, M.: Multi-scale capture of facial geometry and motion. In: Proc. of SIGGRAPH, p. 33 (2007)
6. Botsch, M., Pauly, M., Wicke, M., Gross, M.: Adaptive space deformations based on rigid cells. Comput. Graph. Forum **26**(3), 339–347 (2007)
7. Botsch, M., Sorkine, O.: On linear variational surface deformation methods. IEEE Trans. Visual. Comput. Graph. **14**(1), 213–230 (2008)
8. Bradley, D., Popa, T., Sheffer, A., Heidrich, W., Boubekeur, T.: Markerless garment capture. In: SIGGRAPH '08: ACM SIGGRAPH 2008 Papers, pp. 1–9. ACM (2008)
9. Byrd, R., Lu, P., Nocedal, J., Zhu, C.: A limited memory algorithm for bound constrained optimization. SIAM J. Sci. Comp. **16**(5), 1190–1208 (1995)
10. Carranza, J., Theobalt, C., Magnor, M., Seidel, H.P.: Free-viewpoint video of human actors. In: Proc. SIGGRAPH, pp. 569–577 (2003)
11. de Aguiar, E., Stoll, C., Theobalt, C., Ahmed, N., Seidel, H.P., Thrun, S.: Performance capture from sparse multi-view video. In: SIGGRAPH '08: ACM SIGGRAPH 2008 papers, pp. 1–10. ACM (2008)
12. de Aguiar, E., Theobalt, C., Stoll, C., Seidel, H.: Marker-less 3d feature tracking for mesh-based human motion capture. In: Proc. ICCV Workshop on Human Motion, pp. 1–15 (2007)
13. de Aguiar, E., Theobalt, C., Stoll, C., Seidel, H.P.: Marker-less deformable mesh tracking for human shape and motion capture. In: Proc. CVPR, pp. 1–8. IEEE (2007)
14. de Aguiar, E., Theobalt, C., Thrun, S., Seidel, H.P.: Automatic conversion of mesh animations into skeleton-based animations. Comput. Graph. Forum (Proc. Eurographics EG'08) **27**(2), 389–397 (2008)
15. Einarsson, P., Chabert, C.F., Jones, A., Ma, W.C., Lamond, B., Hawkins, T., Bolas, M., Sylwan, S., Debevec, P.: Relighting human locomotion with flowed reflectance fields. In: Proc. Eurographics Symposium on Rendering, pp. 183–194 (2006)
16. Exluna, Inc.: Entropy 3.1 Technical Reference (2002)
17. Fedkiw, R., Stam, J., Jensen, H.W.: Visual simulation of smoke. In: Fiume, E. (ed.) Proceedings of SIGGRAPH, pp. 15–22. ACM (2001)
18. Goesele, M., Curless, B., Seitz, S.M.: Multi-view stereo revisited. In: Proc. CVPR, pp. 2402–2409 (2006)
19. Gross, M., Würmlin, S., Näf, M., Lamboray, E., Spagno, C., Kunz, A., Koller-Meier, E., Svoboda, T., Gool, L.V., Lang, S., Strehlke, K., Moere, A.V., Staadt, O.: Blue-c: a spatially immersive display and 3D video portal for telepresence. ACM Trans. Graph. **22**(3), 819–827 (2003)
20. Jobson, D.J., Rahman, Z., Woodell, G.A.: Retinex image processing: improved fidelity to direct visual observation. In: Proceedings of the IS&T Fourth Color Imaging Conference: Color Science, Systems, and Applications, vol. 4, pp. 124–125 (1995)
21. Kanade, T., Rander, P., Narayanan, P.J.: Virtualized reality: constructing virtual worlds from real scenes. Proc. IEEE MultiMedia **4**(1), 34–47 (1997)
22. Kartch, D.: Efficient rendering and compression for full-parallax computer-generated holographic stereograms. Ph.D. thesis, Cornell University (2000)

23. Kazhdan, M., Bolitho, M., Hoppe, H.: Poisson surface reconstruction. In: Proc. SGP, pp. 61–70 (2006)
24. Landis, H.: Global illumination in production. In: ACM SIGGRAPH 2002 Course #16 Notes (2002)
25. Leordeanu, M., Hebert, M.: A spectral technique for correspondence problems using pairwise constraints. In: Proc. ICCV (2005)
26. Levoy, M., Pulli, K., Curless, B., Rusinkiewicz, S., Koller, D., Pereira, L., Ginzton, M., Anderson, S., Davis, J., Ginsberg, J., Shade, J., Fulk, D.: The digital michelangelo project. In: Akeley, K. (ed.) Proceedings of SIGGRAPH, pp. 131–144 (2000)
27. Lowe, D.G.: Object recognition from local scale-invariant features. In: Proc. ICCV, vol. 2, p. 1150ff (1999)
28. Matusik, W., Buehler, C., Raskar, R., Gortler, S., McMillan, L.: Image-based visual hulls. In: Proc. SIGGRAPH, pp. 369–374 (2000)
29. Menache, A.: Understanding Motion Capture for Computer Animation and Video Games. Morgan Kaufmann, San Francisco (1999)
30. Mitra, N.J., Flory, S., Ovsjanikov, M., Gelfand, N., AS, L.G., Pottmann, H.: Dynamic geometry registration. In: Proc. Symposium on Geometry Processing, pp. 173–182 (2007)
31. Moeslund, T.B., Hilton, A., Krüger, V.: A survey of advances in vision-based human motion capture and analysis. Comput. Vis. Image Understand. **104**(2), 90–126 (2006)
32. Müller, M., Dorsey, J., McMillan, L., Jagnow, R., Cutler, B.: Stable real-time deformations. In: Proc. of SCA, pp. 49–54. ACM (2002)
33. Nobuhara, S., Matsuyama, T.: Deformable mesh model for complex multi-object 3D motion estimation from multi-viewpoint video. In: 3DPVT06, pp. 264–271 (2006)
34. Paramount: Beowulf movie page. http://www.beowulfmovie.com/ (2007)
35. Park, S.I., Hodgins, J.K.: Capturing and animating skin deformation in human motion. ACM Trans. Graph. (SIGGRAPH 2006) **25**(3) (2006)
36. Park, S.W., Linsen, L., Kreylos, O., Owens, J.D., Hamann, B.: Discrete sibson interpolation. IEEE Trans. Visual. Comput. Graph. **12**(2), 243–253 (2006)
37. Parke, F.I., Waters, K.: Computer Facial Animation. A. K. Peters, Natick (1996)
38. Pellacini, F., Vidimče, K., Lefohn, A., Mohr, A., Leone, M., Warren, J.: Lpics: a hybrid hardware-accelerated relighting engine for computer cinematography. ACM Trans. Graph. **24**(3), 464–470 (2005)
39. Poppe, R.: Vision-based human motion analysis: an overview. Comput. Vis. Image Understand. **108**(1–2), 4–18 (2007)
40. Rosenhahn, B., Kersting, U., Powel, K., Seidel, H.P.: Cloth x-ray: Mocap of people wearing textiles. In: LNCS 4174: Proc. DAGM, pp. 495–504 (2006)
41. Sako, Y., Fujimura, K.: Shape similarity by homotropic deformation. Vis. Comput. **16**(1), 47–61 (2000)
42. Sand, P., McMillan, L., Popović, J.: Continuous capture of skin deformation. ACM Trans. Graph. **22**(3) (2003)
43. Scholz, V., Stich, T., Keckeisen, M., Wacker, M., Magnor, M.: Garment motion capture using color-coded patterns. Comput. Graph. Forum (Proc. Eurographics EG'05) **24**(3), 439–448 (2005)
44. Shinya, M.: Unifying measured point sequences of deforming objects. In: Proc. of 3DPVT, pp. 904–911 (2004)
45. Sorkine, O., Alexa, M.: As-rigid-as-possible surface modeling. In: Proc. SGP, pp. 109–116 (2007)
46. Starck, J., Hilton, A.: Spherical matching for temporal correspondence of non-rigid surfaces. In: IEEE Int. Conf. Computer Vision, pp. 1387–1394 (2005)
47. Starck, J., Hilton, A.: Correspondence labelling for wide-timeframe free-form surface matching. In: Proc. ICCV , pp. 1–8 (2007)
48. Starck, J., Hilton, A.: Surface capture for performance based animation. IEEE Comput. Graph. Appl. **27**(3), 21–31 (2007)
49. Stoll, C., de Aguiar, E., Theobalt, C., Seidel, H.P.: A volumetric approach to interactive shape editing. Research Report MPI-I-2007-4-004, Max-Planck-Institut für Informatik (2007)

50. Stoll, C., Karni, Z., Rössl, C., Yamauchi, H., Seidel, H.P.: Template deformation for point cloud fitting. In: Proc. SGP, pp. 27–35 (2006)
51. Sumner, R.W., Popović, J.: Deformation transfer for triangle meshes. In: SIGGRAPH '04, pp. 399–405 (2004)
52. Vedula, S., Baker, S., Kanade, T.: Image-based spatio-temporal modeling and view interpolation of dynamic events. ACM Trans. Graph. **24**(2), 240–261 (2005)
53. Vlasic, D., Baran, I., Matusik, W., Popović, J.: Articulated mesh animation from multi-view silhouettes. ACM Trans. Graph. **27**(3), 1–9 (2008)
54. Wand, M., Jenke, P., Huang, Q., Bokeloh, M., Guibas, L., Schilling, A.: Reconstruction of deforming geometry from time-varying point clouds. In: Proc. SGP, pp. 49–58 (2007)
55. Waschbüsch, M., Würmlin, S., Cotting, D., Sadlo, F., Gross, M.: Scalable 3D video of dynamic scenes. In: Proc. Pacific Graphics, pp. 629–638 (2005)
56. White, R., Crane, K., Forsyth, D.: Capturing and animating occluded cloth. In: ACM TOG (Proc. SIGGRAPH) (2007)
57. Wilburn, B., Joshi, N., Vaish, V., Talvala, E., Antunez, E., Barth, A., Adams, A., Horowitz, M., Levoy, M.: High performance imaging using large camera arrays. ACM Trans. Graph. **24**(3), 765–776 (2005)
58. Xu, W., Zhou, K., Yu, Y., Tan, Q., Peng, Q., Guo, B.: Gradient domain editing of deforming mesh sequences. In: Proc. SIGGRAPH, p. 84ff. ACM (2007)
59. Yamauchi, H., Gumhold, S., Zayer, R., Seidel, H.P.: Mesh segmentation driven by gaussian curvature. Vis. Comput. **21**(8–10), 649–658 (2005)
60. Yee, Y.L.H.: Spatiotemporal sensistivity and visual attention for efficient rendering of dynamic environments. Master's thesis, Cornell University (2000)
61. Zitnick, C.L., Kang, S.B., Uyttendaele, M., Winder, S., Szeliski, R.: High-quality video view interpolation using a layered representation. ACM Trans. Graph. **23**(3), 600–608 (2004)

Combining Multi-view Stereo and Bundle Adjustment for Accurate Camera Calibration

Yasutaka Furukawa and Jean Ponce

Abstract This article presents an algorithm to achieve accurate camera calibration for 3D reconstruction/visualization systems observing static scenes. The advent of high-resolution digital cameras, and sophisticated 3D reconstruction algorithms such as multi-view stereo offer the promise of unprecedented geometric fidelity in image-based modeling tasks, but it also puts unprecedented demands on camera calibration to fulfill these promises. Camera calibration is an essential step of most such systems involving multiple cameras. While there exist several standard procedure for the task, it is not easy to ensure accurate calibration. In this article, we talk about existing popular camera calibration procedure together with their problems and potential sources of errors, then provide a solution to these problems with an algorithm that produces accurate camera calibration starting from an initial guess possibly containing some errors. More concretely, the algorithm uses a multi-view stereo system on scaled-down input images to reconstruct rough 3D geometry of a scene from initial camera parameters, which is used to effectively guide the search for additional image correspondences. A standard bundle-adjustment algorithm is used with the obtained image correspondences to tighten-up camera calibration. The proposed method has been tested on various real datasets to prove its effectiveness.

1 Introduction

Automated acquisition of 3D geometric models from images has been an important research problem. One of the most successful methods (for static scenes) is *multi-view stereo* (*MVS*) that takes calibrated photographs, namely, images and their

---

Y. Furukawa (✉)
Computer Science and Engineering, University of Washington, Box 352350, Seattle,
WA 98195-2350, USA
e-mail: furukawa@cs.washington.edu

J. Ponce
Willow Team, LIENS (CNRS/ENS/INRIA UMR 8548), Ecole Normale Supérieure, Paris, France
e-mail: Jean.Ponce@ens.fr

R. Ronfard and G. Taubin (eds.), *Image and Geometry Processing for 3-D Cinematography*, Geometry and Computing 5, DOI 10.1007/978-3-642-12392-4_7,
© Springer-Verlag Berlin Heidelberg 2010

associated camera parameters, then produces a dense 3D model. Modern MVS systems are capable of capturing dense and accurate surface models of static complex objects or scenes from a moderate number of calibrated images. More specifically, according to a recent survey by Seitz et al. [20], several algorithms achieve surface coverage of about 95% and depth accuracy of about 0.5 mm for an object 10 cm in diameter observed by 16 low-resolution (640 × 480) cameras. Combined with the emergence of affordable, high-resolution (10 megapixel and higher) consumer-grade cameras, this technology promises even higher, unprecedented geometric fidelity in image-based modeling tasks, but puts tremendous demands on the calibration procedure used to estimate the intrinsic and extrinsic camera parameters, lens distortion coefficients, etc.

1.1 Existing Approaches and Their Problems

There are two main approaches to the calibration problem: The first one, dubbed *chart-based calibration* (or *CBC*) in the rest of this article, assumes that an object with precisely known geometry (the chart) is present in all input images, and computes the camera parameters consistent with a set of correspondences between the features defining the chart and their observed image projections [2, 26]. It is often used in conjunction with positioning systems such as a robot arm [20] or a turntable [12] that can repeat the same motion with high accuracy, so that object and calibration chart pictures can be taken separately but under the same viewing conditions. The second approach to calibration is *structure from motion* (*SFM*), where both the scene shape (structure) and the camera parameters (motion) consistent with a set of correspondences between scene and image features are estimated [11, 19]. In this process, the *intrinsic* camera parameters are often supposed to be known a priori [18], or recovered a posteriori through *auto-calibration* [19, 25]. A final *bundle adjustment* (*BA*) stage is then typically used to fine tune the positions of the scene points and the entire set of camera parameters (including the intrinsic ones and possibly the distortion coefficients) in a single non-linear optimization [16, 24]. A key ingredient of both approaches to calibration is the *selection of feature correspondences* (*SFC*), procedure that may be manual or (partially or totally) automated, and is often intertwined with the calibration process: In a typical SFM system for example [19], features may first be found as "interest points" in all input images, before a robust matching technique such as *RANSAC* [4] is used to simultaneously estimate a set of consistent feature correspondences *and* camera parameters. Some approaches propose to improve feature correspondences for robust camera calibration [17]. However, reliable automated SFC/SFM systems are hard to come by, and they may fail for scenes composed mostly of objects with weak textures (e.g., human faces). In this case, manual feature selection and/or CBC are the only viable alternatives.

Today, despite decades of work and a mature technology, putting together a complete and reliable calibration pipeline thus remains non-trivial procedure requiring much know-how, with various pitfalls and sources of inaccuracy. Automated SFC/SFM methods tend to work well for close-by cameras in controlled environments – though errors tend to accumulate for long-range motions, and they may be ineffective for poorly textured scenes and widely separated input images. CBC systems can be used regardless of scene texture and view separation, but it is difficult to design and build accurate calibration charts with patterns clearly visible from all views. This is particularly true for 3D charts (which are desirable for uniform accuracy over the visible field), but remains a problem even for printed planar grids (the plates the paper is laid on may not be quite flat, laser printers are surprisingly inaccurate, etc.). In addition, the robot arms or turntables used in many experimental setups may not be exactly repetitive. In fact, even a camera attached to a sturdy tripod may be affected during experiments by vibrations from the floor, thermal effects, etc. These seemingly minor factors may not be negligible for modern high-resolution cameras,[1] and they limit the effectiveness of classical chart-based calibration. Of course, sophisticated setups that are less sensitive to these difficulties have been developed by photogrammeters [27], but they typically require special equipment and software that are unfortunately not available in many academic and industrial settings. Our goal, in this article, is to develop a flexible but high-accuracy calibration system that is affordable and accessible to everyone. To this end, a few researchers have proposed using scene information to refine camera calibration parameters: Lavest et al. [15] propose to compensate for the inaccuracy of a calibration chart by adjusting the 3D position of the markers that make it up, but this requires special markers and software for locating them with sufficient sub-pixel precision. The calibration algorithms proposed in Hernández Esteban et al. [13] and Wong and Cipolla [28] exploit silhouette information instead. They work for objects without any texture and are effective in wide-baseline situations, but are limited to circular camera motions.

1.2 Overview of the Proposed Approach

In this article, we propose a very simple and efficient BA algorithm that does not suffer from these limitations and exploits top-down information provided by a rough surface reconstruction to establish image correspondences. Concretely, given a set of input images, possibly inaccurate camera parameters that may have been obtained by an SFM or CBC system, and some conservative estimate of the corresponding reprojection errors, the input images are first scaled down so these errors become

---

[1] For example, the robot arm (Stanford spherical gantry) used in the multi-view stereo evaluation of [20] has an accuracy of $0.01°$ for a 1 m radius sphere observing an object about 15 cm in diameter, which yields approximately $1.0\,[m] \times 0.01 \times \pi/180 = 0.175\,[mm]$ errors near an object. Even with the low-resolution 640×480 cameras used in [20], where a pixel covers roughly 0.25 mm on the surface of an object, this error corresponds to $0.175/0.25 = 0.7$ pixels, which is not negligible. If one used a high-resolution $4,000 \times 3,000$ camera, the positioning error would increase to $0.7 \times 4,000/640 = 4.4$ pixels.

small enough to successfully run a patch-based multi-view stereo algorithm (PMVS [8]) that reconstructs a set of oriented points (points plus normals) densely covering the surface of the observed scene, and identifies the images where they are visible. The core component of the approach proposed in this paper is essentially guided-matching procedure in its second stage, where image features are matched across multiple views using the estimated surface geometry and visibility information. Finally, matched features are input to the SBA bundle adjustment software [16] to tighten up camera parameters.[2] Besides improving camera calibration, the proposed method can significantly speed up SFM systems by running the SFM software on scaled-down input images, then using the proposed algorithm on full-resolution images to tighten-up camera calibration. The left diagram of Fig. 1 illustrates the relationship of the three steps in our algorithm, namely, MVS, BA, and the proposed guided-matching procedure: (1) MVS uses camera parameters (*motion*) to generate a dense 3D model (*structure*); (2) BA uses feature correspondence information (*observation*) to refine camera parameters (*motion*) and 3D coordinates of matched feature points (*structure*); (3) The guided-matching uses rough *structure* information to generate feature correspondence (*observation*). Note that the proposed guided matching procedure closes the *loop* in this diagram, which enables us to iterate the three stages of our algorithm to improve three types of information (i.e., *structure*, *motion*, and *observation*) one by one, and finally achieves high-fidelity camera calibration. The proposed method has been tested on various real datasets, including objects without salient features for which image correspondences cannot be found in a purely bottom-up fashion, and objects with high-curvature and thin structures that are lost in the construction of visual hulls without our bundle adjustment procedure (Sect. 4). In summary, the contributions of the proposed approach can be described as follows:

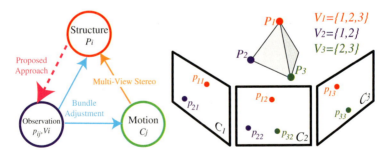

Fig. 1 *Left*: Relationship of multi-view stereo, bundle-adjustment, and the proposed guided-matching approach. *Right* (notations): Three points P_1, P_2, P_3 are observed by three cameras C_1, C_2, C_3. P_{ij} is the image projection of P_i in C_j. V_i is a set of indexes of cameras in which P_i is visible

---

[2] The spirit of guided-matching is also used in [6] to match features of objects with weak textures, although approximate geometry of an object must be known in advance and manual feature correspondences are required in their work.

Combining Multi-view Stereo and Bundle Adjustment for Accurate Camera Calibration   155

- Better feature localization by taking into account surface geometry estimations
- Better coverage and dense feature correspondences by exploiting surface geometry and visibility information
- An ability to handle objects with very weak textures and resolve accumulation errors, two difficult issues for existing SFM and BA algorithms

The rest of this article is organized as follows. Section 2 presents our imaging model together with some notations, and briefly introduce the MVS algorithm used in the article [10]. Section 3 details the proposed algorithm. Experimental results and discussions are given in Sect. 4, with a conclusion given in Sect. 5. Note that implementations of MVS (PMVS by [10]), BA (SBA by [16]), and SFM (Bundler by [22]) algorithms, and several CBC systems such as [2] are publicly available. Bundled with our software, which is also available online at [9], they make a complete software suite for high-accuracy camera calibration.

2   Imaging Model and Preliminaries

Our approach to camera calibration accommodates in principle any parametric projection model of the form $p = f(P, C)$, where P denotes both a scene point and its position in some fixed world coordinate system, C denotes both an image and the corresponding vector of camera parameters, and p denotes the projection of P into the image. In practice, our implementation is currently limited to a standard perspective projection model where C records five intrinsic parameters and six extrinsic ones. Distortion is thus supposed to be negligible, or already corrected, for example by software such as DxO Optics Pro [3]. Standard BA algorithms take the following three data as inputs: a set of n 3D point positions $P_1, P_2, \ldots, P_n$, m camera parameters $C_1, \ldots, C_m$, and the positions of the projections p_{ij} of the points P_i in the images C_j where they are visible (Fig. 1). They optimize both the scene P_i and camera parameters C_j by minimizing, for example, the sum of squared reprojection errors:

$$\sum_{i=1}^{n} \sum_{j \in V_i} (p_{ij} - f(P_i, C_j))^2, \tag{1}$$

where V_i encodes visibility information as the set of indices of images where P_i is visible. Unlike BA algorithms, multi-view stereo algorithms are aimed at recovering scene information alone given fixed camera parameters. In our implementation, we use the PMVS software [8, 10] that generates a set of *oriented* points P_i, together with the corresponding visibility information V_i. We have chosen PMVS because (1) it is one of the best MVS algorithms to date according to the Middlebury benchmarks [20], (2) our method does not require a 3D mesh model but just a set of oriented points, which is the output of PMVS, and (3) as noted earlier, PMVS is freely available [10]. This is also one of the reasons for choosing the SBA software [16] for bundle adjustment, the others being its flexibility and efficiency.

3 Algorithm

The overall algorithm is given in Fig. 2. We first use the oriented points P_i ($i = 1, \ldots, n$) and the corresponding visibility information V_i output by PMVS to form initial image correspondences p_{ij}, then refine these parameters p_{ij} and V_i by simple local image texture comparison in the second step. Given the refined image correspondences, it is possible to rely on SBA to improve the camera parameters. The entire process is repeated a couple of times to tighten up the camera calibration. In this section, we will explain how to initialize and refine feature correspondences.

3.1 Initializing Feature Correspondences

In practice, we have found PMVS to be robust to errors in camera parameters *as long as the image resolution matches the corresponding reprojection errors* – that is, when features to be matched are roughly within two pixels of the corresponding 3D points. Given an initial set of camera parameters, it is usually possible to obtain a conservative estimate of the expected reprojection error E_r by hand (e.g., by visually inspecting a number of epipolar lines) or automatically (e.g., by directly measuring reprojection errors associated with the features matched by a

---

Input: Cameras parameters $\{K_j, R_j, t_j\}$ and
   expected reprojection error E_r.
Output: Refined cameras parameters $\{K_j, R_j, t_j\}$.

Build image pyramids for all the images.
Compute a level L to run PMVS: $L \leftarrow \max(0, \lfloor \log_2 E_r \rfloor)$.
Repeat four times
- Run PMVS on level L of the pyramids to obtain patches $\{P_i\}$ and their visibility information $\{V_i\}$.
- Initialize feature locations: $p_{ij} \leftarrow F(P_i, \{K_j, R_j, t_j\})$.
- Sub-sample feature correspondences.
- *For* each feature correspondence $\{p_{ij} | j \in V_i\}$
  - Identify a *reference* camera C_{j_0} in V_i with the minimum foreshortening factor.
  - *For* each non-reference feature p_{ij} ($j \in V_i, j \neq j_0$)
    - For $L^* \leftarrow L$ down to 0
      - Use level L^* of image pyramids to refine p_{ij}:
        $p_{ij} \leftarrow argmax_{p_{ij}} \mathrm{NCC}(q_{ij}, q_{ij_0})$.
  - Filter out features that have moved too much.
- Refine $\{P_i, K_j, R_j, t_j\}$ by a standard BA with $\{p_{ij}\}$.
- Update E_r by the *mean* and *std* of reprojection errors.

Fig. 2 Overall algorithm description. Starting from initial camera parameters possibly containing some errors, we iterate MVS, the guided-matching, and BA procedure to tighten-up camera calibration

SFM system).[3] Thus, we first build image pyramids for all the input images, then run PMVS on the level $L = \lceil \log_2 E_r \rceil$ of the pyramids. At this level, images are 2^L times smaller than the originals, with reprojection errors of at most about two pixels. We then project the points P_i output by this program into the images where they are visible to obtain an initial set of image correspondences $p_{ij} = f(P_i, C_j)$, with j in V_i. Depending on the value of L and the choice of the PMVS parameter ζ that controls the density of oriented points it constructs, the number of these points, and thus, the number of feature correspondences may become quite large. Dense reconstruction is not necessary for bundle adjustment, and we sub-sample feature correspondences for efficiency.[4] More concretely, we first divide each image into 10×10 uniform blocks, and randomly select within each block at most ϵ features. A feature correspondence will be used in the next refinement step if at least one of its associated image features p_{ij} was sampled in the above procedure. In practice, ϵ is chosen so that the number of feature correspondences becomes 10–20% of the original one after this sampling step. Note that sub-sampling is performed in each block (as opposed to each image) in order to ensure uniformly distributed feature correspondences.

3.2 Refining Feature Correspondences

Due to the use of low-resolution images in PMVS and errors in camera parameters, the initial values of p_{ij} are not accurate. Therefore, the second step of the algorithm is to optimize the feature locations p_{ij} by comparing local image textures. Concretely, since we have an estimate of the surface normal at each point P_i, we consider a small 3D rectangular patch Q_i centered at P_i and construct its projection q_{ij} in the set V_i of images where P_i is visible (Fig. 3). We automatically determine the extent of Q_i so its largest projection covers an image area of about $\delta \times \delta$ pixels (we have used $\delta = 7$ throughout our experiments). In practice, as in [8], a patch Q_i is represented by a $\delta \times \delta$ grid of 3D points and the local image texture inside q_{ij} is, in turn, represented by a set of pixel colors at their image projections that are computed by a bilinear interpolation method.

Next, our problem is to refine feature locations by matching local image textures q_{ij}. For efficiency, we fix the shapes of the image patches q_{ij} and only allow the positions of their centers to change. This is not a problem because, as explained later, we iterate the whole procedure a couple of times and the shapes of the image patches improve over iterations. Note that this image patch optimization is fundamentally

---

[3] In practice, reprojection errors reported by a SFM system tend to be small even when camera parameters contain errors due to poor coverage of matched features. Since E_r is just a conservative estimate of reprojection errors, it is advisable to over-approximate the value.

[4] We could increase the value of ζ to obtain a sparser set of patches without sub-sampling, but, as detailed in [8], a dense reconstruction is necessary for this algorithm to work well and determine visibility information accurately.

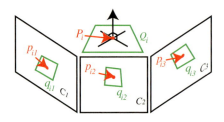

Fig. 3 Given a patch (P_i, Q_i) and the visibility information V_i, we initialize matching images patches (p_{ij}, q_{ij})

different from 3D patch optimization procedure performed by PMVS in that the optimization is carried out as 2D feature matching and does not enforce epipolar geometry constraints that come from possibly erroneous camera parameters. Let us call the camera with the minimum foreshortening factor with respect to P_i the *reference* camera of P_i, and use j_0 to denote its index. We fix the location p_{ij_0} in the reference camera and optimize every other element $p_{ij}, j \neq j_0$ one by one by maximizing the consistency between q_{ij_0} and q_{ij} in a multi-scale fashion. More concretely, starting from the level L of the image pyramids where PMVS was used, a conjugate gradient method is used to optimize p_{ij} by maximizing the normalized cross correlation between q_{ij_0} and q_{ij}. The process is repeated after convergence at the next lower level. After the optimization is complete at the bottom level, we check whether p_{ij} has not moved too much during the optimization. In particular, if p_{ij} has moved more than E_r pixels from its original location, it is removed as an outlier and V_i is updated accordingly. Having refined feature correspondences, we then use the SBA bundle adjustment software [16] to update the camera parameters. In practice, we repeat the whole procedure (PMVS, multi-view feature matching, and SBA) four times to tighten up the camera calibration, while E_r is updated to be the mean plus three times the standard deviation of reprojection errors computed in the last step. Note that L is fixed across iterations instead of recomputed from E_r. This is for efficiency, since PMVS runs slowly with a small value of L.

4 Experimental Results and Discussions

4.1 Datasets

The proposed algorithm has been implemented in C++ and tested on six real datasets, with sample input images shown in Fig. 4, and the number of images and their (approximate) resolution listed in Table 1. The *vase* and *step* datasets have been calibrated by a local implementation of a standard automated SFC/SFM/BA suite as described in [11]. For the *step* dataset, the input images are scaled-down by a factor of five to speed up the execution of the SFM software, but the full-resolution images are used for our refinement algorithm. Our SFM implementation fails on all other datasets except for *predator*, for which 14 out of the 24 images have been calibrated successfully. It is of course possible that a different implementation would

Fig. 4 Sample pictures for the six datasets used in the experiments

Table 1 The number of images and their approximate resolution (in megapixels) are listed for each dataset. E_r is the expected reprojection error in pixels, L is the level of image pyramids used by PMVS, N_p is the number of patches reconstructed by PMVS, and N_t is the number of patches that have successfully generated feature correspondences after sub-sampling

	# of images	# of pixels	E_r	L	N_p	N_t
vase	21	3M	12	3	9,926	1,310
dino	16	0.3M	7	2	5,912	1,763
face	13	1.5M	8	3	7,347	1,997
spiderman	16	1M	7	2	3,344	840
predator	24	2M	7	2	12,760	3,587
step	7	6M	5	2	106,806	9,500

have given better results, but we believe that this is rather typical of practical situations when different views are widely separated and/or textures are not prominent, and this is a good setting to exercise our algorithm. The *spiderman* dataset has been calibrated using a planar checkerboard pattern and a turntable with the calibration software from [2], and the same setup has been used to obtain a second set of camera parameters for the *predator* dataset. The *face* dataset was acquired outdoors, without a calibration chart, and textures are too weak for typical automated SFC/SFM algorithms to work. This is a typical case where, in post-production environments for example, feature correspondences would be manually inserted to calibrate cameras. This is what we have actually done for this dataset. The *dino* dataset is part of the Middlebury MVS evaluation project, and it has been carefully calibrated by the authors of [20]. Nonetheless, this is a very interesting object lacking in salient features and a good example to test our algorithm. Therefore, we have artificially added Gaussian noise to the camera parameters so that reprojection errors become approximately six pixels, yielding a challenging dataset.

Probably due to the use of a rather inaccurate planar calibration board, and a turntable that may not be exactly repetitive, careful visual inspection reveals that *spiderman* and *predator* contain some errors, in particular, for points far away from the turntable where the calibration board was placed. The calibration of *face* is not tight either, because of the sparse manual feature correspondences (at most a few dozens among close-by views) used to calibrate the cameras. The *vase* dataset has relatively small reprojection errors with many close-by images for which SFM algorithms work well, but some images contain large reprojection errors because of the use of a flash and the limited depth of field, and errors do accumulate. The *step* data set does not have these problems, but since scaled-down images are used for the SFC/SFM/BA system, it contains some errors in full resolution images. Note that since silhouette information is used both by the PMVS software and the visual hull computations described in the next section, object silhouettes have been manually extracted using PhotoShop for all datasets except *dino*, where background pixels are close to black, and thresholding followed by morphological operations is sufficient to obtain the silhouettes. Note that the silhouette extraction is not essential for our algorithm, although it helps the system to run and converge more quickly. Furthermore, the use of PMVS is not essential either and this software can be replaced by any other multi-view stereo system.

4.2 Experiments

The two main parameters of PMVS are a correlation window size γ, and a parameter ζ controlling the density of the reconstruction: PMVS tries to reconstruct at least one patch in every $\zeta \times \zeta$ image window. We use $\gamma = 7$ or 9 and $\zeta = 2$ or 4 in all our experiments. Figure 5 shows for each dataset a set of patches reconstructed by PMVS (top row), and its subset that have successfully generated feature correspondences after sub-sampling (bottom row). Table 1 gives some statistics on the matching procedure. E_r denotes a conservative estimate of the expected reprojection errors in pixels, and L denotes the level of image pyramids used by PMVS to reconstruct a set of patches. The number of patches reconstructed by PMVS is denoted by N_p, and the number of patches that successfully generated feature correspondences after sub-sampling is denoted by N_t. Examples of matched 2D features for each dataset are shown in Fig. 6. The histograms of the numbers of images where features are matched by the proposed algorithm are given in Fig. 7. By taking into account the surface orientation and the visibility information estimated by PMVS, the proposed method has been able to match features in many views taken from quite different angles even when image textures are very weak, and hence, producing strong constraints for the BA step. This is also clear from Fig. 8 that shows histograms for feature correspondences obtained by standard SFC/SFM/BA procedure for the *vase* and *step* datasets,[5] and illustrates the fact that features are matched in fewer images compared to the proposed method.

It is impossible to give a full quantitative evaluation of our results given the lack of ground truth 3D data, because constructing such dataset is difficult and expensive, which is beyond the scope of this paper. We can, however, demonstrate that

---

[5] Histograms are shown only for *vase* and *step* in Fig. 8, because the SFC/SFM/BA software fails on the other datasets due to the problems mentioned earlier.

Fig. 5 *Top*: Patches reconstructed by PMVS at level L of the pyramid. *Bottom*: Subsets of these patches that have successfully generated feature correspondences after sub-sampling

our camera calibration procedure does its job as far as improving the reprojection errors of the patches associated with the established feature correspondences. Figure 9 shows matched images features for each dataset, while their colors represent the amounts of the associated final reprojection errors: Red, green, and blue corresponds to two, one, and zero pixels, respectively. Figure 10 shows the mean and standard deviation of these reprojection errors at each iteration of our algorithm for every dataset. The bottom-left graph shows the number of 2D features matched and used to refine camera parameters for the six iterations. The mean reprojection error decreases from 2–3 pixels before refinement to about 0.25–0.5 pixels for most datasets. As described earlier, the process is repeated for four iterations in practice to obtain the final camera parameters, as the two extra iterations in Fig. 10 show a decrease in error but do not seem to affect the quality of our reconstructions much. Note that the following assessment is performed after the fourth iteration of our algorithm.

We have used a couple of different methods to qualitatively assess the accuracy of the estimated camera parameters. First, epipolar geometry has been used to check the consistency between pairs of images (Fig. 2). More concretely, for a pair of images, we draw pairs of epipolar lines in different colors to see if corresponding epipolar lines of the same color pass through the same feature points in the two images. Several images in the *vase* dataset contained large errors before refinement (approximately six pixels in some places) because of the limited depth of field and an exposure difference due to the use of a flash. The *spiderman* and *predator* datasets also contain very large errors, up to seven (or possibly more) pixels for points far from the ground plane where the calibration chart is located. In each case, the proposed method has been able to refine camera parameters to sub-pixel level precision. Inconsistencies in the *dino* dataset introduced by the added noise have also been corrected by our method despite its weak texture.

Next, we have tested the ability of our algorithm to recover camera parameters that are highly consistent across widely separated views. We use the *spiderman*

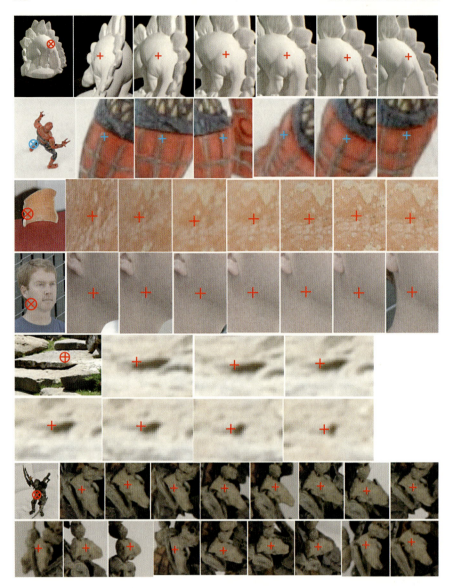

Fig. 6 A set of matching 2D features is shown for each dataset. The proposed method is able to match features in many images even without salient textures due to the use of surface geometry and visibility information estimated by the multi-view stereo algorithm

and *predator* datasets in this experiment (Fig. 12) since parts of these objects are as thin as a few pixels in many images. Recovering such intricate structures normally requires exploiting silhouette information in the form of a visual hull [1] or a hybrid model combining silhouette and texture information [7, 12, 21, 23]. In turn, this requires a high degree of geometric consistency over the cameras, and provides

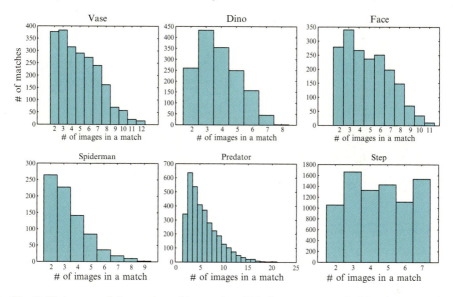

Fig. 7 Histograms of the number of images in which features are matched by the proposed algorithm

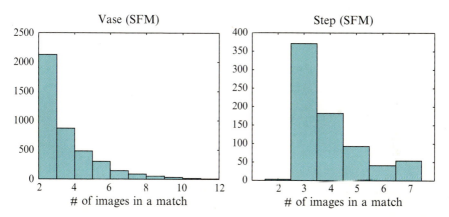

Fig. 8 Histograms of the number of images in which features are matched with a standard SFC/SFM/BA software for *vase* and *step* datasets. In comparison to the proposed algorithm whose results are presented in Fig. 7, features are matched in fewer images

a good testing ground for our algorithm. We have used the EPVH software of Franco and Boyer [5] to construct polyhedral visual hulls in our experiments, and Fig. 12 shows that thin, intricate details such as the fingers of *spiderman* and the blades of *predator* are successfully recovered with refined camera parameters, and completely lost otherwise.

For *dino* and *face*, we have used PMVS to reconstruct a set of patches that are then converted into a 3D mesh model using the method described in [14] (Fig. 12,

Fig. 9 Matched image features are shown for each data set. The *colors* represent the associated reprojection errors computed after the last bundle adjustment step. See text for more details

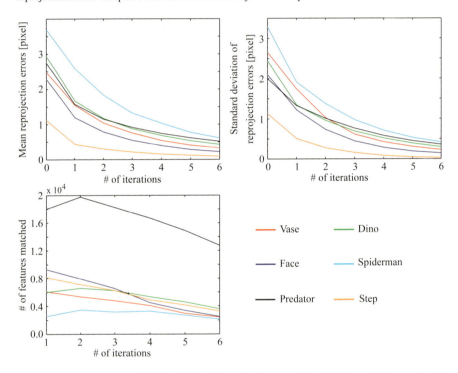

Fig. 10 The mean and standard deviation of reprojection errors in pixel for each dataset at each iteration. The *bottom-left graph* shows the total number of matched 2D features per iteration

bottom right). The large artifacts at the neck and the chin of the shaded *face* reconstruction before refinement are mainly side effects of the use of visual hull constraints in PMVS (patches are not reconstructed outside the visual hull [8]), exacerbated by the fact that the meshing method of [14] extrapolates the surface in areas where data is not present. Ignoring these artifacts, the difference in quality between

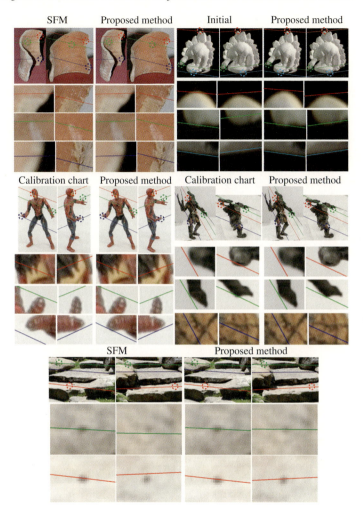

Fig. 11 Epipolar lines are used to assess the improvements in camera parameters. A pair of epipolar lines of the same color must pass through the same feature points

the reconstructions before and after refinement is still obvious in Fig. 12, near the fins of the dinosaur, or the nose and mouth of the face for example. In general, however, the accumulation of errors due to geometric inconsistencies among widely separated cameras is not always visually recognizable in 3D models reconstructed by multi-view stereo, because detailed local reconstructions can be obtained from a set of close cameras, and wide-baseline inconsistencies turn out as low-frequency errors. In order to assess the effectiveness of our algorithm in handling this issue, we pick a pair of widely separated cameras C_1 and C_2, map a texture from one camera C_1 onto the reconstructed model, render it as seen from C_2, and compare the rendered model with the input image associated with C_2. The two images should

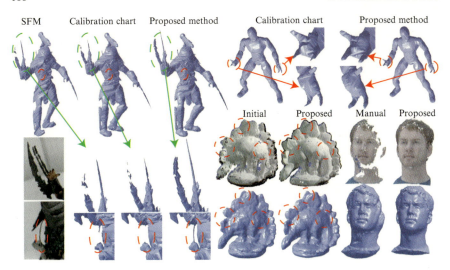

Fig. 12 Visual hull models are used to assess the accuracy of camera parameters for *spiderman* and *predator*. Intricate structures are reconstructed only from the camera parameters refined by the proposed method. For *dino* and *face*, a set of patches reconstructed by PMVS and a 3D mesh model extracted from these patches are used for the assessment. See text for more details

look the same (besides exposure differences) when the camera parameters and the 3D model are accurate. Figure 13 illustrates this on the *vase* and *face* datasets: Mesh models obtained again by combining PMVS [8] and the surface extraction algorithm of [14] are shown for both the initial and refined camera parameters. Although the reconstructed *vase* models do not look very different, the amount of *drifting* between rendered and input images is approximately six pixels for initial camera parameters. Similarly, for the *face* model, the reconstructed surfaces at the left cheek just beside the nose look detailed and similar to each other, while the rendered image is off by approximately six pixels as well. In both cases, the error decreases to sub-pixel levels after refinement. Note that reducing low-frequency errors may not necessarily improve the appearance of 3D models, but is essential in obtaining *accuracy* in applications where the actual model geometry, and not just their overall appearance, is important (e.g., engineering data analysis or high-fidelity surface modeling in the game and movie industries).

Finally, the running time in minutes per iteration of the three steps (PMVS, feature matching, bundle adjustment) of the proposed algorithm on a Dual Xeon 3.2 GHz PC is given in Table 2. As shown by the table, the proposed algorithm is efficient and takes at most a few minutes per iteration to refine camera parameters. Note that the running time of typical CBC systems is also in an order of a few minutes for these data sets. SFC/SFM/BA systems are more computationally expensive, and in particular, a local implementation (Matlab) of a standard SFC/SFM/BA software takes several hours to calibrate the *step* data set with full resolution images. As explained before, the proposed approach reduces such computational expenses

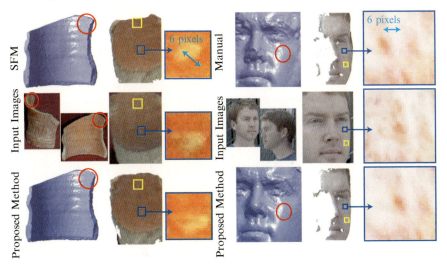

Fig. 13 Inconsistencies in widely separated cameras (accumulation errors) are often not recognizable from 3D mesh models reconstructed by a MVS algorithm. For further assessments, we pick a pair of separated cameras shown in the *middle row*, texture-map the surface from the *right* image, render it to the left, and compare the rendered model with the *left* image. The rendered and the input images look the same only if camera parameters and the reconstructed model are accurate. The *top and the bottom rows* show rendered images and the reconstructed 3D mesh model before and after the refinement, respectively. The amount of errors with the initial camera parameters (calibrated by SFM for *vase* and manual feature correspondences for *face*) is roughly six pixels for both datasets, which are very large

Table 2 Running time in minutes of the three steps of the proposed algorithm for the first iteration

	vase	dino	face	spiderman	predator	step
PMVS	1.9	0.40	0.65	0.34	1.9	4.0
Match	1.1	0.66	0.96	0.24	1.6	0.39
BA	0.17	0.13	0.17	0.03	0.38	1.2

for the step data set by first running a SFC/SFM/BA system with scaled-down input images, which takes only a few minutes, then using the proposed method to tighten up camera calibration.

5 Conclusion

Despite the development and the sophistication of camera calibration software and algorithms, it is, by no means, trivial to obtain camera parameters with high-fidelity for various reasons. We have proposed a novel approach for camera calibration where top-down information from rough camera parameter estimates and the output of a multi-view stereo system on scaled-down input images is used to effectively establish feature correspondences. By taking into account the surface orientation and the visibility information estimated by a multi-view stereo system, the proposed method has been able to match features in many views taken from quite different angles even when image textures are very weak. We have performed three different ways to qualitatively assess the accuracy of refined camera calibration, which shows that the proposed method has successfully reduced calibration errors significantly. Future work will focus on the analysis of remaining errors and influences of various factors that have been ignored in the current framework, such as the second order effects in the camera projection model (distortions) or surface reflectance properties that are assumed to be Lambertian. The implementation of the proposed algorithm is publicly available at [9]. Also see [2, 10, 16, 22] for other useful software that are publicly available.

Acknowledgements This work was supported in part by the National Science Foundation under grant IIS-0535152, the INRIA associated team Thetys, and the Agence Nationale de la Recherch under grants Hfimbr and Triangles. We thank S. Sullivan, A. Suter, and Industrial Light and Magic for the face data set and support of this work. We also thank Jerome Courchay for the SfM software, and Jean-Baptiste Houal for the vase dataset.

References

1. Baumgart, B.: Geometric modeling for computer vision. Ph.D. thesis, Stanford University (1974)
2. Bouguet, J.Y.: Camera calibration toolbox for matlab (2008). http://www.vision.caltech.edu/bouguetj/calib_doc
3. DxO Labs. DxO Optics Pro (2008). http://www.dxo.com
4. Fischler, M., Bolles, R.: Random sample consensus: a paradigm for model fitting with applications to image analysis and automated cartography. Commun. ACM **24**(6), 381–395 (1981)
5. Franco, J.B., Boyer, E.: Exact polyhedral visual hulls. In: British Machine Vision Conference (2003)
6. Fua, P.: Regularized bundle-adjustment to model heads from image sequences without calibration data. Int. J. Comput. Vis. **38**(2), 153–171 (2000)
7. Furukawa, Y., Ponce, J.: Carved visual hulls for image-based modeling. In: European Conference on Computer Vision, pp. 564–577 (2006)
8. Furukawa, Y., Ponce, J.: Accurate, dense, and robust multi-view stereopsis. In: Computer Vision and Pattern Recognition (2007)
9. Furukawa, Y., Ponce, J.: PBA (2008). http://www.cs.washington.edu/homes/furukawa/research/pba
10. Furukawa, Y., Ponce, J.: PMVS (2008). http://www.cs.washington.edu/homes/furukawa/research/pmvs
11. Hartley, R.I., Zisserman, A.: Multiple View Geometry in Computer Vision. Cambridge University Press, Cambridge (2004)
12. Hernández Esteban, C., Schmitt, F.: Silhouette and stereo fusion for 3D object modeling. Comput. Vis. Image Underst. **96**(3), 367–392 (2004)
13. Hernández Esteban, C., Schmitt, F., Cipolla, R.: Silhouette coherence for camera calibration under circular motion. IEEE Trans. Pattern Anal. Mach. Intell. **29**, 343–349 (2007)
14. Kazhdan, M., Bolitho, M., Hoppe, H.: Poisson surface reconstruction. In: Symposium on Geometry Processing (2006)
15. Lavest, J.M., Viala, M., Dhome, M.: Do we really need an accurate calibration pattern to achieve a reliable camera calibration? In: European Conference on Computer Vision (1998)

16. Lourakis, M., Argyros, A.: SBA: a generic sparse bundle adjustment C/C++ package based on the Levenberg–Marquardt algorithm (2008). http://www.ics.forth.gr/~lourakis/sba/
17. Martinec, D., Pajdla, T.: Robust rotation and translation estimation in multiview reconstruction. In: Computer Vision and Pattern Recognition, pp. 1–8 (2007)
18. Nister, D.: An efficient solution to the five-point relative pose problem. IEEE Trans. Pattern Anal. Mach. Intell. **26**(6), 756–777 (2004)
19. Pollefeys, M., Gool, L.V., Vergauwen, M., Verbiest, F., Cornelis, K., Tops, J., Koch, R.: Visual modeling with a hand-held camera. Int. J. Comput. Vis. **59**(3), 207–232 (2004)
20. Seitz, S.M., Curless, B., Diebel, J., Scharstein, D., Szeliski, R.: A comparison and evaluation of multi-view stereo reconstruction algorithms. In: Computer Vision and Pattern Recognition (2006)
21. Sinha, S., Pollefeys, M.: Multi-view reconstruction using photo-consistency and exact silhouette constraints: a maximum-flow formulation. In: International Conference on Computer Vision (2005)
22. Snavely, N.: Bundler: SfM for unordered image collections (2008). http://phototour.cs.washington.edu/bundler/
23. Tran, S., Davis, L.: 3D surface reconstruction using graph cuts with surface constraints. In: European Conference on Computer Vision (2006)
24. Triggs, B., McLauchlan, P., Hartley, R., Fitzgibbon, A.: Bundle adjustment – a modern synthesis. In: W. Triggs, A. Zisserman, R. Szeliski (eds.) Vision Algorithms: Theory and Practice, LNCS, pp. 298–375. Springer, Berlin (2000)
25. Triggs, W.: Auto-calibration and the absolute quadric. In: Computer Vision and Pattern Recognition (1997)
26. Tsai, R.: A versatile camera calibration technique for high-accuracy 3D machine vision metrology using off-the-shelf TV cameras. J. Robot. Automat. **3**(4), 323–344 (1987)
27. Uffenkamp, V.: State of the art of high precision industrial photogrammetry. In: Third International Workshop on Accelerator Alignment. Annecy, France (1993)
28. Wong, K.K., Cipolla, R.: Reconstruction of sculpture from its profiles with unknown camera positions. IEEE Trans. Image Process. **13**(3), 381–389 (2004)

Cell-Based 3D Video Capture Method with Active Cameras

Tatsuhisa Yamaguchi, Hiromasa Yoshimoto, and Takashi Matsuyama

Abstract This paper proposes a 3D video capture method with active cameras, which enables us to produce 3D video of a moving object in a widespread area. Most existing capture methods use fixed cameras and have strong restrictions on allowable object motion; an object cannot move in a wide area. To solve this problem, our method partitions a studio space into a set of subspaces named "cells", and conducts the camera calibration and control for object tracking based on the cells. We first formulate our method as an optimization problem and then propose an algorithm to solve it.

1 Introduction

3D video is a full 3D dynamic shape and texture data generated from multi-view video. A number of studies have proposed 3D video generation methods [1, 2, 5, 9, 10]. These methods first capture target objects as a multi-view video by a set of calibrated video cameras that surrounds the objects, and then generate a 3D video using shape reconstruction algorithms such as Shape-From-Silhouette technique [8]. Namely, these methods are based on the analysis of images using camera geometry and photometry cues that are obtained separately.

In general, the requirements on the multi-view video for 3D video generation from multi-view video are summarized as follows:

Req. 1 Camera calibration: For computing the 3D shape of the objects, all the cameras must be calibrated accurately.

Req. 2 Visual coverage: The cameras must capture the whole surface of the objects.

Req. 3 Spatial resolution: The cameras must capture object texture with high spatial resolution.

T. Matsuyama (✉), H. Yoshimoto, and T. Yamaguchi
Kyoto University, Yoshida-Honmachi, Sakyo-ku, Kyoto, Japan
e-mail: yamaguti@vision.kuee.kyoto-u.ac.jp, yosimoto@vision.kuee.kyoto-u.ac.jp,
tm@vision.kuee.kyoto-u.ac.jp

R. Ronfard and G. Taubin (eds.), *Image and Geometry Processing for 3-D Cinematography*, Geometry and Computing 5, DOI 10.1007/978-3-642-12392-4_8,
© Springer-Verlag Berlin Heidelberg 2010

3D video can be generated in the space where these three requirements are satisfied. We call such space "capture space".

In existing 3D video capture methods, the size of the capture space is small and it is difficult to generate 3D videos of moving objects in a widespread area, such as a walking person. One reason is the use of a fixed set of cameras. As mentioned in [11], camera configurations – camera positions, directions, and focal lengths – are adjusted first in most existing 3D video capture methods. Then, the cameras are calibrated accurately to satisfy Req. 1. However, as the camera configurations are fixed, a trade-off problem between Reqs. 2 and 3 arises. It comes from the trade-off between the angle of view and the spatial resolution of each camera; if we choose a wider angle of view to satisfy Req. 2, the spatial resolution becomes lower and Req. 3 will not be satisfied.

One possible approach to extend the capture space without affecting the resolution is to increase the number of cameras proportionally to the desired capture space size. Nevertheless there is a limitation on the number of cameras due to their cost or spatial constraints of the studio. Moreover, this approach is not desirable from the standpoint of effective use of the cameras because not all cameras can contribute to the reconstruction of each 3D video frame. This issue is also discussed in the chapter by Bennett Wilburn.

Another approach is to track the objects with active cameras. This approach dynamically moves the capture space by controlling directions and zooms of all the active cameras. Hence it can virtually extend the capture space without increasing the number of cameras. In other words, it can use the limited number of cameras more effectively. We adopt this approach.

However, two problems arise when active cameras are used for 3D video production. One concerns the active camera calibration for Req. 1. Although several methods have been proposed to calibrate active cameras [14], it is still difficult to accurately and robustly calibrate active cameras that largely change focal lengths and lens distortions. The other one concerns the real-time camera control for 3D video capture. The 3D position of target objects must be computed from multi-view images and the cameras must be controlled so that Reqs. 2 and 3 are satisfied. Consequently, we derive the fourth requirement:

Req. 4 Track target objects in real-time while satisfying Reqs. 2 and 3.

Ukita et al. have proposed a real-time cooperative active camera control method [13]. They realize real-time tracking of multiple objects with multiple active cameras. Kanade et al. have developed a tracking system with multiple active cameras [6]. These methods cannot be used for 3D video production, because they do not satisfy Reqs. 1 and 2. They produce camera parameters based on the active camera calibration, but there is a limitation in the accuracy. Thus Req. 1 is not satisfied. As for Req. 2, they do not guarantee to satisfy the visual coverage.

We propose a cell-based tracking method for 3D video capture. Our method can be applied to the following situation:

1. There is only one target object.
2. The object moves along a given path and its maximum velocity is given.

3. The resolution requirement is specified by the lowest allowable resolution.
4. The cameras are active PTZ cameras. A camera control value consists of pan, tilt, zoom and focus, and their projection centers are almost fixed.
5. The cameras are surrounding the desired capture space to view the objects there from varying directions.

Our method consists of four steps: The first step divides the space along the given path into N_k subsets named "cells". The second step adjusts a camera control value per camera and per cell, with which every camera can "watch the cell". In this paper, the term "watch a cell" means to capture the whole cell with allowable resolution with a single camera control value. We then calibrate the cameras for each cell; we apply the camera control values associated with the cell and apply existing camera calibration methods for static cameras such as [15] and [12]. By these two steps, Reqs. 1 and 3 are satisfied while each camera is watching one of the cells. The third step makes a camera control schedule by assigning three roles to each camera at every point of the given path in an off-line process: watching the cell where the object is in, switching its view to the next cell, or watching the next cell to anticipate the object movement. The final step is on-line tracking and active camera control. Our method computes the object position from images and controls the cameras based on the schedule. Then, it generates 3D video using images captured by calibrated cameras that were watching a cell. In other words, while a camera is switching its view from one cell to another, the captured images do not satisfy Req. 1. Thus we discard such images.

The rest of this paper formulates the cell-based tracking method as an optimization problem, proposes an algorithm to solve it, and then shows the experimental results and evaluates the performance of our method.

2 Problem Formulation

We formulate each step of our cell based 3D video capture based on the four requirements described in Sect. 1. The processes to satisfy Reqs. 1, 3 and 4 can be formulated as cell-based processes. Req. 1 can be satisfied by camera calibration on each cell. Req. 3 can be satisfied by adjusting zooms for each cell. Req. 4 can be satisfied by on-line tracking and camera control based on cells. In contrast, Req. 2 cannot be always satisfied in 3D video generation from multi-view video. This is because the visual coverage depends on the shape of captured objects, as well as the relative position of objects to cameras. In extreme cases, some types of object shape hinder complete visual coverage by self-occlusion.

This paper proposes a practical method that guarantees to satisfy Reqs. 1, 3 and 4 exactly while achieving best effort to satisfy Req. 2. For this purpose, we first define an evaluation function that estimates how well Req. 2 is satisfied and formulate the process to satisfy Req. 2 as an optimization problem of camera control for maximizing the evaluation function. As for the trade-off problem between Reqs. 2 and 3, mentioned in Sect. 1, our algorithm gives higher priority to Req. 2; it guarantees the

Table 1 Notations for defining the visual coverage

$\mathbb{S} \subseteq \mathbb{R}^3$	3D video studio space.
N_E	Number of active cameras to be used.
$\mathbf{E}_i (i = 1, \cdots, N_\mathrm{E})$	Camera parameters (intrinsic and extrinsic, geometry and photometry parameters.)
r [mm/pixel]	Lowest allowable spatial resolution.
$\mathbb{F}_i \subseteq \mathbb{S}$	View frustum: The space where objects can be captured by camera i with higher resolution than r. This is bounded by the distance from the camera, the depth of field, and the angle of view.
$\mathbf{v} \subseteq \mathbb{S}$	Volume occupied by the object.
$\mathbf{p}_j (j = 1, 2, \ldots, N_j)$	Points on the object surface.
$\mathbf{n}_j$	Surface normal vectors on $\mathbf{p}_j$.
O_i	Projection center of active camera i.
θ_i^j	Angle formed by n_j and $(O_i - \mathbf{p}_j)$
$\mathtt{visible}(\mathbf{p}, (\mathbf{E}_i, \mathbb{F}_i))$	Visibility function: Binary function that returns 1 when $\mathbf{p}$ is visible by a camera with state $(\mathbf{E}_i, \mathbb{F}_i)$ and 0 when unobservable.
$\mathbb{V}(t) \subseteq \mathbb{S}$	Space occupied by the object at time t.
$\mathbf{e}_i(t)$	Camera control value for camera i at time t.
$\bar{\mathbf{E}}_i(\mathbf{e}_i(t))$	Camera parameters of camera i with control value $\mathbf{e}_i(t)$.
$\bar{\mathbb{F}}_i(\mathbf{e}_i(t)) \subseteq \mathbb{S}$	View frustum of camera i with control value $\mathbf{e}_i(t)$

spatial resolution to be higher than the lowest allowable resolution, but no longer optimizes the active camera control for the resolution.

We define the symbols to formulate the problem in Table 1.

Figure 1 shows the geometric relations of some symbols. In order to generate a 3D video, the following two requirements must be satisfied for each point $\mathbf{p}_j$.

- In order to get texture information, $\mathbf{p}_j$ must be observed from the viewpoint directly facing the surface.
- In order to capture good images for Shape-From-Silhouette, $\mathbf{p}_j$ must be observed from the orthogonal direction to the surface normal.

As mentioned, it is sometimes impossible to satisfy these requirements for all of the surface points, considering the shape complexity or self-occlusion of target objects. We first quantify these two requirements for a given pair of a point $\mathbf{p}_j$ and camera i with parameter $\mathbf{E}_i$ and view frustum $\mathbb{F}_i$.

$$q_\mathrm{t}(\mathbf{p}_j, (\mathbf{E}_i, \mathbb{F}_i)) = \mathtt{visible}(\mathbf{p}_j, (\mathbf{E}_i, \mathbb{F}_i)) \cos \theta_i^j$$
$$q_\mathrm{s}(\mathbf{p}_j, (\mathbf{E}_i, \mathbb{F}_i)) = \mathtt{visible}(\mathbf{p}_j, (\mathbf{E}_i, \mathbb{F}_i)) |\sin \theta_i^j|$$

The larger the values of these functions are, the more each requirement is satisfied. These functions imply that it is impossible to satisfy the two requirements by a single camera. However, each requirement can be satisfied by different cameras in the 3D video production with multiple cameras. Thus we quantify the two requirements on each point $\mathbf{p}_j$ as follows:

Cell-Based 3D Video Capture Method with Active Cameras

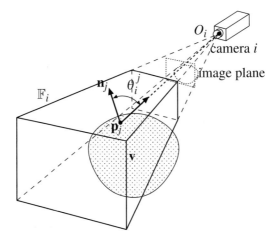

Fig. 1 Geometric relations of the symbols that define the visual coverage

$$q'_t(\mathbf{p}_j, \{(\mathbf{E}_i, \mathbb{F}_i)\}_{i=1,\cdots,N_E}) = \max_i q_t(\mathbf{p}_j, (\mathbf{E}_i, \mathbb{F}_i))$$

$$q'_s(\mathbf{p}_j, \{(\mathbf{E}_i, \mathbb{F}_i)\}_{i=1,\cdots,N_E}) = \max_i q_s(\mathbf{p}_j, (\mathbf{E}_i, \mathbb{F}_i))$$

Then, we choose the least observable point by (1) and (2).

$$Q_t(\mathbf{v}, \{(\mathbf{E}_i, \mathbb{F}_i)\}_{i=1,\cdots,N_E}) = \min_j q'_t(\mathbf{p}_j, \{(\mathbf{E}_i, \mathbb{F}_i)\}_{i=1,\cdots,N_E}) \quad (1)$$

$$Q_s(\mathbf{v}, \{(\mathbf{E}_i, \mathbb{F}_i)\}_{i=1,\cdots,N_E}) = \min_j q'_s(\mathbf{p}_j, \{(\mathbf{E}_i, \mathbb{F}_i)\}_{i=1,\cdots,N_E}) \quad (2)$$

The larger the values of (1) and (2) are, the lager area on the object is observed well. Finally, we define an evaluation function that estimates how well Req. 2 is satisfied by (3).

$$Q(\mathbf{v}, \{(\mathbf{E}_i, \mathbb{F}_i)\}_{i=1,\cdots,N_E}) = Q_t(\mathbf{v}, \{(\mathbf{E}_i, \mathbb{F}_i)\}_{i=1,\cdots,N_E}) Q_s(\mathbf{v}, \{(\mathbf{E}_i, \mathbb{F}_i)\}_{i=1,\cdots,N_E}) \quad (3)$$

Using (3), Req. 2 can be quantitatively evaluated for a given pair of $\mathbf{v}$ and $\{(\mathbf{E}_i, \mathbb{F}_i)\}_{i=1,\cdots,N_E}$. Next, to deal with the object movements, let $\mathbb{V}(t) \subseteq \mathbb{S}$ be the space occupied by the object at time t. The active cameras are controlled via camera control values $\mathbf{e}_i(t)$. Therefore, we denote the camera parameters and view frustums at each time as the functions of $\mathbf{e}_i(t)$; $\bar{\mathbf{E}}_i(\mathbf{e}_i(t))$ and $\overline{\mathbb{F}}_i(\mathbf{e}_i(t))$.

Based on these definitions, the process to capture the whole object surface can be formulated as follows: when given $\mathbb{V}(t)$, optimize $\{\mathbf{e}_i(t)\}_{i=1,2,\cdots,N_E}$ for maximizing $Q(\mathbb{V}(t), \{\bar{\mathbf{E}}_i(\mathbf{e}_i(t)), \overline{\mathbb{F}}_i(\mathbf{e}_i(t))\}_{i=1,\cdots,N_E})$.

3 Cell-Based Active Tracking Algorithm

As described in Sect. 2, cell-based 3D video capture can be formulated as cell-based computations regarding Reqs. 1, 3 and 4, and an optimization for Req. 2 using (3). Our algorithm solves this problem by two steps. The first step divides a studio space into cells and sets up camera control values for each cell in order to satisfy Reqs. 1 and 3. The second step optimizes the camera control timings for Reqs. 2 and 4.

Our algorithm cannot always produce the optimal solution, but just a pseudo-optimal one. This is because the optimization problem mentioned in Sect. 2 is computationally complex. The variables have dependencies on each other; moreover, (3) cannot be computed without setting all the camera control values. These characteristics of the function makes it difficult to apply efficient algorithms such as dynamic programming. Our algorithm is a practical way to find one acceptable solution as it guarantees the multi-view video to satisfy Reqs. 1, 3, and 4.

The inputs to our algorithm consists of scenario and resources described in Table 2 and 3, and the output is a 3D video of a moving object in a widespread area. The algorithm consists of five processes.

1. Cell formation
2. Camera calibration
3. Camera control scheduling
4. Real-time object tracking and camera control
5. 3D video generation

We use the symbols listed in Tables 4 and 5 in the following descriptions.

Table 2 Scenario and its notations

$\{\mathbf{L}(l) \in \mathbb{S} \| l \in [0, L]\}$	Path of the object motion to capture, expressed as a curve with length L, and parameterized by arc length.
L	Length of the path.
$\tilde{\mathbb{V}}(\mathbf{x}) \subseteq \mathbb{S}$	Object shape model. When the object is at $\mathbf{x}(t)$, $\mathbb{V}(t) \subseteq \tilde{\mathbb{V}}(\mathbf{x}(t))$ must be satisfied.
$V_{\max}$	Maximum allowable velocity of the object movement.
r	Lowest allowable spatial resolution.

Table 3 Resources and its notations

τ_{cam}	Video capture interval.
$K_i(\mathbf{v})$	Function that computes an $\mathbf{e}_i$ for camera i to capture all the points in $\mathbf{v}$ with higher resolution than r. If there is no such $\mathbf{e}_i$, $K_i(\mathbf{v}) = \phi$. This function must be designed reflecting the structure of the active camera.
$\tau_i(\mathbf{e}_i, \mathbf{e}_i')$	Length of time needed for camera i to change its state from $\mathbf{e}_i$ into $\mathbf{e}_i'$.
τ_{proc}	Length of time needed to measure the 3D position of the object.
$\text{visible}(\mathbf{p}, (\mathbf{E}_i, \mathbb{F}_i))$	Visibility function (cf. Sect. 2). This must be given depending on the studio setup.

Table 4 Symbols for the cell formation algorithm

$\Delta l = V_{max}\tau_{cam}$	Length of each path fragment.
$\mathbf{f}(n)(n = 0, \ldots, N_f - 1)$	Path fragments.
N_f	Total number of fragments generated.
$\mathbb{W}(n)$	Unit space: space along the nth fragment.
$\mathbb{C}_k \subseteq \mathbb{S}(k = 1, \ldots, N_k)$	Cells.
N_k	Total number of the cells generated.
c_k	Cell border fragment indices. Fragments from $\mathbf{f}(c_k)$ to $\mathbf{f}(c_{k+1})$ belongs to cell k.
$\bar{\mathbf{e}}_i^k$	Camera control value for camera i to watch the kth cell.

Table 5 Symbols for camera control scheduling and real-time tracking

m_i^n	Camera mode. The index of the cell that camera i should watch when the object is on fragment n.
$g^n \subseteq \{1, 2, \ldots, N_E\}$	Subset of cameras that watches a cell when the object is at fragment n.
A_k	Fragment indices where the scheduling problem is divided.
$\mathbf{x}(t)$	Object position.
$\mathbf{I}_i(t)$	Image captured by camera i at time t.

3.1 Cell Formation

This subsection first describes the conditions that cells should satisfy, the definition of cells, and then states the cell formation algorithm.

As mentioned in Sect. 2, our method satisfies Req. 1 and 3 by cell-based processes. Our method satisfies Req. 3 by capturing the whole space of a cell with higher resolution than r, with all the cameras. It means that each cell must be created so that $\bar{\mathbf{e}}_i^k$ satisfying (4) can exist.

$$\mathbb{C}_k \subseteq \overline{\mathbb{F}}_i(\bar{\mathbf{e}}_i^k) \qquad (4)$$

On the other hand, considering the camera calibration costs, the number of the cells should be minimized. In other words, each single cell should be as large as possible.

The cell formation algorithm divides the 3D space along the path into cells that satisfy Reqs. 1 and 3. First, we discretize the given path as well as the nearby space. The path is divided into fragments. The length of each fragment is $\Delta l = V_{max}\tau_{cam}$, which is the maximum possible length that the object can advance in one video capture interval. We also define the unit spaces as the parts of the studio space that may be occupied by the object when the object is on each associated fragment. Equations (6) and (7) show the definitions of the fragments and the unit spaces respectively.

$$k \leftarrow 1$$
$$c_k \leftarrow 0$$
while $c_k < N_\mathrm{f} - 1$ **do**
  for i = 1,2,...,N_E **do**
    $n_i \leftarrow c_k$
    while $n_i < N_\mathrm{f} - 1$ and $K_i \left(\bigcup_{n=c_k}^{n_i+1} \mathbb{W}(n) \right) \neq \phi$ **do**
      $n_i \leftarrow n_i + 1$
    end while
    if $n_i = c_k$ **then**
      return Φ
    end if
    $\bar{\mathbf{e}}_i^k \leftarrow K_i \left(\bigcup_{n=c_k}^{n_i} \mathbb{W}(n) \right)$
  end for
  $c_{k+1} \leftarrow \min\{n_i\}_{i=1,2,...,N_\mathrm{E}}$
  $\mathbb{C}_k \leftarrow \bigcup_{n=c_k}^{c_{k+1}} \mathbb{W}(n)$
  $k \leftarrow k + 1$
end while
$N_k \leftarrow k - 1$
return $\{N_k, \{c_k\}_{k=1,2,...,N_k+1}, \{\mathbb{C}_k\}_{k=1,2,...,N_k}, \{\bar{\mathbf{e}}_i^k\}_{i=1,\cdots,N_\mathrm{E},k=1,2,...,N_k}\}$

Fig. 2 Cell formation algorithm

$$N_\mathrm{f} = \lceil \frac{L}{\Delta l} \rceil \tag{5}$$

$$\mathbf{f}(n) = \{\mathbf{L}(l)|l \in [\Delta l n, \max(L, \Delta l(n+1))]\}(n = 0, \dots, N_\mathrm{f} - 1) \tag{6}$$

$$\mathbb{W}(n) = \bigcup_{\mathbf{x}\in\mathbf{f}(n)} \tilde{V}(\mathbf{x}) \tag{7}$$

Second, we define each cell $\mathbb{C}_k$ as a union of some consecutive $\mathbb{W}(n)$ by the cell formation algorithm presented in Fig. 2. This algorithm outputs N_k, a list of N_k cells and camera control values for each cell that satisfies (4). When there is no such solution, it returns the empty list Φ, meaning that our method cannot satisfy Req. 3 for given scenario with given resources. In this case, our algorithm terminates at this step.

Note that a camera watching $\mathbb{C}_k$ can also satisfy Reqs. 1 and 3 in some part of neighbor cells, i.e., $\mathbb{C}_{k-1}$ and $\mathbb{C}_{k+1}$. These parts of camera views are not necessary for 3D video capture in $\mathbb{C}_k$, but we can make use of it in the camera control scheduling algorithm. As described in Sect. 3.3, such spatial redundancy of camera views are useful for improving the visual coverage.

3.2 Camera Calibration

In order to satisfy Req. 1, camera parameters for each cell, $\bar{\mathbf{E}}_i(\bar{\mathbf{e}}_i^k)$, are obtained by camera calibration. All the active cameras can be regarded as fixed cameras, when they are watching one of the cells. Thus any existing camera calibration methods for fixed cameras, e.g., Zhang's [15] and Svoboda's [12], can be applied here.

3.3 Camera Control Scheduling

In order to satisfy Req. 2, camera control values that maximize the evaluation function (3) are needed. Our algorithm performs this computation by path fragments.

We introduce camera modes m_i^n that express the active camera states. They are defined as follows:

$$m_i^n = \begin{cases} k & \text{watching } \mathbb{C}_k \text{ with parameter } \bar{\mathbf{e}}_i^k \\ 0 & \text{switching its view} \end{cases}$$

These camera modes have the following constraint: Assume that camera i begins to switch its view from $\mathbb{C}_{k_0}$ to $\mathbb{C}_{k_1}$ when the object arrives at $\mathbf{f}(n_0)$. There are two kinds of delays before the camera finishes switching its view. The first one is the processing time τ_{proc}, which includes capturing an image, computing the object position, and communications between computers and active cameras. After that, the active camera requires $\tau_i(\bar{\mathbf{e}}_i^{k_0}, \bar{\mathbf{e}}_i^{k_1})$ time before finishing its motion and resume capturing $\mathbb{C}_{k_1}$. Thus, this camera movement requires $\Delta T = \tau_{\text{proc}} + \tau_i(\bar{\mathbf{e}}_i^{k_0}, \bar{\mathbf{e}}_i^{k_1})$ in total. Meanwhile, the object can advance by $V_{\max} \Delta T$ at the worst case. As shown in Fig. 3, the camera is not guaranteed to capture the object when the object is within the fragments, from $\mathbf{f}(n_0)$ to $\mathbf{f}(n_1 = n_0 + \lfloor \frac{V_{\max} \Delta T}{\Delta l} \rfloor)$. As a result, $m_i^{n_0} = 0, \cdots, m_i^{n_1} = 0$ must be assigned.

Based on the camera modes, we first consider the case when the object is within one of the path fragments $\mathbf{f}(n)$. Only the subset of cameras that is gazing at a cell,

$$g^n = \{i \mid m_i^n \neq 0\}$$

can capture the object, satisfying Req. 1. Note that when the object is in $\mathbb{C}_k$, a camera watching another cell $\mathbb{C}_{k'} (k' \neq k)$ can sometimes capture the object as mentioned in Sect. 3.1, therefore such cameras are contained in g^n. We compute the evaluation function for Req. 2 using only the cameras that belong to g^n as follows:

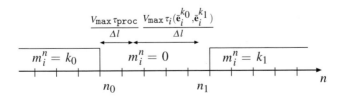

Fig. 3 Assignment of m_i^n reflecting $\tau_i(\mathbf{e}, \mathbf{e}')$ and τ_{proc}

$$Q\left(\mathbb{V}(t), \{(\bar{\mathbf{E}}_i(\bar{\mathbf{e}}_i^{m_i^n}), \mathbb{F}_i(\bar{\mathbf{e}}_i^{m_i^n}))|i \in g^n\}\right)$$

Next, from (7) and the definition of $\tilde{\mathbb{V}}$, $\mathbb{V}(t) \subseteq \mathbb{W}(n)$ can be derived. Therefore we assume that $\{\bar{\mathbf{e}}_i^{m_i^n}|i \in g^n\}$ which maximizes (3) for $\mathbb{W}(n)$ also maximizes the evaluation function for $\mathbb{V}(t)$ as well. Thus we compute the evaluation function for visual coverage for each fragment n using (8).

$$Q(\mathbb{W}(n), \{(\bar{\mathbf{E}}_i(\bar{\mathbf{e}}_i^{m_i^n}), \mathbb{F}_i(\bar{\mathbf{e}}_i^{m_i^n}))|i \in g^n\}) \tag{8}$$

Finally, from the standpoint of guaranteeing 3D video capture, we choose the worst value of (8) in the path as the objective function. In conclusion, the scheduling algorithm solves the following maximization problem.

Variables $\{m_i^n\}_{i=1,\cdots,N_E, n=0,\cdots,N_f}$

Objective Function

$$\min_{n=0,\ldots,N_f-1} Q(\mathbb{W}(n), \{(\bar{\mathbf{E}}_i(\bar{\mathbf{e}}_i^{m_i^n}), \mathbb{F}_i(\bar{\mathbf{e}}_i^{m_i^n}))|i \in g^n\}) \tag{9}$$

Constraints If the gaze of camera i is switched from $\mathbb{C}_{k_0}$ to $\mathbb{C}_{k_1}$ when the object arrives at $\mathbf{f}(n')$, then $m_i^n = 0$ for all n that satisfy $n' \leq n \leq n' + \lfloor \frac{V_{\max}\left(\tau_{\mathrm{proc}}+\tau_i(\bar{\mathbf{e}}_i^k, \bar{\mathbf{e}}_i^{k+1})\right)}{\Delta l} \rfloor$.

This problem requires the full search on the solution space because g^n changes depending on the combinations of the camera mode values. The solution space consists of all the possible combinations of camera mode values, and its size is $O(N_k^{N_f N_E})$. As shown in Sect. 5, N_f exceeds 100 and N_E is more than 20 in our assumed scenarios and resources. Therefore $O(N_k^{N_f N_E})$ is still too large to find the optimal solution by a full search practically. As a reasonable solution, our algorithm divides the problem into $N_k - 1$ independent sub-problems between every pair of adjoining cells, $(\mathbb{C}_k, \mathbb{C}_{k+1})$, by grouping the path fragments into $N_k - 1$ sections $[A_k, A_{k+1}](k = 1, \cdots, N_k - 1)$, where

$$A_k = \begin{cases} c_1 & (k = 1) \\ \lfloor \frac{c_k + c_{k+1}}{2} \rfloor & (2 \leq k \leq N_k - 1) \\ c_{N_k+1} & (k = N_k) \end{cases}$$

and solve each of them. That is, this division puts a restriction on the camera schedule that all the cameras watch $\mathbb{C}_k$ when the object is at A_k and switch its view from $\mathbb{C}_k$ to $\mathbb{C}_{k+1}$ only once when the object moves from $\mathbb{C}_k$ to $\mathbb{C}_{k+1}$. Here, choosing the values for $A_k(k = 2, 3, \ldots, N_k - 1)$ that maximize (9) itself is an optimization problem that is difficult to solve because of the similar reason described above. Instead we choose the center of each cell for A_k based on the following heuristics: In general, the more cameras see the object, the larger the value of (8) is. As to the number of cameras gazing at a cell, it is reduced in the following two cases; While the object is in $[c_k, A_k - 1]$, some of the cameras switch their view from $\mathbb{C}_{k-1}$ to $\mathbb{C}_k$ in order to follow the object. And while the object is in $[A_k, c_{k+1} - 1]$, some of the cameras switch their view from $\mathbb{C}_k$ to $\mathbb{C}_{k+1}$ in order to anticipate the object movement. In both cases, the more frequently those cameras switch their view in the same time, the fewer cameras see the object. Our method reduces such possibilities by making both $[c_k, A_k - 1]$ and $[A_k, c_{k+1} - 1]$ as long as possible. For this reason, we set A_k to the center of each cell, in order to make the value of (9) larger.

Each kth sub-problem is formulated as follows:

Variables $\{m_i^n\}_{i=1,\cdots,N_E,n=0,\cdots,N_f}$

Objective Function

$$\min_{n \in [A_k, A_{k+1}]} Q(\mathbb{W}(n), \{(\bar{\mathbf{E}}_i(\bar{\mathbf{e}}_i^{m_i^n}), \overline{\mathbf{F}}_i(\bar{\mathbf{e}}_i^{m_i^n})) | i \in g^n\}) \tag{10}$$

Constraints Each camera switches its gaze from $\mathbb{C}_k$ to $\mathbb{C}_{k+1}$ only once when the object arrives at fragment $n_i'(A_k < n_i' < A_{k+1} - \lfloor \frac{V_{\max}(\tau_{\mathrm{proc}} + \tau_i(\bar{\mathbf{e}}_i^k, \bar{\mathbf{e}}_i^{k+1}))}{\Delta l} \rfloor)$. m_i^n must satisfy

$$m_i^n = \begin{cases} k & (A_k \le n < n_i') \\ 0 & (n_i' \le n \le n_i' + \lfloor \frac{V_{\max}(\tau_{\mathrm{proc}} + \tau_i(\bar{\mathbf{e}}_i^k, \bar{\mathbf{e}}_i^{k+1}))}{\Delta l} \rfloor \\ k+1 & (n_i' + \lfloor \frac{V_{\max}(\tau_{\mathrm{proc}} + \tau_i(\bar{\mathbf{e}}_i^k, \bar{\mathbf{e}}_i^{k+1}))}{\Delta l} \rfloor < n \le A_{k+1}) \end{cases}$$

The number of solution candidates for each sub-problem is $O((A_k - A_{k-1})^{N_E})$. On the average, $A_k - A_{k-1}$ can be approximated by $\frac{N_f}{N_k}$. Our algorithm solves this problem using genetic algorithm. A sequence of $n_i'(i = 1, \cdots, N_E)$ composes a chromosome, and the fitness function is (10).

3.4 Real-Time Object Tracking and Camera Control

The cell formation and the scheduling processes compute a pseudo-optimal assignment of the cameras as $\{\bar{\mathbf{e}}_i^k\}_{i=1,\cdots,N_E,k=1,\cdots,N_k}$ and $\{m_i^n\}_{i=1,\cdots,N_k,n=0,\ldots,N_f-1}$. Thus the tracking can be performed by measuring the object position $\mathbf{x}(t)$ and controlling the active cameras in parallel.

Figure 4 summarizes the capture process. It is performed by a computer cluster with N_E camera nodes $\pi_i^C(i = 1, \cdots, N_E)$ and one master node π^M. These nodes

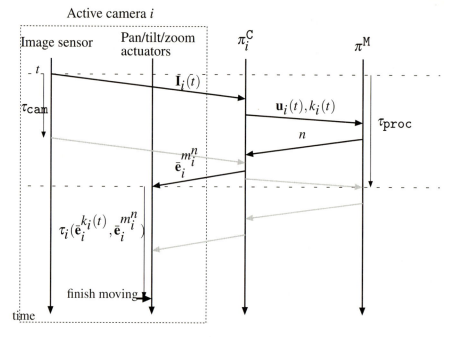

Fig. 4 Overview of capture process

are connected each other to share the object position $\mathbf{x}(t)$. Every camera node has one active camera connected.

In the following descriptions, we denote the time as t and we assume that all the system clocks on the nodes are synchronized. The measurement of the object 3D position is performed as follows:

The 2D Tracking Process on each π_i^C

$\pi_i^C (i = 1, \cdots, N_E)$ repeats the following process in every time interval τ_{cam}.

1. Grab an image $\mathbf{I}_i(t)$.
2. If camera i is gazing at one of the cells,

   a. Store $\{t, \mathbf{I}_i(t), k_i(t)\}$. Here, $k_i(t)$ is the cell number that camera i has been gazing at.
   b. Track the object position on the image and compute its centroid $\mathbf{u}_i(t)$. The tracking is performed by Condensation algorithm [4].
   c. If $\mathbf{u}_i(t)$ is successfully computed, transmit $\{t, \mathbf{u}_i(t), k_i(t)\}$ to π^M.

The 3D Tracking Process on π^M

π^M repeats the following process in every time interval τ_{cam}.

1. When two or more sets out of $\{\{t, \mathbf{u}_i(t), k_i(t)\}|i = 1, \ldots, N_E\}$ have been received, triangulate the 3D position of the object using $\mathbf{u}_i(t), \bar{\mathbf{E}}_i(\bar{\mathbf{e}}_i^{k_i(t)})$.
2. If the 3D position $\mathbf{x}(t)$ is successfully calculated, project $\mathbf{x}(t)$ onto the path and find the corresponding fragment $\mathbf{f}(n)$. Transmit n to all π_i^C.

The camera control is performed by each π_i^C as follows:

The Camera Control Process on each π_i^C

1. Whenever a new n, the fragment number in which the object exists, is received, look up m_i^n and begin switching the active camera state into $\bar{\mathbf{e}}_i^{m_i^n}$.

3.5 3D Video Generation

In the algorithm described above, each π_i^C stores $\{t, \mathbf{I}_i(t), k_i(t)\}$. From these data, a sequence of multi-view images and camera parameters, $\{\mathbf{I}_i(t), \bar{\mathbf{E}}_i(\bar{\mathbf{e}}_i^{k_i(t)})\}$, which satisfy the four requirements can be obtained. It means that our method can generate a 3D video.

4 Experiments and Evaluations

For the experiments described in this section, we arranged 23 active cameras in our 3D video studio, which is about 8 m^2. Each of them is a partially-fixed viewpoint active camera [7] composed of a zoom camera SONY DWF-VL500 and a pan-tilt unit Directed Perception Inc. PTU-46. We set up 23 computers as the camera nodes and 1 computer as the master node.

Figure 5 shows the studio and the active camera arrangement. The hatched area is $\bar{\mathbb{F}}_{16}$, the capturable part of the studio by camera 16, when the camera is directed to a person standing at $(-1,500, 0, 0)$. As shown by this example, in general, camera views that satisfy Req. 3 are limited to part of the studio space and cannot cover all the studio space at one time.

As the resources required by the algorithm, $\tau_i(e_i, e_i')$ was given based on measured dynamic characteristics of each active camera. $K_i(\mathbf{v})$ was made up to find the camera control parameter that captures $\mathbf{v}$ near the center of image was given. As to visible$(\mathbf{p}, \mathbf{n}, (\bar{\mathbf{E}}_i, \bar{\mathbb{F}}_i))$, a function to compute self-occlusion of the object shape was given because there is no other object that occludes the object in our studio.

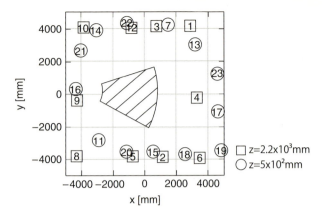

Fig. 5 The studio and the camera arrangement. The numbers represent the camera positions

The following Sects. 4.1, 4.2, and 4.3 describe the details of three experiments conducted under this environment.

4.1 Evaluation of the Visual Coverage Function

First, we show that the evaluation function (3) is effective for estimating the visual coverage of real objects. For this purpose, we compared the evaluation function value Q to the non-observed surface rate R for several different scenes. The lower R is, the better Req. 2 is satisfied.

We used 2 types of digitized human shape model as test data. We generated 1,000 virtual camera configurations for each using the path and camera configuration shown in Fig. 7(a). Each of them was generated by putting an object at a random position on the path, randomly choosing a subset of cameras to be used, and then setting up their control values by K_i. For the object shape model $\tilde{V}$ used for computing (3), a sphere that includes upper half part of the human body was used. The reason why we chose such $\tilde{V}$ is described in Sect. 4.2.

Figure 6 shows the result distribution of (Q, R). Each black dot represents one camera configuration. Because the visual coverage is largely affected by the shape of target object and relative positions of the cameras to the object, there is a large variance in R for any Q. Some parts of the object such as the soles of the feet are physically unobservable and thus R never reaches zero. The maximum value of R for each Q, however, tends to be the lower for the higher Q. It shows that (3) can estimate the worst-case visual coverage for objects that have unknown shape, using the shape model $\tilde{V}$. The results proves that optimizing camera views for (3) leads to better capture of multi-view video from the standpoint of Req. 2.

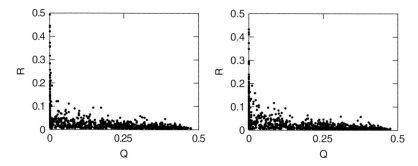

Fig. 6 Distribution of (Q, R) – the visibility function and non-visibility ratio – for two human shape model data

4.2 Tracking Experiments

As an experiment to show the effectiveness of our method, we captured 3D videos of a walking person in a widespread area. The value of (3) changes depending on the relative positions of the cameras to the object. Consequently, the cell formation and camera control scheduling in our algorithm is also affected by them. Hence, we prepared two scenarios with different paths. The arrows in Figs. 7(a) and 7(b) stand for the given path **L** for each scenario. We assumed that the height of the target person is 1.8 m, and that he moves along the given path with varying speed slower than 0.5 m/s. The resolution requirement was set to $r = 8$[mm/pixel].

First, we attempted to capture the whole body of the target. Hence, $\tilde{V}(\mathbf{x})$ was specified as double-stacked spheres at **x** whose radii are 0.45 m. With these scenarios and the resources, our cell formation algorithm detected that it was impossible to capture the target. This is due to our studio setup. Some cameras were too close to the paths. For example, when the object is at $(-1,500, 0, 0)$, camera 16 cannot capture all part of the target because the projected height on its image exceeds the image height. This result shows one of the benefits with our method; it can judge if the requirements can be satisfied before capture.

Second, we attempted to guarantee that the upper half of the body is successfully captured and verified the generated 3D video of it. $\tilde{V}(\mathbf{x})$ was specified by a single 0.45 m-radius sphere 0.9 m above the floor, which includes the upper half part of a standing person. Other resources and scenarios were the same as the first experiment.

We ran the cell formation algorithm with these inputs. The cell formation algorithm was implemented on a computer with Xeon 3.6 GHz CPU. The cells were successfully generated and the running time was less than 5 s in both cases. The hatched areas in Fig. 7 visualize the generated cells $\mathbb{C}_k$ for each scenario. Then, the camera parameters for each cell were calibrated; the intrinsic parameters were estimated by Zhang's method [15], the extrinsic parameters were estimated by the eight-point algorithm [3] for each pair of cameras with 2D-to-2D correspondences

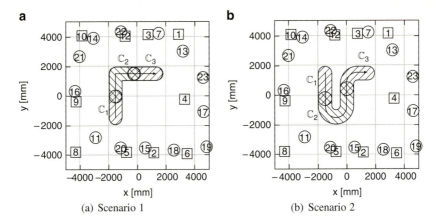

Fig. 7 Path and cells. The arrows represent the given path for each scenario. The hatched parts are generated cells by our algorithm

Table 6 GA optimization details

Number of units per generation	2,000
Crossover method	Uniform crossover
Mutation	Replace one gene, n'_i, with a random value per unit. No mutation for top 200 units.
Generations computed	2,000

of unknown 3D points, and then refined through a bundle adjustment process which minimizes the sum of symmetric epipolar distances of all cameras. After that, camera control was scheduled based on the generated cells. The camera control scheduling algorithm was implemented on a computer with Xeon 3.6 GHz CPU. The parameters for the genetic algorithm in the optimization process is described in Table 6. The computation times were about 80 h. Figure 8 visualizes the result schedules. In these figures, vertical axis represents each active camera and horizontal axis represents target position by fragment number n. Each solid line expresses an interval where the camera is gazing at a cell, and blank parts represents the intervals where the camera is switching view to the next cell. We can see that these schedules are avoiding that too many cameras are switching view to the next cell simultaneously in both cases.

A walking person was tracked for a multi-view video by the cell-based tracking algorithm using these cells and schedules. Then 3D video was generated from the tracking records. Figure 9 shows the captured multi-view video and the generated 3D video. These results indicate that the upper half of the body was successfully captured and 3D video of that part could be generated. Thus it was shown that our method can produce high-resolution 3D video of the specified object.

On the contrary, the lower half was not successfully captured in some frames. This was due to mechanical limitations of the active cameras. This resulted in the lack of texture on the legs, as shown in Fig. 9(c).

Cell-Based 3D Video Capture Method with Active Cameras

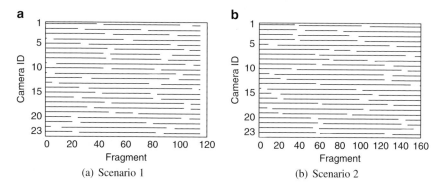

Fig. 8 Optimized schedules. Vertical axis represents each active camera. Horizontal axis represents target position by fragment number. Each solid line expresses an interval where the camera watches a cell, and blank parts represents the intervals where the camera switches its view to the next cell

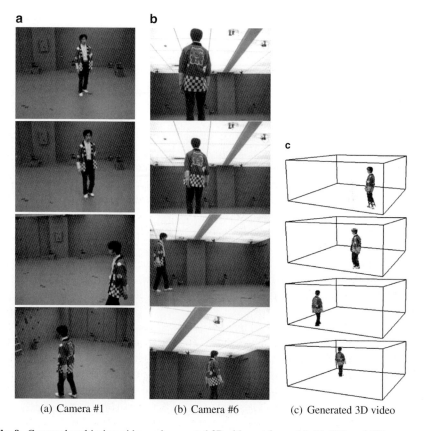

(a) Camera #1   (b) Camera #6   (c) Generated 3D video

Fig. 9 Captured multi-view video and generated 3D video, at frame 15, 54, 104, and 151

In summary, our method has realized these two functions. First, it has realized 3D video production of an object moving in a widespread area. Second, our algorithm can detect that the four requirements cannot be satisfied before tracking.

4.3 Performance Evaluation

For the quantitative evaluation of our method, we compared it with a pair of possible methods with static cameras, from a viewpoint of the effectiveness of the camera usage. We define the following two indices.

$$\text{Viewpoint Usage} = \frac{|G|}{N_{\text{E}}} \tag{11}$$

$$\text{Pixel Usage} = \frac{1}{|G|} \sum_{i \in G} \frac{N_i^{\text{p}}}{N_i^{\text{I}}} \tag{12}$$

G $\{i \,|\, \text{camera } i \text{ is capturing the object}\}$
N_i^{p} Number of the pixels occupied by the object in image i
N_i^{I} Number of pixels in image i. e.g. $307{,}200(= 640 \times 480)$ for VGA.

The larger these indices are, the more information can be obtained for 3D video generation from images. Thus, it leads to the high fidelity of 3D video. As mentioned in Sect. 1, there is a trade-off between the two indices when methods with static cameras are used, especially when capturing an object moving in a widespread area. As mentioned in Sect. 1, we can think of two methods using static cameras: (1) the "view-optimized" method, which gives higher priority to the viewpoint usage and (2) the "resolution-optimized" method, which gives higher priority to the pixel usage.

We used the same scenario as the experiment in Sect. 4.2. In the fixed camera settings for the view-optimized method, views of every camera were adjusted in order to include the entire volume where the object passes. Hence, lenses with very wide angle of view were virtually generated. For the resolution-optimized method, the cameras were divided into 3 groups and assigned to watch one of the three cells generated by the cell formation algorithm. There are 3^{23} combinations to assign 23 cameras to one of the 3 cells independently. Thus we randomly generated 10^{12} combinations and chose the best one that maximizes (3).

The two indices are also sensitive to the shape of the object, as well as the camera configurations. In order to evaluate different camera control methods using the same dynamic scene, we first generated a 3D video using our method and then simulated the other two methods by synthesizing virtual images from the 3D video data.

Figures 10 and 11 shows the two indices for each method. These results show the trade-off problem between viewpoint usage and pixel usage in the methods using static cameras. For example, as shown in Fig. 10, if higher priority is given to viewpoint usage, it is high through the sequences but the pixel usage is lower, since

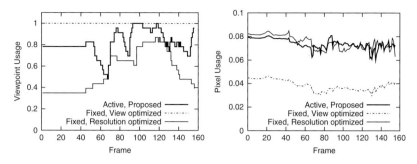

Fig. 10 Viewpoint and pixel usage in scenario 1

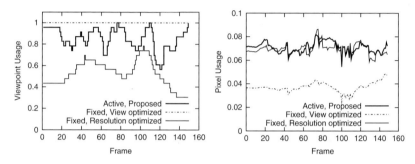

Fig. 11 Viewpoint and pixel usage in scenario 2

the object is projected small into those camera images. The same trade-off is also shown in Fig. 11. As shown, the methods using static cameras cannot capture without losing one of them. On the contrary, our method can improve viewpoint usage while keeping the same pixel usage with the resolution-optimized method. As to the viewpoint usage, though it changes largely depending on the object position and the path, our method has improved the worst values in the sequences compared to the resolution-optimized method in both case.

Finally, we discuss the minimal number of static cameras that would be necessary to achieve the same quality of 3D video with our methods, with these scenarios. Resolution-optimized method retained the same pixel usage with our method. Therefore, the average viewpoint usage gives a rough estimation of the 3D video quality. The average viewpoint usage is 83% with our method whereas it is 54% with the resolution-optimized method. From these figures, we estimate that our method can attain the viewpoint usage about 1.6 times higher than the resolution-optimized method, while keeping the same pixel usage. Consequently, we estimate that roughly 36 fixed cameras are required.

To summarize, our method is effective from the standpoint of viewpoint usage and pixel usage as well.

5 Conclusion and Future Work

Existing 3D video capture methods cannot capture moving objects in a widespread area with high resolution. It is because they use a static camera array with fixed camera views and a trade-off problem arises between the capture space size and the spatial resolution. In order to overcome this limitation, we first summarized the requirements for 3D video capture using active cameras and then formulated the active camera control problem for 3D video capture. Then we proposed a cell-based capture method as a practical solution and showed that our method can capture high-resolution 3D video of an object that moves along a given path.

However, our method can only find a pseudo-optimal camera control but not the optimal one. Moreover, it puts a strong assumption that the target object's path is given in advance. In the near future, we would like to address situations where the target object's path is unknown and would also like to invent the algorithm that can find the optimal camera control, based on the formulations that we have discussed in this paper.

Acknowledgements This research was supported by "Foundation of Technology Supporting the Creation of Digital Media Contents" project (CREST, JST) and Ministry of Education, Culture, Sports, Science and Technology under the Leading Project: "Development of High Fidelity Digitization Software for Large-Scale and Intangible Cultural Assets". We would like to thank Dr. Lyndon Hill and Dr. Tony Tung for their valuable writing corrections.

References

1. Esteban, C., Schmitt, F.: Silhouette and stereo fusion for 3D object modeling. Comput. Vis. Image Underst. **96**(3), 367–392 (2004)
2. Furukawa, Y., Ponce, J.: Carved visual hulls for image-based modeling. In: European Conference on Computer Vision (ECCV), pp. 564–577 (2006)
3. Hartley, R., Zisserman, A.: Multiple View Geometry in Computer Vision, 2nd edn. Cambridge University Press (2004)
4. Isard, M., Blake, A.: Condensation – conditional density propagation for visual tracking. Int. J. Comput. Vis. **29**(1), 5–28 (1998). URL citeseer.ist.psu.edu/isard98condensation.html
5. Kanade, T., Rander, P., Narayanan, P.: Virtualized reality: constructing virtual worlds from real scenes. IEEE Multimed. **4**(1), 33–47 (1997)
6. Kitahara, I., Saito, H., Akimichi, S., Onno, T., Ohta, Y., Kanade, T.: Large-scale virtualized reality. In: Proceedings of IEEE Conference on CVPR Technical Sketches (2001)
7. Kondou, J., Wu, X., Matsuyama, T.: Calibration of partially-fixed viewpoint active camera. IPSJ SIG Notes. CVIM **2003-CVIM-137 No.36**(137-19), 149–156 (20030327). URL http://ci. URL nii.ac.jp/naid/110002664037/
8. Laurentini, A.: The visual hull concept for silhouette based image understanding. IEEE Trans. Pattern Anal. Mach. Intell. **16**(2), 150–162 (1994)
9. Matsuyama, T., Wu, X., Takai, T., Nobuhara, S.: Real-time 3D shape reconstruction, dynamic 3D mesh deformation, and high fidelity visualization for 3D video. Comput. Vis. Image Underst. **96**(3), 393–434 (2004). doi:http://dx.doi.org/10.1016/j.cviu.2004.03.012
10. Starck, J., Hilton, A.: Virtual view synthesis of people from multiple view video sequences. Graph. Models **67**(6), 600–620 (2005)

11. Starck, J., Maki, A., Nobuhara, S., Hilton, A., Matsuyama, T.: The multiple-camera 3-D production studio. IEEE Trans. Circuits Syst. Video Technol. **19**(6), 856–869 (2009)
12. Svoboda, T., Martinec, D., Pajdla, T.: A convenient multicamera self-calibration for virtual environments. Presence: Teleoper. Virtual Environ. **14**(4), 407–422 (2005). doi:http://dx.doi.org/10.1162/105474605774785325
13. Ukita, N., Matsuyama, T.: Real-time cooperative multi-target tracking by communicating active visionagents. Comput. Vis. Image Underst. **97**(2), 137–179 (2005)
14. Wada, T., Matsuyama, T.: Appearance sphere: background model for pan-tilt-zoom camera. In: The 13th International Conference on Pattern Recognition, vol. A, pp. 718–722 (1996)
15. Zhang, Z.: A flexible new technique for camera calibration. IEEE Trans. Pattern Anal. Mach. Intell. **22**(11), 1330–1334 (2000)

Dense 3D Motion Capture from Synchronized Video Streams

Yasutaka Furukawa and Jean Ponce

Abstract This article proposes a novel approach to nonrigid, markerless motion capture from synchronized video streams acquired by calibrated cameras. The instantaneous geometry of the observed scene is represented by a polyhedral mesh with fixed topology. The initial mesh is constructed in the first frame using the publicly available PMVS software for multi-view stereo (Furukawa and Ponce, PMVS, 2008). Its deformation is captured by tracking its vertices over time, using two optimization processes at each frame: a local one using a rigid motion model in the neighborhood of each vertex, and a global one using a regularized nonrigid model for the whole mesh. Qualitative and quantitative experiments using seven real datasets show that our algorithm effectively handles complex nonrigid motions and severe occlusions.

1 Introduction

The most popular approach to motion capture today is to attach distinctive markers to the body and/or face of an actor, and track these markers in images acquired by multiple calibrated video cameras [24]. The marker tracks are then matched, and triangulation is used to reconstruct the corresponding position and velocity information. The accuracy of any motion capture system is limited by the temporal and spatial resolution of the cameras. In the case of marker-based technology, it is also limited by the number of markers available: Although relatively few (say, 50) markers may be sufficient to recover skeletal body configurations, thousands

---

Y. Furukawa (✉)
Computer Science and Engineering, University of Washington, Box 352350, Seattle, WA 98195-2350, USA
e-mail: furukawa@cs.washington.edu

J. Ponce
Willow Team, LIENS (CNRS/ENS/INRIA UMR 8548), Ecole Normale Supérieure, Paris, France
e-mail: Jean.Ponce@ens.fr

R. Ronfard and G. Taubin (eds.), *Image and Geometry Processing for 3-D Cinematography*, Geometry and Computing 5, DOI 10.1007/978-3-642-12392-4_9,
© Springer-Verlag Berlin Heidelberg 2010

may be needed to accurately recover the complex changes in the fold structure of cloth during body motions [25], or model subtle facial motions and skin deformations [17, 18], a problem exacerbated by the fact that people are very good at picking unnatural motions and "wooden" expressions in animated characters. Markerless motion capture methods based on computer vision technology offer an attractive alternative, since they can (in principle) exploit the dynamic texture of the observed surfaces themselves to provide reconstructions with fine surface details[1] and dense estimates of nonrigid motion. Markerless technology using special make-up is indeed emerging in the entertainment industry [15], and several approaches to local *scene flow* estimation have also been proposed to handle less constrained settings [5, 13, 16, 19, 23]. Typically, these methods do not fully exploit global spatiotemporal consistency constraints. They have been mostly limited to relatively simple and slow motions without much occlusion, and may be susceptible to error accumulation. We propose a different approach to motion capture as a 3D tracking problem and show that it effectively overcomes these limitations.

1.1 Related Work

Three-dimensional *active appearance models* (AAMs) are often used for facial motion capture [11, 14]. In this approach, parametric models encoding both facial shape and appearance are fitted to one or several image sequences. AAMs require an a priori parametric face model and are, by design, aimed at tracking relatively coarse facial motions rather than recovering fine surface detail and subtle expressions. *Active sensing* approaches to motion capture use a projected pattern to independently estimate the scene structure in each frame, then use optical flow and/or surface matches between adjacent frames to recover the three-dimensional motion field, or *scene flow* [10, 26]. Although qualitative results are impressive, these methods typically do not exploit the redundancy of the spatiotemporal information, and may be susceptible to error accumulation over time. Several *passive* approaches to scene flow computation have also been proposed [5, 13, 16, 19, 23]. Some start by estimating the optical flow in each image independently, then extract the 3D motion from the recovered flows [13, 23]. Others directly estimate both 3D shape and motion [5, 16, 19]: A variational formulation is proposed in [19], the motion being estimated in a level-set framework, and the shape being refined by the multi-view stereo component of the algorithm (see [6] for related work). A subdivision surface model is used in [16], the shape and motion of an object being initialized independently, then refined simultaneously. In contrast, visible surfaces are

---

[1] This has been demonstrated for static scenes, since, as reported in [21], modern multi-view stereo algorithms now rival laser range scanners with sub-millimeter accuracy and essentially full surface coverage from relatively few low-resolution cameras. Of course, instantaneous shape recovery is not sufficient for motion capture, since nonrigid motion cannot (easily) be recovered from a sequence of instantaneous reconstructions.

represented in [5] as collections of *surfels* – that is, small patches encoding shape, appearance, and motion. In this case, shape is first estimated in each frame independently by a multi-view stereo algorithm, then the 3D motion of each surfel from one frame to the next is estimated.

Existing scene flow algorithms suffer from two limitations: First, they have so far mostly been restricted to simple motions with little occlusion. Second, local motions are typically estimated independently between adjacent frames, then concatenated into long trajectories, causing accumulating drift [20] which may pose problems in applications such as body and face motion capture, or facial expression transfer from human actors to imaginary creatures [3, 15]. A strategy aptly called "track to first" in [4] solves the accumulation problem, and it is exploited in our approach (see [1, 22] for approaches free from accumulation drifts).

1.2 Problem Statement and Proposed Approach

This article addresses motion capture from synchronized, calibrated video streams as a *3D tracking* problem, as opposed to scene flow estimation. The instantaneous geometry of the observed scene is represented by a polyhedral mesh with fixed topology. An initial mesh is constructed in the first frame using the publicly available PMVS software for multi-view stereo [7, 8], and its deformation is captured by tracking its vertices over time with two successive optimization processes at each frame: a local one using a rigid motion model in the neighborhood of each vertex, and a global one using a regularized nonrigid deformation model for the whole mesh. Erroneous motion estimates at vertices with high deformation energy are filtered out as outliers, and the optimization process is repeated without them (see Fig. 1). As demonstrated by our experiments (Sect. 4), the main contributions of this article are in three areas:

- *Handling complex, long-range motions:* Our approach to motion capture as a 3D tracking problem allows us to handle fast, complex, and highly nonrigid motions with limited error accumulation over a large number of frames. This involves several key ingredients: (a) an effective mixture of locally rigid and globally nonrigid, regularized motion models; (b) the decomposition of the former into normal and tangential components, which allows us to use the mature machinery of multi-view stereopsis for shape estimation; and (c) a simple expansion procedure that allows us to propagate to a given vertex the shape and motion parameters inherited from its neighboring vertices.
- *Handling gross errors and heavy occlusion.* Our approach is capable of detecting and recovering from gross matching errors and tracks lost due to partial occlusion, thanks to a second set of key ingredients: (d) an effective representation of surface texture and image photoconsistency that allows us to easily spot outliers; (e) a global representation of shape by an evolving mesh that allows us to stop or restart tracking vertices as they become occluded or once again visible; and (f) an effective means for associating with a surface patch a reference frame and the

Fig. 1 In our approach to motion capture, a polyhedral mesh deforms as its vertices are continuously tracked under locally rigid and globally nonrigid motion models. This is illustrated here with a mesh extracted from real data consisting of eight synchronized video streams 155 frames long [25]. The mesh is shown from two different viewpoints in states 30 frames apart, along with the trajectories of a subset of its vertices (the translational motion is exaggerated for better visualization). See Sect. 4 for details

corresponding texture adaptively during the sequence, which frees us from the need for a perfect initialization.
- *Quantitative validation.* This issue has been mostly ignored in scene flow research, in part because ground truth is usually not available. We have quantitatively as well as qualitatively evaluated the proposed algorithm.

The rest of this article is organized as follows. We first describe our local surface model together with the core tracking procedure to estimate its parameters in Sect. 2, then provide an overall algorithm description in Sect. 3. Experimental results and their discussion are given in Sect. 4, while Sect. 5 concludes this article with some future work.

2 Spatiotemporal Surface Model

We model the surface being tracked as a polyhedral mesh model with a fixed topology and moving vertices $v_1, \ldots, v_n$. As will become clear in the rest of this section, each vertex may or may not be tracked at a given frame, including the first one, allowing us to handle occlusion, fast motion, and parts of the surface that are not visible initially. The core computational task of our algorithm is to estimate in each frame f the position v_i^f of each vertex v_i. The rest of this section presents our local geometric and photometric models of the surface area s_i in the vicinity of v_i (Fig. 2), as well as the core tracking procedure used by our algorithm.

Dense 3D Motion Capture from Synchronized Video Streams

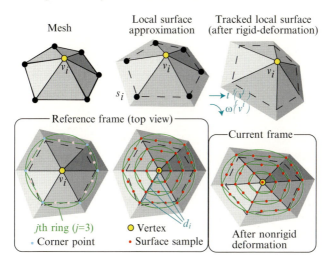

Fig. 2 Local geometric (*top*) and photometric (*bottom*) surface models. The surface region s_i associated with the vertex v_i is simply the union of the incident triangles. The motion of the local surface region around each vertex v_i is represented by a 3D rigid transformation, i.e., translational $t^f(v_i)$ and rotational velocities $\omega^f(v_i)$

2.1 Local Surface Model

2.1.1 Local Geometric Model

We represent the surface in the vicinity of a vertex v_i by the union s_i of the incident triangles, and assume *local* rigid motion at each frame (the mesh *globally* moves in a nonrigid manner with the iteration of the local/global motion steps as explained in Sect. 3.2). Concretely, we attach a coordinate system to s_i with an origin at v_i and a z axis along the surface normal at v_i (the x axis is arbitrarily in the tangent plane), and represent its rigid motion by translational and rotational velocities $t^f(v_i)$ and $\omega^f(v_i)$ (Fig. 2, top).

2.1.2 Local Photometric Model

Some model of spatial texture distribution is needed to measure the *photoconsistency* of different projections of the surface region s_i. We assume that a surface is Lambertian and represent the appearance of s_i in an image by a finite number of pixel samples, which are computed as follows: We construct a discrete set of points sampled at regular intervals in concentric rings around v_i on s_i (Fig. 2, bottom). The spacing d_i between rings is chosen so images of two consecutive rings are (roughly) separated by one pixel in the image where s_i is visible with minimum foreshortening. There are τ rings around each vertex ($\tau = 4$ or 5 in all our experiments), and

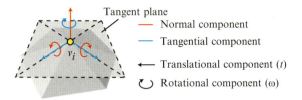

Fig. 3 A local rigid motion consists of translation and rotation, while it can be also considered as a combination of the *tangential* and *normal* components. The normal component encodes depth as well as surface normal, which amounts to the *shape* information of a surface

the ring points are sampled uniformly between *corner points* located at d_i intervals from each other along the edges incident to v_i, with $i-1$ samples per face for ring number i. Finally, each sample point is assigned the corresponding pixel value from an image by bilinear interpolation of neighboring pixel colors. Note that the sample point positions are computed as above only in the *reference frame* $\hat{f}_i$ attached to v_i (see Sect. 3 for how it is determined), and stored as barycentric coordinates in the affine coordinate systems formed by the vertices of the triangles they lie in. In all the other frames, the barycentric coordinates are used to recompute the sample positions. This provides a simple method for projecting them into new images despite nonrigid motions, and retrieving the corresponding texture patterns.[2]

2.2 Shape and Motion Estimation

The rotational and translational velocities $\omega^f(v_i)$ and $t^f(v_i)$ representing the local rigid motion of the patch s_i can be decomposed into *normal* and *tangential* components (Fig. 3). The normal components essentially encode what amounts to shape information in the form of a "tangent plane" (the first two elements of $\omega^f(v_i)$) and a "depth" (the third element of $t^f(v_i)$) along its "normal", and their tangential components encode an in-plane rotation (the third element of $\omega^f(v_i)$) and a translational motion tangent to the surface (the first two elements of $t^f(v_i)$). Instead of estimating all six parameters at once, which is difficult for complex motions, we first estimate the normal (shape) component, then the full 3D motion.

In the following, we describe how motion parameters are initialized before each optimization by a simple *expansion* strategy, how the two motion estimation routines are executed, and how the visibility information (V_i^f) is estimated for these optimizations.

---

[2] Many reconstruction methods, both in multi-view stereo and scene flow reconstructions, use a tangent plane to approximate a local surface region [5, 7], but we choose instead a set of adjacent mesh triangles, because this approximation is more accurate and provides a natural and easy way to handle nonrigid surface deformations.

2.2.1 Initial Motion Estimation by Expansion

Expansion strategies have recently proven extremely effective in turning a sparse set of matches into a dense one in multi-view stereo applications [7, 12]: Typically, a set of matching image patches is iteratively expanded using the spatial coherence of nearby features to *predict* the approximate position of a (yet) unmatched patch. Here, we propose to use the spatiotemporal coherence of nearby vertices to predict the motion structure of a vertex not tracked yet. Concretely, before applying the optimization procedures described below, the instantaneous motion parameters are simply initialized by taking an average of the values at the adjacent vertices that have already been tracked in the current frame. When no adjacent vertex has been tracked (yet), motion parameters are initialized by the values estimated at the vertex itself in the previous frame.

2.2.2 Shape Optimization

Optimizing the normal component of motion is very similar to optimizing depth and surface normal in multi-view stereo [7, 9]. Concretely, we maximize the sum of a *shape* photoconsistency function and a smoothness term

$$\left[\sum_{j \in V_i^f} \sum_{k \in V_i^f, j \neq k} \frac{N(Q_{ij}^f, Q_{ik}^f)}{|V_i^f|(|V_i^f| - 1)/2} \right] - \mu_v^f |\bar{v}_i^f - v_i^f|^2 / \epsilon^2 \tag{1}$$

using a conjugate gradient (CG) method. The first term simply compares sampled local textures in multiple images of the *current* frame to compute an average pairwise correlation score (Fig. 4). In this term, V_i^f denotes the set of indexes of the cameras in which v_i is visible in frame f; Q_{ij}^f is the set of sampled pixels colors for v_i in the image I_j^f acquired by camera number j; and $N(Q, Q')$ denotes the normalized cross correlation between Q and Q'. Note that Q_{ij}^f is determined by the normal components of the velocity field: This is how these parameters enter in our energy function. The second (smoothness) term prevents the vertex from moving too far from its initial position. In this term, μ_v^f is the number of nearby vertices used to initialize the motion parameters, which increases the effect of the smoothness term in the presence of many tracked neighbors, $\bar{v}_i^f$ denotes the position of the vertex at initialization, and ϵ is the average edge length in the mesh for normalization.

2.2.3 Motion Optimization

After optimizing the normal component, the local velocity parameters are all refined by maximizing the sum of a *full motion* photoconsistency function and the same

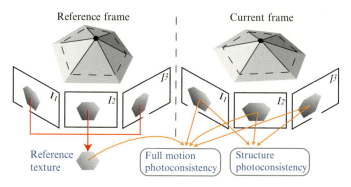

Fig. 4 Structure photoconsistency is computed by comparing images in the current frame, which is the same function used in multi-view stereo algorithms. Full motion photoconsistency is computed from images in the current frame as well as in the reference frame, since it essentially provides tracking information

smoothness term as before:

$$\left[\sum_{j\in V_i^f}\sum_{k\in V_i^{\hat{f}_i}}\frac{N(Q_{ij}^f,Q_{ik}^{\hat{f}_i})}{|V_i^f||V_i^{\hat{f}_i}|}\right]-\mu_v^f|\bar{v}_i^f-v_i^f|^2/\epsilon^2, \qquad(2)$$

using again a CG method. Here, $\hat{f}_i$ is the reference frame of v_i (see Sect. 3 for the method used to determine it), and the first term simply compares reference textures with the image textures in the current frame (this is an example of the "track to first" [4] strategy mentioned earlier). In practice, both the shape and the full motion optimization steps are performed in a multi-scale, coarse-to-fine fashion using a three-level pyramid for each input image.

2.2.4 Visibility Estimation

The computation of the photoconsistency functions ((1) and (2)) requires the visibility information V_i^f, which is estimated as follows: We use the current mesh model to initialize V_i^f, then perform a simple photoconsistency check to filter out images containing unforeseen obstacles or occluders. Concretely, for each image in V_i^f, we compute an average normalized cross correlation score of sampled pixel colors with the remaining visible images. If the average score is below a certain threshold ψ_1, the image is filtered out as outlier. Specific values for this threshold as well as all other parameters are given in Sect. 4 (Table 1).

Dense 3D Motion Capture from Synchronized Video Streams

Table 1 *Top*: Characteristics of the seven datasets: N, F and M are the numbers of cameras, frames and vertices on the mesh; w and h are the width and the height of input images in pixels; and s is the approximate size in pixels of the projection of mesh edges in frontal views. *Bottom*: Parameter values for our algorithm: (η_1, η_2) are the regularization parameters for the first deformation step (they are four times smaller after the filtering); (ψ_1, ψ_2, ψ_3) are thresholds on photoconsistency functions; and ρ is the minimum number of images in which a vertex has to be visible

	flag	*shirt*	*neck*	*face1*	*pants1*	*pants2*	*face2*
N	7	7	7	22	8	8	10
F	37	12	69	90	100	155	325
M	4,828	10,347	5,593	9,035	8,652	8,652	39,612
w	722	722	722	644	480	480	1,000
h	482	482	482	484	640	640	1,002
s	10	6	6	10	6	6	3
η_1	16	16	32	80	20	20	20
η_2	8	200	64	32	40	40	40
ψ_1	0.3	0.3	0.3	0.3	0.1	0.1	0.3
ψ_2	0.5	0.5	0.5	0.5	0.3	0.3	0.5
ψ_3	0.4	0.4	0.4	0.4	0.2	0.2	0.4
ρ	3	3	3	3	2	2	3

3 Algorithm

This section presents the three main steps – local optimization, mesh deformation, and filtering – of our tracking procedure. In practice, these steps are repeated four times at each frame to improve the accuracy of the results. See Fig. 5 at the end of this section for the overall algorithm.

3.1 Local Tracking

Let us now explain how the optimization procedures presented in Sect. 2.2 can be used to estimate the velocity of each vertex in the mesh and identify as needed the corresponding reference frame and reference texture. Vertices to be tracked are stored in a priority queue Z, where pairwise priority is determined by the following rules in order: (1) if a vertex has already been assigned a reference frame, and another one has not, the first one has higher priority; (2) the vertex with most neighbors already tracked in the current frame has higher priority; (3) the vertex with smaller translational motion in the previous frame has higher priority. At the beginning of each frame, we compute a set of visible images for each vertex as described in Sect. 2.2.4, then push onto Z all the vertices with (yet) unknown motion parameters that are visible in at least ρ images. While the queue is not empty, we pop a vertex v_i from the queue, and initialize its instantaneous motion parameters $\omega^f(v_i)$ and $t^f(v_i)$ by the expansion procedure of Sect. 2.2.1. If the vertex has already been

> *Input:* Vertices v_i^{f-1} in the previous frame.
>
> *Output:* Vertices v_i^f in the current frame.
>
> ---
>
> *Repeat* four times
>   Update V_i^f for each vertex v_i (Sect. 2.2.4).
>   Push vertices with unknown motion parameters that
>   are visible in at least ρ images onto a queue Z.
>   *While* Z is not empty
>   Pop a vertex v_i from Z.
>   *If* v_i does not have a reference texture
>   Perform the shape optimization (Sect. 2.2.2).
>   *If* the optimization succeeds
>   Remember the reference texture and sampling points.
>   Update priorities of its adjacent vertices in Z.
>   *else*
>   Perform the shape optimization (Sect. 2.2.2).
>   Perform the full motion optimization (Sect. 2.2.3).
>   *If* the optimization succeeds
>   Update priorities of its adjacent vertices in Z.
>   Deform the mesh by estimated motions (Sect. 3.2).
>   Filter out erroneous motion estimates (Sect. 3.3).
>   Deform the mesh without the erroneous motions (Sect. 3.2).

Fig. 5 Overall tracking algorithm. In each frame, it iterates local tracking, global mesh deformation and filtering four times to deform a mesh to track a surface of interests

assigned a reference frame, the shape optimization and full motion optimization (Sect. 2.2) are performed. At this point, tracking is deemed a failure if the shape photoconsistency term in (1) is below ψ_2 or the full motion photoconsistency term in (2) is below ψ_3, and a success otherwise. If the vertex has not been assigned a reference frame yet, we first compute barycentric coordinates of sample points as described in Sect. 2.1, then perform shape optimization only (the full motion optimization cannot be performed due to the lack of a reference frame). At this point, if the shape photoconsistency in (1) is below ψ_2, we reject the estimated motion. Otherwise, the shape optimization is deemed a success, f becomes the reference frame $\hat{f}_i$, and the corresponding texture is computed by averaging the pixel values in Q_{ij}^f over the images j in V_i^f. In all cases, when tracking succeeds, we update the priority of the vertices adjacent to v_i and their positions in the queue.

3.2 Mesh Deformation

The local tracking step may contain erroneous motion estimates due to its rather greedy approach and the lack of regularization. Therefore, instead of just moving each vertex independently according to the estimated motion, we deform the mesh as a whole by minimizing an energy function that is a weighted sum of *data-attachment*, *smoothness*, and *local rigidity* terms over all the vertices:

$$\sum_i |v_i^f - \hat{v}_i^f|^2 + \eta_1 |[\zeta_2 \Delta^2 - \zeta_1 \Delta] v_i^f|^2 + \eta_2 [\epsilon(v_i^f) - \epsilon(v_i^{\hat{f}_i})]^2.$$

The first (data-attachment) term simply measures the deviation between the actual position v_i^f of v_i in frame f and the position $\hat{v}_i^f$ predicted by the local optimization process. The second term uses the (discrete) Laplacian operator Δ of a local parameterization of the surface in v_i to enforce smoothness ($\zeta_1 = 0.6$ and $\zeta_2 = 0.4$ in all our experiments) [7]. The third (local rigidity) term prevents too much stretching or shrinking of the surface in the neighborhood of v_i by measuring the discrepancy between the mean $\epsilon(v_i^f)$ of the edge lengths around v_i in frame f and its counterpart $\epsilon(v_i^{\hat{f}_i})$ in the reference frame $\hat{f}_i$. The total energy is minimized with respect to the 3D positions of all the vertices again by a CG method. Note that the data-attachment term is used only for vertices that have been successfully tracked.

3.3 Filtering

After surface deformation, we use the residuals $r_d(v_i)$ and $r_l(v_i)$ of the data-attachment and local rigidity terms to filter out erroneous motion estimates.[3] Concretely, we smooth the values of $r_d(v_i)$ and $r_l(v_i)$ at each vertex by replacing each of them by its average over v_i and its neighbors, which process is repeated ten times. After smoothing, a motion estimate is detected as an outlier if $r_d(v_i)$ is more than $\epsilon^2(v_i^{\hat{f}_i})$ or $r_l(v_i)$ is more than $\epsilon^2(v_i^{\hat{f}_i})/4$. Having filtered out the erroneous motions, the mesh is deformed again. We decrease the two regularization parameters η_1 and η_2 by a factor of 4 after the filtering, since the main purpose of the first deformation is to act as a filter, while the second one is used to estimate an accurate surface model.

4 Experimental Results and Discussion

- *Implementation and datasets.* The proposed algorithm has been implemented in C++. A 3D mesh model for each dataset is obtained in the first frame by using the publicly available PMVS software [8] that implements [7], one of the best multi-view stereo algorithms to date according to the Middlebury benchmarks [21].

---

[3] The smoothness residual is not used for filtering, since we want to keep sharp features of the mesh. On the other hand, we want to avoid too much stretching or shrinking for materials such as cloth.

This program outputs a set of *oriented points* (points plus normals), then fits a closed polyhedral mesh over these points, deforming it under the influence of photoconsistency and regularization energy terms. The resulting mesh smoothly extrapolates the reconstructed data in places occluded from the cameras, an important point in practice, since it allows us to start tracking the (extrapolated) vertices when the corresponding surface area becomes visible. Seven real datasets are used for the experiments (Fig. 6): *flag*, *shirt*, *neck* (courtesy of [5]); *face1* (courtesy of [2]); *pants1*, *pants2* (courtesy of [25]), and *face2* (courtesy of ImageMoversDigital). The characteristics of these datasets and the parameter values used in our experiments are given in Table 1. The motions in *neck* and *face1* are very slow, but the textures are weak compared to the other datasets, and the mouth and eye motions in *face1* are challenging. Motions are fast in *flag* and *shirt*, but still relatively simple. On the other hand, *pants1* and *pants2*, although heavily textured, are quite challenging datasets involving fast and complex motions of cloth and its folds, with occlusions in various parts of the videos: In *pants1*, the actor picks up the cloth with his hands, causing severe occlusions and, in *pants2*, he dances very fast, yielding very complex motion and severe self-occlusions due to cloth folding, with image velocities greater than twenty pixels per frame in some image regions. Motions are relatively slow for *face2* throughout the sequence, but occasionally become very fast when the actress speaks.

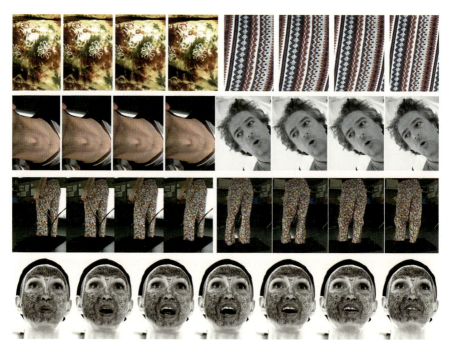

Fig. 6 Sample input images of several contiguous frames from one camera for each dataset. *From left to right* and *top to bottom*, *flag*, *shirt*, *neck*, *face1*, *pants1*, *pants2*, and *face2* datasets

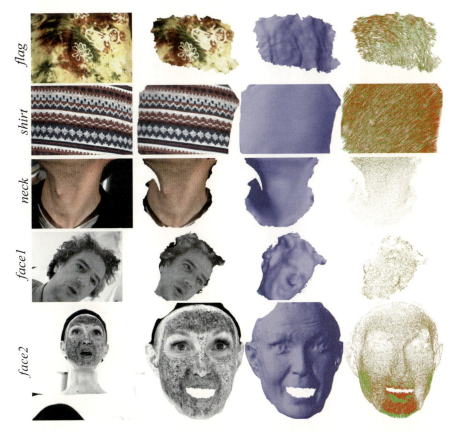

Fig. 7 *From left to right*: an input image, a tracked mesh with and without texture-mapping, and the corresponding motion field. See Fig. 7 for the same results on *pants1* and *pants2* datasets

For *shirt* and *pants1*, we have reversed the original sequences: the motion was too fast otherwise at the beginning of the *shirt* sequence for tracking to succeed. The *pants1* sequence has been reversed in order not to track the hand and arm of the actor, that occlude large portions of the pants in the first frame.

- *Qualitative motion capture experiments.* Figures 7 and 7 show, for each dataset, a sample input image from one frame in the sequence, the corresponding mesh with and without texture mapping, and the estimated motion field, rendered by line segments connecting the positions of sample vertices in the previous frame (red) to the current ones (green). Textures are mapped onto the mesh by averaging the back-projected textures from every visible image in every tracked frame. This is a good way to visually assess the quality of the results, since textures will only look sharp and clear when the estimated shape and motion parameters are accurate throughout the sequence. As shown by the figure, this is indeed the case in our experiments, with sharp images looking very close to the originals. Of course, there are discrepancies in some places. The eyes of *face1* and *face2*

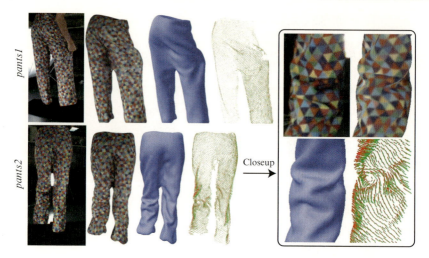

Fig. 8 *From left to right*: an input image, a tracked mesh with and without texture-mapping, and the corresponding motion field. In the close-ups of *pants2*, our texture-mapped model is indeed very close to the corresponding input image, but there are moderate discrepancies in some places, in particular in the middle of the complex fold structure where a surface region not visible by a sufficient number of cameras has not been tracked

provide an example, with a motion different from the other parts of the face and strongly conflicting with our local rigidity term $r_l(v_i)$. A second example is given for *pants2* (see closeups of Fig. 7), corresponding to a case where part of the fold structure of the cloth is not clearly visible by several of the eight cameras. Overall however, our algorithm has been able to accurately capture the cloth's very complicated shape and motion. Given the absence of ground truth data, it is difficult to compare our results to other experiments on the same datasets. The most obvious difference is that our method captures much denser information than [5, 25] for the *pants, flag, shirt*, and *neck* datasets. Indeed White et al. only track the vertices (about 2,400 total) of the triangular pattern printed on the pants in [25], as opposed to the 7,000 mesh vertices or so that we track throughout the two *pants* sequences (see Fig. 11 for more details), without of course exploiting the known structure of the triangular pattern in our case. Likewise, Carceroni and Kutulakos track about 120 *surfels* for *neck*, and 200 for *shirt* and *flag* in [5], whereas the number of tracked vertices varies from about 4,000 to 8,000 for our method.

- *Qualitative evaluation of different key components of the proposed algorithm.* Two experiments have been used for this evaluation. First, we have run the proposed algorithm without the expansion procedure on *pants2* (Fig. 9, top) simply copying motion parameters estimated at the previous frame instead of interpolating the motion of nearby vertices already tracked. As shown in Fig. 9, the cloth motion cannot be captured in frame 124 without the expansion procedure. Our second experiment assesses the contribution of the proposed motion

Dense 3D Motion Capture from Synchronized Video Streams

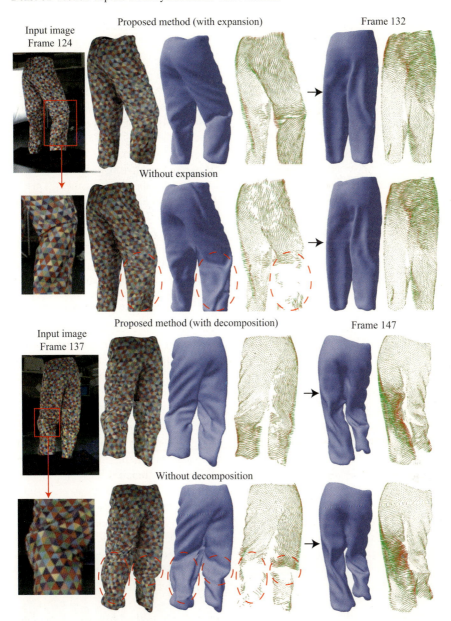

Fig. 9 Qualitative assessments on two key procedure in our algorithm. *Top*: Tracking results with/without the expansion procedure. *Bottom*: Tracking results with/without instantaneous motion decomposition

decomposition. This time, we have run our algorithm by directly applying the full motion optimization step without shape optimization. The bottom half of Fig. 9 shows that tracking without the decomposition fails in recovering details at the back side of both legs. One interesting observation regarding these two experiments is that tracking fails in frame 124 without the expansion scheme, and in frame 137 without motion decomposition, but the algorithm quickly recovers and recaptures the correct shape and motion in frames 132 and 147 of the two sequences (Fig. 9). Thus, even when our basic tracking procedure (local optimization, mesh deformation, and filtering) fails locally in certain frames due to overly complex or fast motions, it is capable of recovering from gross errors, *even when deprived of two key ingredients that further enhance its robustness*. This is a very appealing property for motion capture in practice, because one needs not reinitialize the model every time the tracker fails, and users can just work later on frames where automatic tracking is difficult.

- *Quantitative experiments.* Our last experiments demonstrate the robustness of the proposed method against drift (accumulating errors) and occlusion. Let us first show how our "track to first" strategy limits drift. We have chosen the *flag*, *pants1*, and *pants2* sequences for this experiment, since the corresponding motions are relatively complex. We run the proposed algorithm with and without using a reference frame, updating the reference texture in every frame when a vertex is tracked successfully in the latter case (this resembles the approach followed by most scene flow algorithms). In order to quantitatively measure accuracy, we have appended to each sequence of F frames in the three datasets its reversed copy (without its first frame) to form a new sequence consisting of $2F-1$ frames. Images at frames $F - x$ and $F + x$ are the same, hence the corresponding two meshes should be very close to each other (see [20] for similar experiments for assessing drift in 2D tracking). Let d denote the distance between the positions of the same vertex in frames $F - x$ and $F + x$, divided by the mean edge length of the mesh for normalization. The leftmost graph in Fig. 10 plots, for each frame

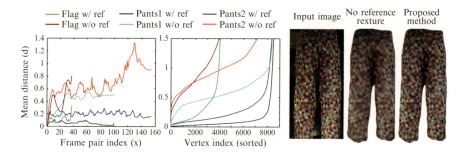

Fig. 10 A synthetic dataset is used to assess the amounts of accumulation errors with and without the use of a reference frame. The *left graph* shows an average distance between a tracked surface and the *ground truth* model over vertices throughout the sequence. The *graph in the center* shows the distribution of errors associated with all the vertices in one frame. *Right*: The accumulation errors can also be assessed qualitatively by comparing texture-mapped models, where the model is blurred if no reference frame is used

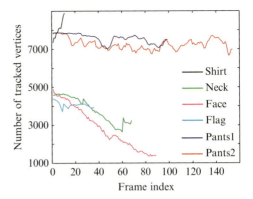

Fig. 11 The number of vertices that have been successfully tracked in each frame. Our algorithm can start tracking new surface regions as shown in the graph for *shirt* dataset

and each dataset, the value of d averaged over all vertices with and without the use of a reference frame.[4] As shown by this figure, the mean distance is consistently three to five times larger for each dataset when reference frames are not used. The value of d for frames 1 and $2F - 1$ ($x = F - 1$) is plotted for every vertex in the next graph (the vertices being sorted in an increasing order of the values of d), showing a similar contrast between the two variants for long-term drift. The added value of reference frames is also (qualitatively) clear from the texture-mapped models for *pants2* shown in the right of Fig. 10, where texture is blurred for the model not using reference frames.

Figure 11 shows the number of vertices that have been successfully tracked in each frame. The number keeps decreasing for *face1*, because a large surface region faces away from most cameras in the middle of the sequence. On the other hand, the number keeps increasing for *shirt* as the cloth moves away from the camera and more surface regions become visible, which illustrates the fact that our method is able to start tracking new vertices in surface areas that have been extrapolated by PMVS but are not visible from the cameras in the first frame with the topology of the mesh being fixed. Of course, a portion of the surface completely hidden from all cameras in the first frame cannot be reconstructed by the multi-view stereo at the beginning for *shirt*, but a surface model larger than the visible portion can be output by interpolation and regularization. Finally, Fig. 12 shows how occlusions are handled by our algorithm for the *pants1* dataset. In frame 51, vertices at the left side of the pants are not tracked due to the severe occlusions caused by a hand, but our algorithm restarts tracking these vertices once they become visible again as shown in our results for frame 77. Note that the right side of the pants is also occluded by the actor's right hand in frame 77, but it is visible from two other cameras, and thus has been successfully tracked by our algorithm.

---

[4] We only retain the "best" vertices with the smallest distances to construct this graph. This is to exclude "outliers" such as vertices that do not correspond to actual parts of the surface (e.g., the top and bottom portions of the mesh in the pants sequences). In practice, 30% and 20% of the vertices in the *flag* and the *pants* sequences are excluded, respectively. See the next graph for the full distance distribution including "outliers".

Fig. 12 At the *left side of the pants*, the tracking has not been performed due to occlusion in frame 55, but our algorithm starts tracking in frame 77 once it becomes visible. In frame 77, the *right side of the pants* is occluded by a hand in one image but visible in other two, and hence, is tracked

Finally, the running time of the proposed method depends on the dataset, the mesh resolution, and the number of input images, but it takes about one to two minutes per frame on a dual Xeon 3.2 GHz PC. It should be possible to speed up the whole computation quite a bit, for example by replacing the numerical derivatives currently used by our conjugate gradient implementation by analytical ones.

5 Conclusion and Future Work

This article proposes a markerless non-rigid motion capture algorithm that tracks a surface by deforming a polyhedral mesh with fixed topology. The proposed algorithm is able to capture dense 3D motion information, which is impossible with existing marker-based technology. One of our future work is to analyze the relative contributions of the key components of our algorithm more thoroughly to reduce the amount of redundant computations. It also seems wasteful to compute angular velocities during local tracking, then discard them during global surface deformation, so we will seek more effective uses for this local information. More importantly perhaps, our current approach to the appearance of new surface regions over time is somewhat ad hoc, relying on PMVS to extrapolate the mesh in regions that are not matched in the first frame without allowing the topology of a mesh to change. A key part of our future work will be to address this problem in a more principled fashion.

Acknowledgements This work was supported in part by the National Science Foundation under grant IIS-0535152, the INRIA associated team Thetys, and the Agence Nationale de la Recherche under grants Hfimbr and Triangles. We thank R.L. Carceroni and K. Kutulakos for *flag*, *shirt* and *neck* datasets, and thank P. Baker and J. Neumann for *face1* dataset. We also thank Hiromi Ono, Doug Epps and ImageMoversDigital for *face2* dataset.

References

1. de Aguiar, E., Theobalt, C., Stoll, C., Seidel, H.P.: Marker-less deformable mesh tracking for human shape and motion capture. In: Computer Vision and Pattern Recognition (2007)
2. Baker, P., Neumann, J.: 3D-photography challenge (2006). http://www.3d-photography.org

3. Bickel, B., Botsch, M., Angst, R., Matusik, W., Otaduy, M., Pfister, H., Gross, M.: Multi-scale capture of facial geometry and motion. ACM Trans. Graph. **26**(3), 33 (2007)
4. Buchanan, A., Fitzgibbon, A.: Interactive feature tracking using k-d trees and dynamic programming. In: Computer Vision and Pattern Recognition (2006)
5. Carceroni, R.L., Kutulakos, K.N.: Multi-view scene capture by surfel sampling: from video streams to non-rigid 3D motion, shape and reflectance. Int. J. Comput. Vis. **49**(2-3), 175–214 (2002)
6. Frederic Huguet, F.D.: A variational method for scene flow estimation from stereo sequences. In: International Conference on Computer Vision (2007)
7. Furukawa, Y., Ponce, J.: Accurate, dense, and robust multi-view stereopsis. In: Computer Vision and Pattern Recognition (2007)
8. Furukawa, Y., Ponce, J.: PMVS (2008). http://www-cvr.ai.uiuc.edu/~yfurukaw/research/pmvs
9. Goesele, M., Snavely, N., Curless, B., Hoppe, H., Seitz, S.M.: Multi-view stereo for community photo collections. In: International Conference on Computer Vision (2007)
10. Hernández Esteban, C., Vogiatzis, G., Brostow, G., Stenger, B., Cipolla, R.: Non-rigid photometric stereo with colored lights. In: International Conference on Computer Vision (2007)
11. Koterba, S.C., Baker, S., Matthews, I., Hu, C., Xiao, J., Cohn, J., Kanade, T.: Multi-view AAM fitting and camera calibration. In: International Conference on Computer Vision (2005)
12. Lhuillier, M., Quan, L.: A quasi-dense approach to surface reconstruction from uncalibrated images. IEEE Trans. Pattern Anal. Mach. Intell. **27**(3), 418–433 (2005)
13. Li, R., Sclaroff, S.: Multi-scale 3D scene flow from binocular stereo sequences. In: IEEE Workshop on Motion and Video Computing (2005)
14. Matthews, I., Baker, S.: Active appearance models revisited. Int. J. Comput. Vis. **60**(2), 135–164 (2004)
15. Mova: LLC. Mova contour reality capture (2009). http://www.mova.com
16. Neumann, J., Aloimonos, Y.: Spatio-temporal stereo using multi-resolution subdivision surfaces. Int. J. Comput. Vis. **47**(1-3), 181–193 (2002)
17. Odisio, M., Bailly, G.: Shape and appearance models of talking faces for model-based tracking. In: IEEE International Workshop on Analysis and Modeling of Faces and Gestures (2003)
18. Park, S.I., Hodgins, J.K.: Capturing and animating skin deformation in human motion. ACM Trans. Graph. **25**(3), 881–889 (2006)
19. Pons, J.P., Keriven, R., Faugeras, O.: Multi-view stereo reconstruction and scene flow estimation with a global image-based matching score. Int. J. Comput. Vis. **72**(2) (2007)
20. Sand, P., Teller, S.: Particle video: long-range motion estimation using point trajectories. In: Computer Vision and Pattern Recognition (2006)
21. Seitz, S.M., Curless, B., Diebel, J., Scharstein, D., Szeliski, R.: A comparison and evaluation of multi-view stereo reconstruction algorithms. In: Computer Vision and Pattern Recognition (2006)
22. Starck, J., Hilton, A.: Correspondence labelling for wide-timeframe free-form surface matching. In: International Conference on Computer Vision (2007)
23. Vedula, S., Baker, S., Kanade, T.: Image-based spatio-temporal modeling and view interpolation of dynamic events. ACM Trans. Graph. **24**(2), 240–261 (2005)
24. Vicon: Vicon motion systems (2009). http://www.vicon.com
25. White, R., Crane, K., Forsyth, D.: Capturing and animating occluded cloth. ACM Trans. Graph. **26**(3), 34 (2007)
26. Zhang, L., Snavely, N., Curless, B., Seitz, S.M.: Spacetime faces: high resolution capture for modeling and animation. ACM Trans. Graph. **23**(3), 548–558 (2004)

Part III
Recent Developments in Image Processing

Wavelet-Based Inverse Light and Reflectance from Images of a Known Object

Dana Cobzas, Cameron Upright, and Martin Jagersand

Abstract Having an accurate model of the lights in a scene is important for many applications. Augmented reality requires light information for seamlessly composing computer generated objects into a real scene. Similarly, computer vision algorithms such as photometric stereo and shape from shading require accurate lighting. Inverse lighting allows such algorithms to be used in real world environments, instead of strictly controlled laboratory setups. While algorithms that recover light as individual point light sources work for simple illumination environments, it has been shown that a basis representation achieves better results for complex illumination. We propose a light model that uses Daubechies wavelets and a method for recovering light from cast shadows and specular highlights in images. Experimentally, we tested our method for difficult cases of both uniform and textured objects and under complex geometry and light conditions. We evaluate the stability of estimation and quality of the relight scene using our smooth wavelet representation compared to a non-smooth Haar basis and two other popular light representations (a discrete set of infinite light sources and a global spherical harmonics basis). We show good results using the proposed Daubechies basis on both synthetic and real datasets.

1 Introduction

Light representation and recovery is an important but less explored problem in computer vision. A practical solution is of value both as input to other computer vision algorithms and in computer graphics rendering. As an example, photometric stereo and shape from shading require known light. In most situations, simplified assumptions about light or object reflectance are made in order to solve the shape problem. With estimated light, such techniques could be applied in everyday environments

---

D. Cobzas (✉), C. Upright, and M. Jagersand
Computer Science, University of Alberta, Edmonton, Canada
e-mail: dana@cs.ualberta.ca, upright@cs.ualberta.ca, jag@cs.ualberta.ca

R. Ronfard and G. Taubin (eds.), *Image and Geometry Processing for 3-D Cinematography*, Geometry and Computing 5, DOI 10.1007/978-3-642-12392-4_10,
© Springer-Verlag Berlin Heidelberg 2010

outside of controlled lab conditions. Light estimated from images is also important in augmented reality applications in order to consistently relight an artificially introduced object. Therefore light reconstruction (referred as inverse light) has significant importance among vision problems.

In this work we consider the problem of estimating light from images given some 3D shape (i.e. shape of an object) but not necessarily the whole scene. One advantage of capturing the lighting on an existing object as opposed to introducing a light probe is that we not only recover the scene's lighting but also the reflectance parameters of the object. This allows us to render the scene under the same or different lighting conditions as well as to introduce synthetic objects into an augmented reality application. Another advantage of using existing shapes in a scene instead of a light probe is that the recovered light minimizes the error in the current scene, allowing a potentially higher quality rendering for moderate view changes. Furthermore, reflectance probes are not always practical to use, and they cannot be inserted afterwards into existing images. By contrast, many images contain some man-made object for which either a CAD model can be obtained, or an identical copy of the object can be scanned post-hoc to obtain the required shape information. Hence a technique that recovers light from known shape has more general applicability.

A variety of techniques have been applied to the inverse light problem, which can be classified mainly into two categories: (1) techniques that estimate a small number of point or directional lights and (2) techniques that recover light as projected on a set of basis functions across the entire light hemisphere (usually spherical harmonics basis [1, 20]). Methods in the first category are designed for recovering sharp light effects while the ones from the second category work best only for diffuse (Lambertian) scenes.

To model both diffuse and sharp light effects, people have proposed bases that provide local support in both the spatial domain (here image dimension) and the frequency domain. Wavelets are an example of such a basis. Few works [16, 17] used a Haar wavelet basis for light representation. Our work proposes a new light representation based on Daubechies wavelets, and a general method for estimating the light basis from images. We investigate the advantages of using this smooth basis compared to the non-smooth Haar basis and show the superiority of the smooth basis for both synthetic and real datasets.

Our inverse light method uses shadows and specular highlights on textured objects. Sato et al. [23] previously used shadows on diffuse objects for recovering light as a discrete illumination hemisphere. We extend the method to incorporate specular highlights. Other works [13, 29] incorporate multiple cues in the light estimation but without giving insights on which cue most helps the light reconstruction. Here we study the influence of shadows vs. specular highlights on the quality and stability of the reconstruction.

In addition to proposing a new light representation and a complete system to estimate light from images, we also study the stability, efficiency and quality of the light reconstruction when comparing four representations: a smooth Daubechies and non-smooth Haar wavelet basis and two other popular choices for the basis – a discrete set of infinite light sources and spherical harmonics basis. We provide upper bounds on error propagation using classical Wilkinson condition number numerical analysis [8], and experimentally, we give real-world practical results for simple and complex scenes, for diffuse and specular objects, for different illumination conditions, and for different levels of noise in real and synthetic images. This study brings an important contribution and insight to practical aspects of the inverse light problem which is known be inherently ill-conditioned. For example, Marschner and Greenberg [15] empirically observed that under diffuse reflectance inverse light tends to be ill-conditioned. For a global basis, Ramamoorthi and Hanrahan theoretically proved that only the low-frequency light components can be reconstructed from diffuse scenes [20]. They later showed that for a specular scene, estimating the spherical harmonics light coefficients is well conditioned only up to an order related to the surface roughness [21]. Okabe et al. [17] showed that the inverse problem becomes well conditioned when using shadows and a spherical harmonics light representation under diffuse light conditions. However no study has been made for a basis with local support, like the wavelet, for both high and low frequency lights as well as different perturbations in the input data.

The remainder of this chapter is organized as follows. The next section summarizes most relevant works in inverse light. Section 3 presents the proposed wavelet-based light representation and a method that estimates light either from one image of a uniformly colored object or from multiple images of a textured object. The complete inverse light system and few implementation details are described in Sect. 4. Section 5 presents comparative experimental results and Sect. 6 a short discussion and future directions.

2 Related Works in Inverse Light

A variety of inverse light methods have been proposed over the last 20 years, that can be divided in two main categories. The first category, presented in Sect. 2.1, involves using a small number of point lights to represent the illumination environment. While this is useful in simple laboratory setups, in general the lighting in a real scene is far more complex. The second category of inverse lighting approaches, detailed in Sect. 2.2, uses a set of basis functions over the entire sphere (representing all incoming directions). This representation allows more complex illumination conditions when compared to the individual point-light representation, especially in scenes with large area-lights or many point-lights. One work that doesn't fall in either of these two categories is the work of Debevec [5] that ray traced the reflections on a spherical mirror ball (an illumination probe) to directly measure the lighting hemisphere. Sato et al. [22] proposed a similar approach where the radiance distribution of the scene is estimated from a set of omnidirectional images. The difference is that the radiance map is projected onto the estimated scene geometry and not on an infinite illumination hemisphere.

2.1 Point Light Representations

Much of the literature about the inverse light problem involves estimating the position of a small number of point lights. A point light representation may seem limited, but Debevec [6] shows how a small number of point lights can be used to approximate a complex radiance map (as few as 64 point lights can accurately represent a complex environment).

Early methods recover light using a *Lambertian white sphere* to detect *critical points* [2,31,32]. Critical points are points where the surface normal is perpendicular to light source direction. The critical points form critical lines which divide the object into regions illuminated by a single virtual light source. The real light sources corresponding to a critical line can be detected using the difference of virtual lights of the two adjacent regions. Different methods can be employed for finding the critical points. Zhang and Yang [32] detect critical points by analyzing intensity along an arc of the sphere, while Bouganis uses a moving window to find critical points [2]. Takai et al. [25] used the intensity difference between two spheres to estimate location and radiance for near point light sources together with directional light sources and ambient light.

More general methods of inverse light use a *Lambertian object* instead of a sphere. One of the first approaches to the inverse light problem was that of Yang and Yuille [31]. They use image intensity of a Lambertian object along the silhouette to determine light directions. A different approach using a Lambertian object is taken by Wang and Samaras [28]. Building on the work of Zhang and Yang [32], they detect critical points as borders between large difference in intensities. Once critical points have been detected, they are grouped into critical lines which are used to determine the positions of virtual and real lights (similar to the case when the object is a sphere).

An interesting analysis of the inverse light problem that also uses a Lambertian object is that of Luong, Fua and Leclerc [14]. They solved for multiple lights, camera response and surface albedo using multiple images. Pixel values are formulated as an affine transformation of actual surface radiance resulting in a linear solution for the lighting. They also demonstrate many similarities between the radiometric reconstruction problem and that of geometric recovery with multiple views. A nonlinear radiometric bundle adjustment algorithm is presented which is analogous to the geometric bundle adjustment algorithm.

Better and more stable light reconstruction methods use specular objects. This is due to the smoothing effect of a Lambertian BRDF, which makes inverse light ill-posed. Highlights on two *specular spheres* with known position can be used to triangulate the light position [18]. Similarly two images of a specular calibrated sphere are used by Zhou and Kambhamettu [33] for detecting discrete light positions. The intensity difference between corresponding points in the two images is used to detect specular and diffuse pixels. The illuminant direction is then found by tracing the reflection at a specular highlight, which is averaged between images. A similar approach was recently proposed by Wong, Schnieders and Li [30] to recover both the light directions and camera calibration from multiple images of a specular sphere. Unlike other methods, their work doesn't require knowledge of the sphere's radius. The image of the sphere is represented as a conic, and a single parameter family of solutions are found for the sphere's position based on the radius. They show that the light directions recovered by tracing the reflections of specular pixels are independent of the sphere's radius.

Specular objects were used as an alternative to diffuse objects in inverse light algorithms. Methods make the assumption that the object geometry is known and use a parametric or non-parametric model for object reflectance. Hara and Ikeuchi [9] propose a method to estimate both the position of a finite light source along with a Torrance–Sparrow [26] reflectance model of the surface. Specular highlights are used to determine light direction, while the diffuse parts of the object are used to calibrate the distance of the light. To separate out the specular from diffuse reflectance components they use polarization filters. Once the light position has been determined, the diffuse parameters can be solved for analytically. Another approach based on the Torrance–Sparrow reflectance model is that of Hara, Nishino and Ikeuchi [10] that uses the von Mises–Fisher distribution from directional statistics to approximate the Torrance–Sparrow model. Light is estimated by backprojecting rays from specular highlights on a discrete illumination sphere, represented by a geodesic dome. This resulting illumination sphere is equivalent to a mixture of von Mises–Fisher distributions that are estimated using a modified version of the Expectation Maximization algorithm.

Strong constraints between shape, reflectance and illumination can be formed using *frontier points* [27]. Frontier points are points on the object where the epipolar plane is tangential to the object's surface, giving both the point and the surface normal and allowing calculation of BRDF and illumination in that point (using deconvolution from multiple images).

The vast majority of inverse light approaches use objects without texture due to the fact that it is difficult to distinguish the effect of lights from varying albedo. Nevertheless, a few methods have been proposed to solve the inverse light problem in the presence of *texture*. Li, Lin, Lu and Shum [13] proposed a method that makes use of several image cues (critical points, specular highlights and shadows cast by the object) to disambiguate light and object texture. Pixels with similar albedo (color) along critical lines are used to determine light color. Exploiting consistency between cues helps prevent spurious light detection. Another method that uses specular highlights on a textured object to detect multiple lights is proposed by Lagger and Fua [12]. They detect specular highlights as local maxima in image intensity and filtered out texture maxima by ignoring maxima which don't change position between images. These remaining specularities are then traced back onto the illumination sphere to detect both the number of lights and their directions.

2.2 Basis Function Representations

While a small number of point lights can be enough to accurately illuminate a Lambertian object (due to the low-pass filtering effect of the Lambertian BRDF), a more complete representation is required to properly reconstruct shadows and specular highlights. To deal with this problem the entire lighting sphere is discretized with a set of point lights infinitely far away (directional lights). A set of basis functions is defined over the discretized sphere, and the inverse light problem reduces to the problem of finding the best set of basis coefficients.

Marschner and Greenberg [15] were the first to represent lighting as a set of basis functions over the illumination sphere. From an arbitrary set of lighting basis functions, they generate basis images of the object illuminated under those basis functions. An image of the object under arbitrary illumination is therefore obtained as a linear combination of those basis images.

While Marschner and Greenberg observed poor conditioning in the inverse light of a Lambertian scene, it was not until few years later when Ramamoorthi and Hanrahan [20] and Basri and Jacobs [1] formally proved that light deconvolution is ill-posed. They represented the reflectance and the lighting in terms of *Spherical Harmonics* and show that the Lambertian BRDF acts as a low-pass filter in which 99.2% of the energy is present in the first nine spherical harmonics. Ramamoorthi and Hanrahan [21] extended their previous work to use arbitrary BRDF functions. The Phong BRDF is shown to act like a Gaussian filter with a width proportional to the shininess. In this case inverse light is shown to work best when the shininess is high. Similarly, the ability for a Torrance–Sparrow model to reconstruct higher frequency components is limited by the roughness parameter. This analysis allowed them to present a method for estimating both BRDF and lighting simultaneously. The theoretical results provided by [1,20] are used by Simakov et al. [24] to perform dense shape reconstruction. A low-dimensional linear approximation of spherical harmonics are used to create a new consistency measure, which is used in the dense shape reconstruction algorithm of Boykov, Veksler and Zabih [3].

While Spherical Harmonics have been useful for analyzing the inverse light problem, they are not very good at compactly representing lighting distributions that are not spatially localized. Each basis function has an effect over the entire sphere. Representing a point light in spherical harmonics requires a large number of basis functions. If not enough functions are used, ringing artifacts can be noticed around the light (i.e. the Gibbs phenomenon). One alternative to spherical harmonics are *Haar wavelets* which were investigated by Ng, Ramamoorthi and Hanrahan [16]. They showed that Haar was much better at compactly representing radiance maps, and used this to create an algorithm to re-light scenes in real time using graphics hardware.

One way of dealing with the poor conditioning of the inverse light problem is to analyze the *shadows cast by the object* [23]. Cast shadows essentially introduce high frequency components into the otherwise smooth transfer function, allowing high frequency lighting to be robustly estimated. Sato and Ikeuchi [23] use a geodesic dome to represent the lighting that is iteratively refined using constraints from cast shadows. The method was later extended by Okabe and Sato [17] to use a light representation based on Haar wavelets. They also analyzed the relationship between frequency in illumination and rank deficiency of the inverse light problem. They show that as the lighting distribution gains more high frequency components the rank deficiency of the inverse light increases. In addition they noticed that the geometry of the object casting the shadow has a large influence on the stability of inverse light. Finally they compare the performance of spherical harmonics and Haar wavelets and found that spherical harmonics blur shadows given larger reconstruction error.

We made few improvements over previous work in hemisphere-based representations. Our method uses the Daubechies wavelet to represent the lighting. We show that Daubechies is able to accurately represent the lighting with fewer basis functions. Secondly, by using multiple images, the lighting and per-point albedo on the object can be obtained. This allows us to use textured objects in the inverse light problem, and to reconstruct albedo maps that can be used later with the object in augmented reality applications.

3 Theory

This section presents our proposed wavelet-based light representation (Sect. 3.2) and the use of this model in two inverse light methods: one that uses a single calibrated image of a uniformly colored object (Sect. 3.4) and a second one that uses multiple calibrated images of a textured object (Sect. 3.5). Light estimation is interleaved with reflectance estimation (Sect. 3.6). Details on regularization and numerical light estimation are presented in the last subsection. We begin by mathematically formalizing the problem and the image formation process.

3.1 Inverse Light Problem Definition

The light recovery (or the inverse light) problem can be stated as follows: given a set of calibrated images $I_i, i = 1 \ldots N$ of an object with known geometry $S = \{x_j, j = 1 \ldots M\}$ (x_j are 3D points that define the object geometry) and reflectance recover the light that illuminates the scene. This process, as implemented in our system, is illustrated in Fig. 1.

The inverse light problem formulation relies on a chosen light model as well as an image formation (rendering) model. As described in the following subsection, the light is represented as a linear combination of basis functions. Here we propose the use of a Daubechies wavelet basis, but the formulation is valid for any basis. The image formation model involves choosing both a camera projection model that defines how the object geometry is mapped on the image, and an object reflectance model that defines how the object appearance is formed given the light conditions

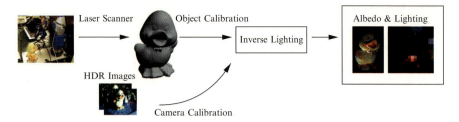

Fig. 1 A brief outline of the inverse light process. High dynamic range input images $I_i, i = 1\ldots N$ and a scanned 3D model $S = \{\mathbf{x}_j, j = 1\ldots M\}$ are used by the inverse light method to reconstruct the lighting

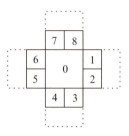

Fig. 2 The *top half* of a cubemap which represents the illumination hemisphere, divided into nine faces. The resolution of one face of the cube is n

and camera position. We chose to use a projective camera model and compare two parametric reflectance models, Lambertian and specular Phong (Sect. 3.3).

It is known that given only one image it is not possible to disambiguate light color from albedo. We therefore assume uniform albedo for the proposed single-view inverse-light method. This is later generalized for objects with general unknown textured albedo by using multiple images from different viewpoints around the scene.

3.2 Wavelet-Based Light Model

We chose to represent light using a basis defined over the illumination hemisphere. The hemisphere is discretized into many point lights, each one representing a light at infinity. A basis is defined over those lights. While a sphere map is natural for representing lighting, a cubemap was chosen due to its simplicity and common use in computer graphics. We limit the lighting representation to only include lights above the ground plane, assuming that the image shading is mostly caused by light coming from above the object. The top half of the cube map shown in Fig. 2 is divided into nine faces, to ensure that each face is square. This allows a basis to be defined over each face, without having to be scaled non-uniformly.

The simplest basis we can define over our cube map is one where each basis corresponds to only one pixel, a per-pixel basis function (*discrete hemisphere*).

Although this representation is simple to implement, it has a few drawbacks. In order to represent all the point lights on the top half of a cubemap with a per pixel basis, $3n^2$ bases are required, where n is the width (resolution) of a single face of the cubemap. Therefore inverse light involves estimating many degrees of freedom and overfitting can become a concern.

Spherical harmonics provide a more compact basis representation for the illumination of a scene. For diffuse objects, only a small number of spherical harmonics are needed to accurately illuminate the object. However, spherical harmonics are not well suited for representing the radiance of specular objects, in which case a large number of functions must be used to accurately model high-frequency variation in the lighting.

The two sets of basis functions discussed so far are on opposite sides of the spectrum: the per-pixel basis is highly localized in space while spherical harmonics have a global influence over the pixels in the cube map. *Wavelets* are localized in both the spatial and frequency domains making them a preferred representation for both diffuse and specular lighting effects. There are a variety of wavelet basis functions, the simplest being Haar. For all wavelet types, the wavelet basis functions, or daughter wavelets, are scaled and translated versions of a single wavelet known as the mother wavelet. In addition to all the wavelet basis functions, there is a single function known as the scaling function. We propose here the use of a smooth wavelet basis (Daubechies) for representing light, and we compare it with the more simple Haar basis.

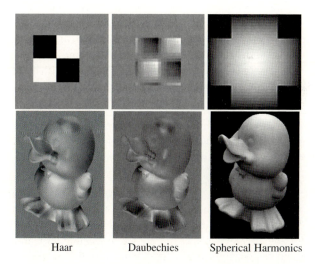

Fig. 3 *Top*: Cubemap corresponding to three basis functions. *Bottom*: scene illuminated under those basis functions; *From left to right*: Diagonal Haar and Daubechies wavelet at lowest scale, and a first order Spherical Harmonic. Each image is normalized to be in a [0:1] range and therefore zero basis entries appear as *gray*

Images of a scene illuminated under a single Haar, Daubechies and Spherical Harmonic basis function can be seen in Fig. 3. Each image is normalized to be in a [0:1] range. Since the wavelet basis functions are not always positive, a value of zero shows up as gray in the images.

All the basis functions given above can be linearly combined to represent the light map. More specifically, the intensity of the each light is given by the following:

$$\mathbf{l}_i = \sum_{k=1}^{K} B_{ik} w_k \tag{1}$$

or in vector form:

$$\mathbf{l} = B\mathbf{w} \tag{2}$$

where a column of matrix B is a basis function $\mathbf{B}_w$, $\mathbf{w}$ a column vector containing corresponding weights, and $\mathbf{l}$ is the vector corresponding to the flattened lightmap.

3.3 Reflectance Models

We next show how the light model is incorporated in the image formation equation. Throughout this chapter we assume that the cameras are calibrated and we denote the image projection of a 3D point $\mathbf{x}_j$ by $I(\mathbf{x}_j) = I_j$. The appearance of a point on a surface depends on the material properties of that surface. The Bidirectional Reflectance Distribution Function (BRDF) defines how light interacts with a surface. A BRDF describes the proportion of incoming radiance that leaves in a given outgoing direction, and is denoted as follows:

$$f(\theta_i, \phi_i, \theta_o, \phi_o)$$

The angles of incoming and outgoing radiance, represented in spherical coordinates, are given respectively as (θ_i, ϕ_i) and (θ_o, ϕ_o). The total amount of radiance leaving a point on a surface at a given outgoing angle is given by:

$$L(\theta_o, \phi_o) = \iint_\Omega f(\theta_i, \phi_i, \theta_o, \phi_o) L(\theta_i, \phi_i) \cos(\theta_i) \sin \theta_i \, d\theta_i \, d\phi_i$$

where $L(\theta_i, \phi_i)$ is the amount of radiance arriving at the point from the angle (θ_i, ϕ_i). $\sin \theta_i \, d\theta_i \, d\phi_i = d\omega$ where ω is the solid angle subtended by an infinitesimal patch at the given point. When estimating a lighting distribution, we use a set of point lights at infinity as opposed to some function over the hemisphere. In the case of a point light with an intensity l at an incoming angle of (θ, ϕ), the term $L(\theta_i, \phi_i)$ becomes $l\delta(\theta_i - \theta, \phi_i - \phi)$, where δ is the Dirac delta function. The double integral then simplifies to:

Wavelet-Based Inverse Lighting

$$L(\theta_o, \phi_o) = f(\theta, \phi, \theta_o, \phi_o) l \cos(\theta) \omega \qquad (3)$$

The simplest reflectance model is the Lambertian model, which has a constant BRDF $f = k_d$. The Lambertian reflectance model is useful for modeling rough or matte surfaces, but is not good for materials with specular reflections. To represent general reflectance we chose a dichromatic model that has a Lambertian term for the diffuse part and a specular term given by the Phong model. While not physically based, it works very well in practice and it is one of the most popular reflectance models used for rendering in the computer graphics community. The BRDF of the Phong model is given by:

$$f(\theta, \phi, \theta_o, \phi_o) = k_d + k_s \frac{\cos(\theta_r)^n}{\cos(\theta)}$$

where θ_r is the angle between the reflected viewing direction and the light direction, θ in the angle of the incoming light and n is the shininess of the material.

For implementation reasons, it is much easier to represent the Phong and Lambertian models in the world's frame of reference. See Fig. 4 for an illustration of light geometry. We define **N**, **R** and **L** as the normalized surface normal, reflected viewing direction and light direction respectively. Rewriting (3) for the intensity of an observed pixel **x** we get:

$$I(\mathbf{x}) = V(\mathbf{x}) S(\mathbf{x}) (k_d \langle \mathbf{N}, \mathbf{L} \rangle + k_s \langle \mathbf{R}, \mathbf{L} \rangle^n) l \omega \qquad (4)$$

where $I(\mathbf{x})$ is the image intensity of the point **x**, $V(\mathbf{x})$ is the visibility of the point from the camera's view, and $S(\mathbf{x})$ is an indication function which is equal to one when the point is not in shadow, and zero otherwise. One commonly used variation of the Phong reflectance model is the Blinn–Phong model. Instead of using the dot product between the normalized reflected viewing direction and light direction $\langle \mathbf{R}, \mathbf{L} \rangle$, Blinn–Phong uses the dot product between the surface normal and half-angle vector **H** between light and view direction $\langle \mathbf{N}, \mathbf{H} \rangle$. With this modification, (4) becomes:

$$I(\mathbf{x}) = V(\mathbf{x}) S(\mathbf{x}) (k_d \langle \mathbf{N}, \mathbf{L} \rangle + k_s \langle \mathbf{N}, \mathbf{H} \rangle^n) l \omega \qquad (5)$$

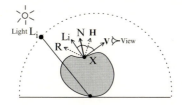

Fig. 4 The vectors used in lighting calculations

3.4 Inverse Light from a Single Image

When given just a single image of a scene we are unable to distinguish between lighting and albedo only the product of the two can be found. The same is true for a specular object. Due to this limitation, with one image we are only able to reconstruct lighting for objects with uniform albedo. Without loss of generality we can assume that the object is white, with $k_d = 1$. We also assume known reflectance, setting $k_s = 1$ and n to a fixed value in the Blinn–Phong model. The intensity of a pixel $\mathbf{x}_j$ illuminated with N lights can be obtained using the Blinn–Phong model (5) with multiple light directions:

$$I(\mathbf{x}_j) = V(\mathbf{x}_j) \sum_i^N S_i(\mathbf{x}_j)(k_d \langle \mathbf{N}_j, \mathbf{L}_i \rangle + k_s \langle \mathbf{H}_{ij}, \mathbf{N}_j \rangle^n) l_i \omega_i \qquad (6)$$

Rewriting in a more simple linear form we get:

$$\mathbf{I} = \mathbf{T}_s \mathbf{l} \qquad (7)$$

where the components of the light transport matrix $\mathbf{T}_s$ are given by:

$$t_{ji} = S_i(\mathbf{x}_j)(k_d \langle \mathbf{N}_j, \mathbf{L}_i \rangle + k_s \langle \mathbf{H}_{ij}, \mathbf{N}_j \rangle^n) \omega_i \qquad (8)$$

The visibility term $V(\mathbf{x}_j)$ is dropped because all the points we use are visible in the image. By combining this equation with the wavelet basis light model from (2) we get:

$$\mathbf{I} = \mathbf{T}_s \mathbf{B} \mathbf{w} \qquad (9)$$

Generalized linear inequality constraints are required to prevent negative lighting and are written compactly as:

$$\mathbf{B} \mathbf{w} >= \mathbf{0}$$

The inverse light problem is therefore formulated as a constrained linear equation system that is solved for the light basis linear weights $\mathbf{w}$. Equation (9) can be interpreted as a sum of "basis images", and can be efficiently implemented on graphics hardware. Examples of these basis images can be seen in Fig. 3. The top row displays the light basis while the second row an example of image basis (scene illuminated with the light basis).

3.5 Inverse Light from a Multiple Images

We now generalize the light and reflectance estimation method for objects with varying albedo. We make the assumption that the lighting does not change and that the object remains stationary with respect to the lighting (only the camera moves).

In this case, when viewing a fixed point on an object from different camera angles, the difference in the image intensity of that point is due to the specular component. By adding a superscript v to (6) to indicate the view, and taking the difference of the two observed pixel values $I^{v_0}(x_j)$ and $I^{v_1}(x_j)$ we get the following system of equations:

$$I^{v_0}(\mathbf{x}_j) - I^{v_1}(\mathbf{x}_j) = \sum_i^N S_i(\mathbf{x}_j)k_s((\langle \mathbf{H}_{ij}^{v_0}, \mathbf{N}_j \rangle^n - \langle \mathbf{H}_{ij}^{v_1}, \mathbf{N}_j \rangle^n)l_i \omega_i \qquad (10)$$

where the visibility term is dropped because we only consider points which are visible in both views. Rewriting in a more compact form we get:

$$I^{v_0}(\mathbf{x}_j) - I^{v_1}(\mathbf{x}_j) = \sum_i^N c_{ji}^{v_0 v_1} l_i \qquad (11)$$

where

$$c_{ji}^{v_0 v_1} = S_i(\mathbf{x}_j)k_s((\langle \mathbf{H}_{ij}^{v_0}, \mathbf{N}_j \rangle^n - \langle \mathbf{H}_{ij}^{v_1}, \mathbf{N}_j \rangle^n)\omega_i \qquad (12)$$

Every pair of views (v_0, v_1) and every point visible in both views adds a new equation. All these equations can be stacked together into a linear system of equations similar to the one from the single-view case,

$$\mathbf{I'} = \mathbf{T}_m \mathbf{l} = \mathbf{T}_m \mathbf{B} \mathbf{w} \qquad (13)$$

and again we have to impose non-negativity constraints $\mathbf{B}\mathbf{w} >= \mathbf{0}$.

After an initial light is calculated from corresponding specular pixels in pairs of images, per-point albedo is recovered from the corresponding diffuse pixels in multiple views v. This can be easily done by linearly solving for k_{dj} in the least square form equations obtained by replacing the constant albedo k_d with varying per-point albedo k_{dj} in (6):

$$I^v(\mathbf{x}_j) = k_{dj}a_j + b_j^v \qquad (14)$$

where

$$a_j = V(\mathbf{x}_j) \sum_i^N S_i(\mathbf{x}_j)\langle \mathbf{N}_j, \mathbf{L}_i \rangle l_i \omega_i$$

$$b_j^v = V(\mathbf{x}_j) \sum_i^N S_i(\mathbf{x}_j)k_s\langle \mathbf{H}_{ij}^v, \mathbf{N}_j \rangle^n l_i \omega_i$$

Next we refine the light using the full equation system (9) (not pairwise) with the estimated per-pixel albedo. Figure 5 shows on the left the reconstructed albedo for the object on the right side.

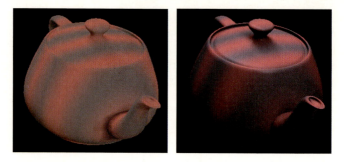

Fig. 5 Example of albedo estimation. Reconstructed albedo (*left*), one of the original images (*right*)

3.6 Reflectance Estimation

Having a good light estimate, we turn our attention to find the reflectance parameters. Initial light estimation is done using approximative values for the specular parameters. We have noticed through experimentation with synthetic data that incorrect specular parameters only seem to influence the overall brightness of the recovered lightmap, not the individual positions of the lights. The inverse light method seems not to be very sensitive to moderately incorrect specular parameters. By interleaving reflectance estimation with lighting estimation, we achieve nearly the same quality reconstruction as the true non-linear global optimum would.

When presented with just a single image, there is no way to separate the recorded values into specular and diffuse parts. The best we can do is try to find the specular parameters that make the reconstructed image look as close as possible to the original. We can rewrite (6) in a simpler form:

$$I(\mathbf{x}_j) = a_j + k_s \sum_i^N b_{ij} c_{ij}^n \qquad (15)$$

with

$$a_j = V(\mathbf{x}_j) \sum_i^N S_i(\mathbf{x}_j) \langle \mathbf{N}, \mathbf{L}_i \rangle l_i \omega_i$$
$$b_{ij} = V(\mathbf{x}_j) S_i(\mathbf{x}_j) l_i \omega_i$$
$$c_{ij} = \langle \mathbf{H}_{ij}, \mathbf{N} \rangle$$

where k_{dj} is omitted since it is assumed to be equal to one. We pose the reflectance estimation problem as a non-linear least squares problem:

$$\{k_s^r, k_s^g, k_s^b, n\} = \operatorname*{argmin} \sum_{j \in S} (I_j - a_j - k_s \sum_i^N b_{ij} c_{ij}{}^n)^2 \qquad (16)$$

Wavelet-Based Inverse Lighting

where S is the set of sample points. In practice we only use points where the specular dot product $\langle \mathbf{R}, \mathbf{L}_i \rangle$ is near one. Those points that have the normal almost aligned with light direction correspond to a strong specular highlight. This way we avoid the problem of fitting specular highlights to noise in regions where there is no highlight.

When using multiple images we estimate specular parameters from pairs of image differences. This way we avoid using albedo values estimated with incorrect specular parameters. Similarly to the single view case we select only visible specular points that have a specular dot product close to unity in one of the views for at least one light. If we define $S_{v_0 v_1}$ as the set of sample points for the pair of views v_0 and v_1, the least squares problem can be posed as:

$$\{k_s^r, k_s^g, k_s^b, n\} = \text{argmin} \sum_{v_0 v_1, v_0 \neq v_1} \sum_{j \in S_{v_0 v_1}} ((I_j^{v_0} - I_j^{v_1}) - k_s \sum_i^N b_{ij} (c_{ij}^{v_0 \, n} - c_{ij}^{v_1 \, n}))^2 \tag{17}$$

where

$$c_{ij}^v = \langle \mathbf{H}_{ij}^v, \mathbf{N_j} \rangle$$

For both single and multiple view cases, the nonlinear minimization is solved with the Levenberg–Marquardt algorithm [11].

3.7 Numerical Light Estimation and Regularization

Inverse light is an inherently ill-conditioned problem, especially when using real images and noisy data. To alleviate this problem, we used a smoothness regularization on the gradient of the light map.

$$E_s = \lambda_s \sum_i \|\nabla l_i\|^2 \tag{18}$$

where λ_s is a positive constant that controls the magnitude of the smoothness penalty. The gradient of the lightmap is computed as:

$$\nabla l_i = \begin{bmatrix} l_{iR} - l_{iL} \\ l_{iT} - l_{iB} \end{bmatrix}$$

where l_{iR}, l_{iL}, l_{iT} and l_{iB} are the right, left, top and bottom neighbors respectively of light l_i.

This penalty term is quadratic in the lighting coefficients. By rewriting (18) as a quadratic form and combining it with (2), we get the following:

$$E_s = \lambda_s \mathbf{l}^T Q_{sl} \mathbf{l} = \lambda_s \mathbf{w}^T B^T Q_{sl} B \mathbf{w} = \lambda_s \mathbf{w}^T Q_s \mathbf{w} \tag{19}$$

The light estimation with known reflectance is therefore a linear least squares problem with linear inequality constraints for both the one image with uniform albedo and the multi-image with varying albedo cases ((9), (13), and (19)). It can be re-formulated as a quadratic programming problem.

$$\min_{\mathbf{w}} \left(\mathbf{w}^T ((TB)^T (TB) + \lambda_s Q_s)\mathbf{w} - \mathbf{I}^T (TB)\mathbf{w} \right) \qquad (20)$$

subject to

$$B\mathbf{w} \geq 0 \qquad (21)$$

where T is the light transport matrix, B is the light basis, $\mathbf{w}$ is the vector of basis function coefficients and $\mathbf{I}$ is a vector of the observed pixel values.

As shown by Ramamoorthi and Hanrahan [19] in the context of efficient rendering, a good approximation of an environmental map can be achieved by zeroing small wavelet coefficients. Here, in the case of the inverse light problem, the areas of the cubemap that need more detail are initially unknown. We start with a coarse basis that is then locally refined based on projected image residual (see next section for details).

4 System and Implementation Details

Algorithm 1: Wavelet-based Lighting Reconstruction

Require: n images $I_1 \ldots I_n$ with camera calibration
    object geometry $\mathbf{x}_j$, $j = 1 \ldots M$
    initial specular params k_s, n
    initial basis subdivision

1: solve for light basis coefficients $\mathbf{w}$ (See Equation 9 or 13 for single- and multi-view cases)
2: project the image residual onto the scene illuminated with the unused basis functions, and add the top 75%
3: solve for $\mathbf{w}$ again using the new basis matrix (Equations 9 or 13)
4: solve for reflectance parameters (See Equation 16 or 17)
5: **if** multi-view reconstruction **then**
6:    solve for albedo
7: **end if**
8: estimate lighting coefficients $\mathbf{w}$ for the final time

---

The lighting reconstruction algorithm outlined above is an iterative method that starts with an initial estimation of the reflectance parameters. In the case of the wavelet basis functions, we start with a sparse initial set of basis functions comprised of the scaling function and lowest level wavelets. For efficiently representing the wavelet basis we use a quad tree. An empty tree has just a single scaling function to represent the image. Every node added into the tree contains the three wavelets needed to add the next level of detail to that region. By using this structure, we are able to focus our representation in regions of the cubemap that need it, while keeping dark or smooth regions sparsely encoded. After solving for the basis coefficients, we need to add new basis functions which are useful in the reconstruction. We calculate the image residual and project it onto the unused basis of the light transport matrix T and add new basis vectors (and their parents in the quad tree) corresponding to the top 75% of residuals. In the case of the Spherical Harmonic and discrete hemisphere basis, no refinement is needed. Once an initial light is estimated, we estimate the specular parameters and, in the multi-view case, we solve for albedo. It is important to do this after performing the reflectance estimation, because errors in specular highlights will work their way into the albedo calculations. With a good estimate of the reflectance (and albedo), we perform one last pass of solving for lighting basis coefficients.

One important detail that is necessary when doing light estimation is related to the quality of input images. We found that we need High Dynamic Range images [7] for reliable light reconstruction. Typical fixed-range images are unable to represent the large range of radiance values in some scenes. Without HDR, a compromise would have to be made between detail in dark regions and prevention of pixel saturation in bright regions. We used the algorithm developed by Debevec and Malik [7] to reconstruct the non-linear camera response to exposure and to create high dynamic range images from multiple images taken at different shutter times.

An extra normalization step was also needed when using multiple images in estimating light. Some of the images come out brighter than the others, causing diffuse regions of the object to show up at different intensities. These errors show up as specular differences when solving for the lighting, creating spurious lights with the same color as the object. This problem is dealt with by scaling the input images to minimize the pixel value differences of points which are diffuse.

5   Experiments

We experimentally compared the proposed smooth Daubechies wavelet light representation (DB4) with the Haar (Haar) [17] and spherical harmonics (SH) [17, 21] based light maps as well as a simple uniform discrete map defined over the cube map (DH) [23]. For Haar and Daubechies representations only 75% of the basis functions are used, which is 75% of the total number of cubemap pixels. In the spherical harmonics basis the first nine low-order basis functions are used. For all experiments, the reconstructed lightmaps have a resolution of 8×8 for the top face and 4×4 for the side faces (as presented in Fig. 2).

We performed two types of experiments. First, in Sect. 5.1, we evaluate the *quality* of the light reconstruction in real and synthetic images. In Sect. 5.2 we evaluate the *stability* of the inverse light method for the different representations in noisy conditions (for both scene and image noise). We also test the impact of shadows and specular highlights on the quality of inverse light. Finally in Sect. 5.3 we show the application of the inverse light method in augmented reality.

5.1 Quality of the Reconstructed Light Basis

The use of *synthetic scenes* allows us to evaluate the quality of reconstructed lightmaps as well as the reconstructed images. We generated synthetic images lit with two environmental maps courtesy of P. Debevec [4]. We chose these two light maps as examples of an environment with few sharp lights (St. Peter) and an environment with a large area light (Ufizzi). For all light representations, we first tested the single-view inverse light method (Sect. 3.4) on a white teapot model. Next, we tested the multiview inverse light method (Sect. 3.5) using a colored teapot with a simple texture based on a sine wave.

We also tested the quality of the proposed inverse light method on two *real scenes*. For the real scenes, we used images of two identical glossy ceramic ducks. One was painted white for the one view case and the other left colored for the multiview case. We obtained the geometric model of the duck using a laser scanner, and we registered it with the image using manually selected corresponding points. The images were calibrated with respect to the camera using the dotted calibration pattern shown in Fig. 7 (third row left). We notice that the inverse light is not sensitive to small changes in surface reflectance parameters, and therefore we used a heuristic approximation of the reflectance parameters for the real images.

Reconstructed lightmaps for the synthetic case are displayed in Fig. 6 (left) for the St. Peter's Basilica lightmap and in Fig. 6 (right) for the Ufizzi lightmap. Results show that the spherical harmonics representation tends to blur the whole light map, and all other three basis show good quality similar results. For St. Peter's light map that has sharp lights Haar works better than the Daubechies while for the Ufizzi

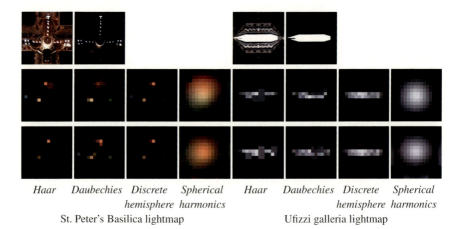

Haar   Daubechies   Discrete   Spherical   Haar   Daubechies   Discrete   Spherical
                    hemisphere harmonics   *hemisphere harmonics*
             St. Peter's Basilica lightmap   Ufizzi galleria lightmap

Fig. 6 Lightmaps reconstructed by the proposed inverse light method using four example basis: Haar wavelet, Daubechies wavelet, simple discrete cubemap and spherical harmonics. All experiments are performed with the teapot object from Fig. 7. The *first row* presents the original lightmap and a scaled version that shows the important lights. The *second* presents the results on the one-view specular object with uniform texture and the *last row* presents the result of the multi-view textured scene

lightmap that has one big area light the smooth basis gives a more accurate reconstruction. This is expected since the smooth basis offers better area support than the non-smooth one. In both cases the discrete hemisphere also shows good performance. Among different objects tested, the white specular one (row 2) gives better results. The multiview case (row 3) is numerically more difficult as the estimated non-uniform albedo creates some artifacts in the light map.

Next we tested the quality of the *reconstructed images*. We chose a view that was not included in the training set used for the light estimation and rendered the image with the recovered light (and albedo for the multiview case). We performed experiments for the two inverse light methods (single and multi-view case) for both synthetic and real test images. In the case of real images, an extra view was taken and not used in the light reconstruction method but kept only for the purpose of evaluation. A selection of the reconstructed images are shown in Fig. 7, while the

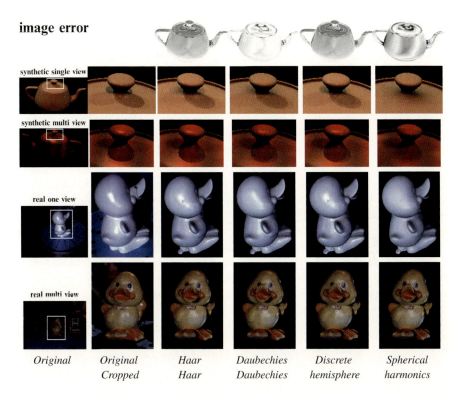

Fig. 7 Results with the four representations (Haar, Daubechies, discrete hemisphere and spherical harmonics) for synthetic and real images. We tested on a novel view, not included in the light reconstruction. The *second and third rows* show the single and multi view reconstruction for a white and colored specular teapot and the *last two rows* are real experiments for a white and colored shinny duck. The corresponding lightmaps for the synthetic case are shown in Fig. 6 (left) third, fourth rows. The numerical errors are presented in Table 1. An illustration of the image errors for the single view case is shown in the *first row*. *White* represents low errors, and *black* represents big errors

Table 1 Results for quality of light reconstruction for different basis. All image errors represent mean intensity error of the original with respect to image rendered with reconstructed light. Errors are normalized to the range [0:255]. The corresponding reconstructed images are shown in Fig. 7

Lightmap	Object	Image error			
		Haar	DB4	SH	DH
St. Peter's	*Teapot diffuse*	8.2	2.6	6.4	8.3
	Teapot specular	6.9	2.3	5.6	7.0
	Teapot multi	3.7	2.0	3.2	3.7
Ufizzi	*Teapot diffuse*	9.0	1.6	8.8	9.0
	Teapot specular	9.2	1.8	9.1	9.0
	Teapot multi	5.6	2.1	3.3	5.7
White duck indoor		21.3	21.3	21.4	21.3
Yellow duck indoor		25.6	25.6	25.6	25.6

numerical errors are shown in Table 1. All image errors are normalized to the range [0:255]. In Fig. 7, the first row displays the error images corresponding to the second row in the same figure as well as the second row in Table 1.

From this experiment we noticed that the Daubechies basis gives the best over all performance. The spherical harmonics give very poor perceptual quality (looking at the images in Fig. 7) but the measured error is surprisingly better than the Haar or discrete hemisphere case (see Table 1 and first row in Fig. 7). This is due to the fact that the spherical harmonics perform better for the diffuse/non-shadowed parts of the object, but it tends to blur any specular highlights or shadows that have large influence on the perceptual appearance. The Haar and discrete hemisphere lightmaps give similar results with good perceptual quality on the reconstructed images but poor image error. The Daubechies basis gives the compromise between visual quality and numerical accuracy.

Despite having bigger numerical errors[1] the real scene reconstruction was visually quite close to the original, with the Daubechies basis giving the best results. This shows that when using real (and thus non-perfect) images, the smooth wavelet representation is more robust and spreads the error more evenly on the objects giving a better appearance in the reconstructed view. In the multi-view case (yellow duck – last row Fig. 7) the results are quite similar.

5.2 Stability of the Inverse Light Method for Different Light Basis

When solving the inverse light problem, a variety of sources of noise can interfere with results. Errors in camera calibration, image noise and object geometry can all

---

[1] The larger numerical errors for the real images are likely due to difficulty in calibration. It is important for the light estimation to get the rays exactly right from light via reflection to camera. This is a much more difficult problem than camera calibration alone.

Wavelet-Based Inverse Lighting

Table 2 Condition number for stability experiments

Test	Condition numb. $\varkappa$ ($*10^5$)			
	Haar	DB4	SH	DH
Geometry error	1.2	1.1	2.5	1.2
Calibration error	1.2	0.82	2.5	1.3
Image noise	1.2	0.83	2.5	1.3
No shadow cues	3.4	1.1	9.6	6.8
No specular cues	6.5	7.4	3.1	10.9
No spec. & no shadows	11,893	5,541	13.5	18,920

have a negative effect on the reconstructed lightmaps and images. In a second set of experiments, we analyzed the stability of the inverse light reconstruction in the presence of noise. We again compared all four basis representations (Haar-Haar, Daubechies-DB4, discrete hemisphere-DH and spherical harmonics-SH). All tests are performed on the single-view method with the white teapot and the St. Peter's Basilica environmental map. We introduced three types of errors: noise in the object geometry (by perturbing the normals), noise in camera calibration and image noise. Finally, we compared the contribution of shadow cues vs. specular highlight cues in the reconstruction method. Numerical results are presented in Table 2.

Given a fixed setup, the image intensities are coupled to the light sources by the light transport matrix (T is (9) and (13)). An error in calibration or geometry manifests itself as a perturbation of the matrix $\tilde{T}_s = T_s + E$. Likewise an image error/noise can be modeled as $\tilde{I} = I + \Delta I$. Classical Wilkinson perturbation analysis [8] gives a bound on the calculated light as

$$\frac{\|\Delta \mathbf{w}\|}{\|\mathbf{w}\|} \leq \varkappa(A_s) \left(\frac{\|E\|}{\|T_s\|} + \frac{\|\Delta I\|}{\|I\|} \right)$$

The above formula assumes that the angle between the residual and the solution is small. The condition number $\varkappa$ thus gives an upper bound on how difficult it is to recover light. A large condition number indicates that small image errors or calibration problems could lead to very large changes in the lightmap. The condition number thus indicates relative "difficulty" of the reconstruction (higher $\varkappa$ = more difficult).

Among the different bases, the spherical harmonics are the most stable. This can be explained by the fact that they uses global functions that smooth any high frequency perturbations. Additionally, we noticed that in practice, the Daubechies and Haar wavelets are influenced more by the errors in the camera position than the errors in object geometry or image noise. This is due to the fact that camera calibration errors result in large consistent image misalignment compared to the noise due to perturbed normals that is more evenly distributed on the object (we noticed this difficulty in the real image experiments from Fig. 7). Looking at the influence of shadows vs. specular reflections, we notice that the specular reflectance makes the inverse light more stable than just the shadows (except for the spherical harmonic basis). As expected and in accordance to previous results [15, 20], when no shadows and no specular reflections are present the light estimation from only shading becomes ill-conditioned. Only the spherical harmonics are able to reconstruct from diffuse light effects.

It should be noted that the number of basis functions used in our experiments differed between each basis. The discrete basis used the most (the same as the number of lights), Haar and Daubechies used 75% of the available basis functions, while Spherical Harmonics only used the top 9. One would expect the discrete basis to outperform the others, since there are more degrees of freedom available. While this basis does appear to outperform the others in Fig. 6, it gives higher reconstructed image errors in Table 1. Another disadvantage of the discrete basis is that it does not scale too well. When using a 16×16 cubemap for example, there are 768 variables to solve for. Even if running time was not a problem in this case, the conditioning is so bad that the constrained least square solver fails.

5.3 Augmented Reality

An application of the inverse light problem is augmented reality. With an accurate knowledge of the lighting in a scene, it is possible to seamlessly composite a computer generated image into a real scene. An experiment was performed to test the visual quality of reconstructed lighting in an augmented reality scene. Refer to Fig. 8 for the following description of the experiment. We used our proposed single-view

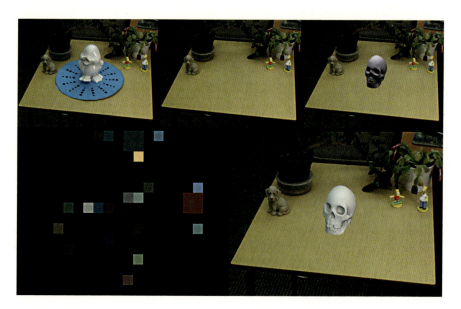

Fig. 8 The *top row* contains from *left* to *right*: the original scene with calibration object, the empty scene and the augmented reality scene with incorrect lighting. The recovered lightmap and augmented reality scene with the correct lighting are shown on the *bottom row*

algorithm and the top left image to reconstruct lighting. Next, without changing light conditions, we took another image with no calibration object (top middle) and inserted a graphically generated skull lit with the recovered light (bottom right). The AR image that uses the recovered light looks realistic, especially compared with the top right image that has incorrect lighting on the skull. The bottom left image shows the recovered lightmap, where one can notice the presence of overhead lights and a bright light coming from the side.

6 Conclusion and Discussion

We have presented a new light representation based on a Daubechies wavelet basis, and we have presented an inverse light method that uses the proposed representation to recover light from calibrated images of a known object using shadows and specular highlights. We showed that when using multiple images, both the light and the varying object albedo can be recovered. To get a more stable solution and overcome the ill-conditioned nature of the inverse light problem, we propose the use of a smoothing penalty over the lightmap. For testing the quality and stability of the proposed Daubechies representation in the context of inverse light, we compared it with three other light bases: spherical harmonics, Haar wavelet and a discrete hemisphere. We have shown that the Daubechies basis gives the most accurate reconstruction for several types of real and synthetic scenes. We also compared the stability on the four representations in the presence of errors (geometric, camera and image noise). While spherical harmonics gave the most stable light reconstruction, Daubechies proved to be comparably stable.

One limitation of the light recovery method is its sensitivity to geometry and calibration noise. Inaccurate surface normals attribute the lighting effects to incorrect areas of the lightmap. We notice the effects of geometry error, especially in the multiple image case because the lighting is reconstructed based on specular highlights. A promising area of future work is to integrate both lighting and surface normal deviation into the minimization framework. This formulation may no longer be represented as a constrained least squares problem, and an iterative method may be required to repeatedly solve for lighting followed by surface normal optimization.

References

1. Basri, R., Jacobs: Lambertian reflectance and linear subspaces. IEEE Trans. Pattern Anal. Machine Intell. **25**(2) (2003)
2. Bouganis, C., Brookes, M.: Multiple light source detection. IEEE Trans. Pattern Anal. Machine Intell. **26**(4), 509–514 (2004)
3. Boykov, Y., Veksler, O., Zabih, R.: Fast approximate energy minimization via graph cuts. IEEE Trans. Pattern Anal. Machine Intell. **23**(11), 1222–1239 (2001)
4. Debevec, P.: Light probe image gallery. http://www.debevec.org/probes/

5. Debevec, P.: Rendering synthetic objects into real scenes: bridging traditional and image-based graphics with global illumination and high dynamic range photography. In: SIGGRAPH (1998)
6. Debevec, P.: A median cut algorithm for light probe sampling. In: Siggraph poster (2005)
7. Debevec, P.E., Malik, J.: Recovering high dynamic range radiance maps from photographs. In: SIGGRAPH (1997)
8. Deuflhard, P., Hohmann, A.: Numerical Analysis in Modern Scientific Computing: An Introduction. Springer, New York (2003)
9. Hara, K., Nishino, K., Ikeuchi, K.: Determining reflectance and light position from a single image without distant illumination assumption. In: ICCV (2003)
10. Hara, K., Nishino, K., Ikeuchi, K.: Multiple light sources and reflectance property estimation based on a mixture of spherical distributions. In: ICCV, pp. 1627–1634 (2005)
11. Heath, M.T.: Scientific Computing. An Introductory Survey. McGraw-Hill, New York (2002)
12. Lagger, P., Fua, P.: Multiple light sources recovery in the presence of specularities and texture. Tech. rep., EPLF (2005)
13. Li, Y., Lin, S., Lu, H., Shum, H.Y.: Multiple-cue illumination estimation in textured scenes. In: ICCV, p. 1366 (2003)
14. Luong, Q., Fua, P., Leclerc, Y.G.: Recovery of reflectances and varying illuminants from multiple views. In: ECCV (2002)
15. Marschner, S.R., Greenberg, D.P.: Inverse lighting for photography. In: Proceedings of the Fifth Color Imaging Conference (1997)
16. Ng, R., Ramamoorthi, R., Hanrahan, P.: All-frequency shadows using non-linear wavelet lighting approximation. ACM Trans. Graph. **22**(3), 376–381 (2003)
17. Okabe, T., Sato, I., Sato, Y.: Spherical harmonics vs. haar wavelets: basis for recovering illumination from cast shadows. In: CVPR, pp. 50–57 (2004)
18. Powell, M., Sarkar, S., Goldgof, D.: A simple strategy for calibrating the geometry of light sources. IEEE Trans. Pattern Anal. Machine Intell. **23**(9), 1022–1027 (2001)
19. Ramamoorthi, R., Hanrahan, P.: An efficient representation for irradiance environment maps. In: SIGGRAPH (2001)
20. Ramamoorthi, R., Hanrahan, P.: On the relationship between radiance and irradiance: determining the illumination from images of a convex lambertian object. J. Opt. Soc. Am. A **18**(10), 2448–2459 (2001)
21. Ramamoorthi, R., Hanrahan, P.: A signal-processing framework for inverse rendering. In: SIGGRAPH (2001)
22. Sato, I., Sato, Y., Ikeuchi, K.: Acquiring a radiance distribution to superimpose virtual objects onto a real scene. IEEE Trans. Visual. Comput. Graph. **5**(1), 1–12 (1999)
23. Sato, I., Sato, Y., Ikeuchi, K.: Illumination from shadows. IEEE Trans. Pattern Anal. Machine Intell. **25**(3), 290–300 (2003)
24. Simakov, D., Frolova, D., Basri, R.: Dense shape reconstruction of a moving object under arbitrary, unknown lighting. In: ICCV (2003)
25. Takai, T., Niinuma, K., Maki, A., Matsuyama, T.: Difference sphere: an approach to near light source estimation. In: CVPR, pp. 98–105 (2004)
26. Torrance, K., Sparrow, E.: Theory for off-specular reflection from roughened surfaces. J. Opt. Soc. Am. **57**(9), 1105–1114 (1967)
27. Vogiatzis, G., Favaro, P., Cipolla, R.: Using frontier points to recover shape, reflectance and illumunation. In: ICCV, pp. 228–235 (2005)
28. Wang, Y., Samaras, D.: Estimation of multiple illuminants from a single image of arbitrary known geometry. In: ECCV, pp. 272–288 (2002)
29. Wang, Y., Samaras, D.: Estimation of multiple directional light sources for synthesis of augmented reality images. Graph. Models **65**(4), 185–205 (2003)
30. Wong, K.Y.K., Schnieders, D., Li., S.: Recovering light directions and camera poses from a single sphere. In: ECCV (2008)
31. Yang, Y., Yuille, A.: Sources from shading. In: CVPR, pp. 534–439 (1991)
32. Zhang, Y., Yang, Y.: Illuminant direction determination for multiple light sources. In: CVPR, vol. I, pp. 269–276 (2000)
33. Zhou, W., Kambhamettu, C.: Estimation of illuminant direction and intensity of multiple light sources. In: ECCV, pp. 206–220 (2002)

3-D Lighting Environment Estimation with Shading and Shadows

Takeshi Takai, Susumu Iino, Atsuto Maki, and Takashi Matsuyama

Abstract We present a novel method for estimating 3-D lighting environments that consist of time-varying 3-D volumetric light sources from a single-viewpoint image. While various approaches for lighting environment estimation have been proposed, most of them assume the lighting environment as a distribution of directional light sources or a small number of near point light sources. Therefore, the estimation of a 3-D lighting environment still remains a challenging problem. In this paper, we propose a framework for estimating 3-D volumetric light sources, e.g. a frame of candles, using shadows cast on surfaces of a reference object by taking into account the geometric structures of the real world. We employ a combination of the *Skeleton Cubes* as a reference object and verify the utilities. We then describe how it works to estimate the 3-D lighting environment stably. We prove its effectiveness with experiments using a real scene under flames.

1 Introduction

Lighting environment estimation is one of the important research topics in computer vision. By recovering the lighting information from an input image of a real scene, we can superimpose virtual objects into the scene with natural shading and shadows by controlling the lighting effects in the generated or captured images accordingly (Fig. 1). The applications of lighting environment estimation are widespread, such as object tracking, face recognition, VR and graphics.

Various methods for lighting environment estimation, either direct or indirect, have been proposed. Pioneering work includes those of estimating illuminant directions in the framework of shape from shading [10, 23]. In direct sensing such

---

T. Takai (✉), S. Iino, A. Maki, and T. Matsuyama
Graduate School of Informatics, Kyoto University, Yoshida-Honmachi, Sakyo-ku, Kyoto, 606–8501 Japan
e-mail: takai@vision.kuee.kyoto-u.ac.jp, s-iino@vision.kuee.kyoto-u.ac.jp,
maki@vision.kuee.kyoto-u.ac.jp, tm@vision.kuee.kyoto-u.ac.jp

R. Ronfard and G. Taubin (eds.), *Image and Geometry Processing for 3-D Cinematography*, Geometry and Computing 5, DOI 10.1007/978-3-642-12392-4_11,
© Springer-Verlag Berlin Heidelberg 2010

Fig. 1 CG image rendered under lighting environment estimated by the proposed method. The candles in the left are real, but the objects in the right are virtual ones lit by the estimated lighting environment including the candles

as in [2, 14, 18], images of the illuminant distribution are directly captured by a camera, and the lighting environment is estimated by analyzing the pixel values of the images. Ihrke and Magnor presented a method for representing a fire with multiple cameras, but not in the context of estimating radiant intensities as a light source [4].

Among indirect sensing approaches some exploit the occluding boundary that puts strong constraints on the light source directions using an object that has a (locally) Lambertian surface [9, 20]. Some other indirect approaches [7, 19, 21, 24], so-called *inverse lighting*, employ images containing a reference object with known shape and reflectance properties, and estimate the lighting environment by analyzing its shading or highlights. Nevertheless, since most of these methods represent the lighting environment as a set of directional light sources, it is difficult to apply them to certain types of real scene, especially to an indoor scene which requires to consider the effects such as attenuation of radiant intensity caused by the positional relationship between the light source and the object. Although a few advances [3, 8, 12] allow estimation of near light source, it is still a challenging problem to solve for parameters of multiple light sources including their radiant intensity. On the other hand, as pointed out in [7] the Lambertian model for surface reflectance is too simple. Theoretically as supported by [1, 13] it performs low-pass filtering of the lighting environment, which results in a coarse resolution estimated lighting distribution.

Accordingly, another clue is required in order to estimate complex lighting environments. Sato et al. have introduced a method which utilizes *shadows* and estimate radiant intensities of a known distributed set of light sources [15]. While this method can estimate the lighting environment with higher stability than the shading-based methods, it only works in a simple case, that is, the lighting environment is represented by directional light sources. For estimating a complex 3-D lighting environment, we have the following difficulties, i.e. the solution space is larger than the measurement space (underdetermined problem), and the ambiguity of matching shadows and light sources.

In this paper, we employ a combination of the Skeleton Cubes [16] as a reference object, and propose a method for solving an *unsolvable problem* by simple linear algebra, utilizing the *geometric structure* of the real world. In the following sections, we first describe the lighting environment estimation with shading and shadows, and verify the utility of the Skeleton Cube. We then describe the difficulties of this problem, and how it can be solved with our geometric-structure based approach. After describing the algorithm of our method, we demonstrate the effectiveness of our method with a real scene, and conclude with discussions.

2 Lighting Environment Estimation with Shading and Shadows

In this section, we describe the framework for estimating lighting environment from shadows. The lighting environment is approximated by a distribution of point light sources so that the problem is to solve for the radiant intensity of each of discretely sampled sources.

Before going to the details, we summarize the following terms.

Reference object	The object with known shape and reflectance.
Shadow surface	The surface of the reference object onto which shadows are cast.
Camera	The camera that captures an image of the reference object, which is calibrated so that we know the relationship of corresponding points in the scene and in the image.
Point light source	The light source that has a position and radiant intensity, which illuminates in all directions (isotropic) as a light bulb. We also consider the effect of attenuation, that is, the energy of the light source decreases in proportion to the squared distance.

2.1 Reflectance Model

We assume that the surface reflection of the reference object is described by the simplified Torrance-Sparrow model [5, 17] which can represent both diffuse and specular reflection, and it especially describes the specular reflection with physical properties. It is in this sense more general than other models such as the Phong reflectance model [11]. In this paper, we presume the influence of interreflection is ignorable and concentrate our discussion on the first reflection.

With the simplified Torrance-Sparrow model, observed radiance, $I(x)$, at the minute area around point x is described as

$$I(x) = (k_d R_d + k_s R_s)L_{\mathcal{L}}, \tag{1}$$

where L is the radiant intensity of light source $\mathcal{L}$. R_d and R_s denote the diffuse and the specular component of a reflectance function, respectively, whereas k_d and

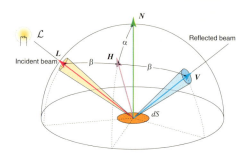

Fig. 2 The geometry of reflection. V is unit vector in direction of the viewer, L is unit vector in direction of the light source, N is unit surface normal, and H is unit angular bisector of V and L

k_s are the weighting coefficients of them. Though $I(x)$ and L are functions with respect to the wavelength, we utilize the functions at three wavelengths for red, green and blue. For the sake of a simple description, we do not denote it explicitly in the following.

Assuming the diffuse component is represented as Lambertian, we have $R_d = N \cdot L / r^2$, where N, L and r denote the surface orientation, the direction of light source L from point x, and the distance between light source L and point x, respectively. The specular component, R_s, is represented as

$$R_s = \frac{1}{r^2} \frac{1}{N \cdot V} \exp\left[-\frac{(\cos^{-1}(N \cdot H))^2}{2\sigma^2}\right], \quad (2)$$

where V, H, and σ denote the viewing direction, the half vector of L and V, and the surface roughness, respectively (Fig. 2).

The observed radiance of point x on the surface can be generally formulated as

$$I(x) = \sum_{i=1}^{N} \mathcal{M}(x, \mathcal{L}_i)\,(k_d R_d + k_s R_s) L_{\mathcal{L}_i}, \quad (3)$$

where N is the number of point light sources, and $\mathcal{M}(x, \mathcal{L}_i)$ is a mask term that encodes the self shadow of the skeleton cube.[1] The mask term, $\mathcal{M}(x, \mathcal{L}_i)$, is binary, indicating whether point x is illuminated by light source $\mathcal{L}_i$ or not, i.e. $\mathcal{M}(x, \mathcal{L}_i) = 1$ if light source $\mathcal{L}_i$ illuminates point x and $\mathcal{M}(x, \mathcal{L}_i) = 0$ otherwise (see Fig. 3).

---

[1] Note that the observed radiance is constant despite the distance between a point and a camera. While irradiance decreases with distance from the surface, the size of the captured area increases instead. Accordingly, these two effects will get balanced out and the incoming energy to one element stays constant. We can regard the energy, i.e. a value of a pixel, as radiance of one certain point since the size of the element is minute.

Fig. 3 Relationship between a light source and a reference object. The green object in the right denotes an occluding object, and the black region at the bottom denotes a shadow. The mask term of the shadow point is set to 0, and the observed intensity at the point becomes 0

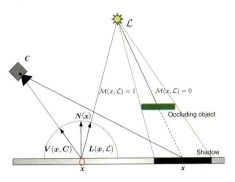

2.2 Computational Algorithm

When we sample M points on the shadow surface for observing the radiance, we have a matrix representation based on (3)

$$\begin{bmatrix} I(x_1) \\ I(x_2) \\ \vdots \\ I(x_M) \end{bmatrix} = \begin{bmatrix} K_{11} & K_{12} & \cdots & K_{1N} \\ K_{21} & K_{22} & \cdots & K_{2N} \\ \vdots & \vdots & \ddots & \vdots \\ K_{M1} & K_{M2} & \cdots & K_{MN} \end{bmatrix} \begin{bmatrix} L_{\mathcal{L}_1} \\ L_{\mathcal{L}_2} \\ \vdots \\ L_{\mathcal{L}_N} \end{bmatrix}, \tag{4}$$

where

$$K_{mn} = \mathcal{M}(x_m, \mathcal{L}_n)(k_d R_d + k_s R_s). \tag{5}$$

We then write (4) simply as

$$I = KL, \tag{6}$$

where $K = (K_{mn})$. Given a sufficient number of surface radiance samples (i.e. $M > N$), it is in principle possible to solve (6) for L by the linear least squares method with non-negative variables (the non-negative least squares; NNLS [6]),

$$\min_{L} \frac{1}{2} \| KL - I \|_2^2 \text{ subject to } L \geq 0. \tag{7}$$

Namely, we can obtain the radiant intensities in vector $[L_{\mathcal{L}_1}, L_{\mathcal{L}_2}, \ldots, L_{\mathcal{L}_N}]$.

3 Skeleton Cube

3.1 3-D geometric Features

In order to estimate the parameters of 3-D distributed light sources, we need to have a 3-D structured object so as to observe shadows caused by the lighting environment. Given that we can utilize a reference object of known shape, we design the skeleton

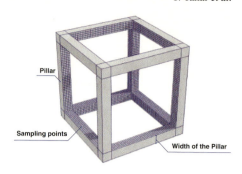

Fig. 4 The skeleton cube. Sampling points are at the lattice positions on the inside surface of the cube

cube, a hollow cube which casts shadows to its inside surface, as shown in Fig. 4. The design is on the basis of the two requirements that are inconsistent to each other. Namely, the shape should be simple while some complexity is desirable:

Simplicity   A large portion of the surface should be observable for sampling the surface intensity, and also a simple shape reduces the computational cost.

Complexity   The shape needs to be complex to a certain extent so that self shadows occur under variable lighting conditions.

We employ the skeleton cube as a reference object that satisfies the above requirements. That is, under light sources at almost any positions in a scene, it casts shadows on its inside surfaces and the shadows can be observed from any viewpoints.

The skeleton cube can also be used as a reference object for geometric calibration of cameras, which is an ordinary method by matching corresponding points in a captured image and the model. Although we do not go into the details of the geometric calibration in this paper.

3.2 Verifying Utilities of Skeleton Cube

In this section, we verify the utilities of the skeleton cube as a reference object for estimating the lighting environment. We first investigate the effectiveness of the lighting environment estimation using shadows by analyzing matrix K that constitutes (4). We then examine the occurrence of shadows for different positions of point light sources.

3.2.1 Evaluating Effectiveness of Lighting Environment Estimation with Shadows

Since the lighting environment estimation based on Lambertian shading of an object is an ill-posed or numerically ill-conditioned problem, we utilize the shadows, i.e.

Fig. 5 Distribution of light sources. The points in the images illustrate point light sources. The skeleton cube and the camera are also placed in the lighting environment. Grid lines are added for the sake of display

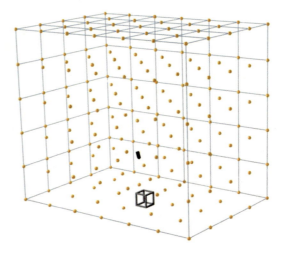

the $\mathcal{M}$ term in (5) to make the problem better-conditioned. Here, we study the relevancy of our lighting environment estimation with shadows by comparing matrix K in (6) for the cases with and without the $\mathcal{M}$ term. We generate matrix K in respective cases with the following conditions.

- Lighting environment: A set of point light sources on 3-D grid points ranging from $(-1500, -1000, 0)$ to $(1500, 1000, 2500)$ with spacing of 500 as shown in Fig. 5. The number of the light sources is 209.
- Camera position: $(450, 450, 450)$.
- Skeleton cube
  - Cube size: 100 on a side. The width of pillar 10.
  - Reflectance properties: $k_d = 1.0E+9$, $k_s = 0.8E+2$ and $\sigma = 0.3$.[2]
  - Number of the sampling points from a viewpoint: 2,293.

We visualize the generated matrix in Fig. 6. The rows of matrix K is sorted according to the categories of surface normals of the skeleton cube whereas the row and the column correspond to sampling points and light sources, respectively. The color denotes the value of entries; blue for low, red for high and white for 0 values.

With the $\mathcal{M}$ term as a consequence of shadows, matrix K contains variations of rows as observed in Fig. 6 because occluded light sources are different depending on the sample points even if the surface normals on those points are identical. Without the $\mathcal{M}$ term, on the other hand, variations of the rows are roughly in three categories, which illustrate that the rank of matrix K without the $\mathcal{M}$ term is degenerated close to rank three. This is because the number of the sorts of surface normals that are viewable from a single viewpoint is at most three. In other words, the rows within each categories are quite similar except for the difference due to specularities or the

---

[2] These are the estimated parameters of a skeleton cube under a controlled lighting environment.

Fig. 6 Visualized matrix K. The row and the column correspond to sampling points and light sources, respectively. The color denotes the value of entries; *blue* for low, *red* for high and *white* for 0 values

(a) With $\mathcal{M}$ term.   (b) Without $\mathcal{M}$ term.

distance to the light sources, which causes rank deficiency. We later discuss the rank of matrix K in detail.

In order to verify the rank of matrix K more qualitatively, we apply the singular value decomposition to then matrix in the cases of using the point light sources with and without the $\mathcal{M}$ term, and investigate the contributions of each singular value to the entire sum.

We compare matrix K in the two cases (with or without the $\mathcal{M}$ term) by analyzing their rank in terms of the contributions of the major singular values. In order to solve (7) stably, a certain rank of matrix K is required. As shown in Fig. 7, however, matrix K without the $\mathcal{M}$ term degenerates close to rank three, which is observable in contributions of the singular values that falls drastically after the third singular value. With the $\mathcal{M}$ term, singular values the third still show relatively high contributions, and it proves that matrix K with the $\mathcal{M}$ term has a higher rank, which is more desirable for solving (7).

Overall we have seen that the shadows effectively reflect the complication of the lighting environment, so that it is beneficial for computing the radiant intensities of the light sources with higher stability.

3.2.2 Evaluating Capability of Shadow Generation of the Skeleton Cube

In order to examine the occurrence of shadows, we count the number of sampling points on the inside surface of the skeleton cube that are self-occluded by another pillar of the hollow cube when viewed from each possible point light source. We consider the space for the point light sources to be placed, ranging from $(-1500, -1500, 0)$ to $(1500, 1500, 3000)$ with spacing 100.

3-D Lighting Environment Estimation with Shading and Shadows

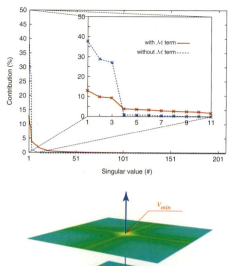

Fig. 7 Contributions of singular values of matrix K. Horizontal axis for the singular values and vertical axis for the corresponding contributions to the entire sum. The part for the higher singular values is enlarged

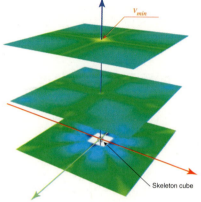

Fig. 8 Self-occlusion of the skeleton cube. We count the number of sampling points on the inside surface of the skeleton cube that are self-occluded by other pillar of the hollow cube when viewed from each possible 3-D grid position for a point light source. The color corresponds to the number of self-occluded points: *red* for few, *blue* for many

When a certain sampling point is self-occluded viewing from a certain grid point, a shadow is cast on the point by a light source on the grid point. Hence, it would be a problem if there were only few points that were self-occluded, since in that case almost all the $\mathcal{M}$ term in (5) would equal to 1, and thus the estimation of radiant intensities might become unstable. In this respect the skeleton cube is desirable of it has shadows to some extent wherever the point light sources are placed.

We show some results for verification in Fig. 8. The color of each position on the planes in Fig. 8 signifies the possible number of self-occlusions by a point light source that is placed on that position. The color denotes the number, i.e. red denotes a low and blue denotes a high number. For the total number of sampling points, 5,300, even the minimum number of occluded points was 704 (when viewed from $V_{\min}$ in Fig. 8) whereas the maximum was 3,540. This indicates that the skeleton cube has shadows of some good extent by light sources at almost all the considered positions, but in some points such as $V_{\min}$ in Fig. 8 it cannot create enough shadows to be observable.

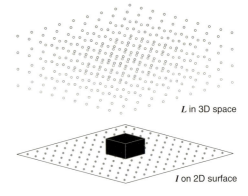

Fig. 9 Example of underdetermined problem

4 Estimation Algorithm for 3-D Lighting Environment

4.1 Problems: Underdeterminedness and Uncertainty

The framework shown in Sect. 2 enables us to estimate a lighting environment with a set of directional light sources (hemispherical lighting) [15], but for a complex 3-D lighting environment, we have the following fundamental problems.

- **Underdeterminedness:** Equation (7) easily becomes underdetermined when we extend it to 3-D distribution estimation, because the dimension of the sampling surface is two and the estimation space is three (see Fig. 9). Accordingly, we cannot have a unique solution of (7).
- **Uncertainty:** Even if we make the problem overdetermined, we have a problem of uncertainty of the non-negative least squares (NNLS). As shown in the name of the NNLS, the solution of it has no negative components and thus it does not fit as well as the least squares method without constraints. Furthermore, it finds a solution only from a viewpoint of minimizing the sum of the errors, it cannot take account of the geometrical structures in the real world. Hence, almost all of the solutions that the NNLS finds is a local optimum, which has a lot of unnatural elements when they are projected in the real world. We call them as *phantoms*, which are caused by the following two reasons, i.e. the structure of a reference object *(Phantom-I)* and the ambiguity of shadow–light source matching *(Phantom-II)*.

    – Phantom-I
      As described in Sect. 3.2.2, the skeleton cube has some regions where shadows are not easily observed. The estimated radiant intensities of light sources in these regions will be uncertain, and there can be phantoms as a result (Fig. 10).

3-D Lighting Environment Estimation with Shading and Shadows

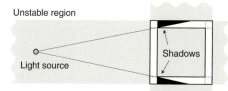

Fig. 10 Unstable region of Skeleton Cube. The gray regions denote the unstable region of the skeleton cube. Estimated radiant intensity of a light source in the region is unstable, because the shadows generated by the light source are small and hard to be observed

Fig. 11 Uncertainty of shadow–light source matching. Suppose we have two pair of light source candidates, i.e. *black pair* and *white pair*. We cannot distinguish which pair the true light sources are based on the shading and shadows on the sampling surface, because both possibilities can represent the same shading and shadows

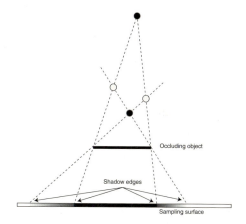

- Phantom-II

  On the other hand, the shadow based method implicitly evaluates shadow–light source matching during the numerical optimization. It is obvious from the results that the estimated light sources appear where the observed shadows can be represented by them. The ambiguity of this shadow–light source matching generates another type of phantoms. Figure 11 shows an example of this uncertainty. We have two pairs of light sources drawn by the white and black circles, and cannot distinguish which pair is the phantom by the shading and shadows on a single surface. Hence, these light sources are in a uncertain situation, and can be phantoms.

As described the above, this is a typical ill-posed problem that often appears in the computer vision problems, which cannot be solved by simple numerical approaches. One of the reasons of this is that the most of these numerical approaches cannot preserve the structure of the real world because the permutation operation of the equations does not affect the solutions. That is, these approaches ignore the relationships between points that are closely connected in the real world, and make it into the *pointwise process*. We need to apply a clue to make the problem well-posed, that is what we propose here, the *geometric structure*.

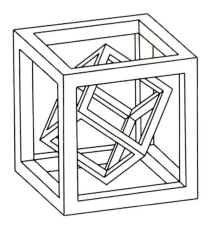

Fig. 12 Combination of skeleton cubes

4.2 Geometric Structure based Approach

As we described above, 3-D lighting environment estimation from shading and shadows is underdetermined and uncertain. For overcoming this, we utilize the geometric structure.

To make the problem overdetermined, we dived the problem into solvable ones and integrate partial solutions by taking account of the real world's structure. We also utilize the geometric structure for eliminating phantoms, which are caused by the structure of the skeleton cube (Phantom-I) and the uncertainty of the coefficient matrix K (Phantom-II). To cancel out the unstable regions and eliminate the phantoms, we employ two skeleton cubes so that the direction of each surface normal is different each other (Fig. 12). By integrating the results that are separately estimated from each skeleton cube,[3] true light sources can be extracted since they will appear in the same space of the both estimation results, but the phantoms will appear in different spaces corresponding to the structure of each skeleton cube. The details of the algorithm is shown in the next section.

4.3 Overall Algorithm

Our basic strategy for estimating light source distribution in the 3-D space is twofold, i.e., directional analysis and distance identification. We first assume a light source distribution on a hemisphere that has a certain radius, and estimate the radiant intensities of the light sources by solving (6). Then, by repeating the estimation of the radiant intensities with different radii of hemispheres, we can identify the direc-

---

[3] Note that we compute matrices K of both skeleton cubes in the same space at the same time, and thus these matrices have shadow information cast from the other cube as well as from itself, which denotes the correct relationship between light sources and skeleton cubes.

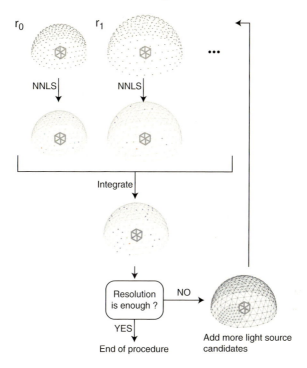

Fig. 13 Diagram of directional search

tions along which the estimated radiant intensities remain high at all the distances as those of major light sources in the scene[4] (Fig. 13). Through this process, light sources with weak radiant intensities such as reflections from the walls are regarded as ambient lighting. In this step, we separately estimate radiant intensities from each skeleton cube, and then integrate the results to eliminate phantoms. The detail of the directional search is shown in the following (Alg. 1).

In order to explicitly identify the light sources in the 3-D space, we also need to know the distances to them, i.e. from the center of the reference object, which is also the origin of the world coordinate system. This is important especially when there are near light sources in the scene. For this purpose, we can investigate possible locations of the light sources on lines along the above estimated directions. In practice, we consider distributions of light source candidates at a certain interval along the lines and then solve (6) for L, analogously as in the directional analysis. In this step, we estimate radiant intensities from both of the skeleton cubes. Note that it is useful to retain the assumption of a light source distribution on the hemisphere for describing the ambient light. After these two steps, we distribute light source candidates in the vicinity of the once identified locations of light sources, and obtain the distribution of light source candidates in the 3-D space.

---

[4] We also utilize the coarse to fine strategy for this step.

Algorithm 2: Directional search.

for *scID* ← 1 **to** *numberSkeletonCubes* **do**
    $L^{init, 0}$ ← Create a recursively subdivided icosahedron as an initial candidate distribution;
    for *itr* ← 1 **to** *maxIterations* **do**
        for *r* ← R_{min} **to** R_{max} **do**
            $L_r^{init, itr}$ ← Generate a distribution of light source candidates with $L^{init, itr-1}$ that has radius, *r*;
            L_r^{est} ← Estimate radiant intensities by the NNLS with $L_r^{init, itr}$;
        $L^{opt, scID}$ ← Integrate the light source candidates that have values of all L_r^{est};
        if *resolution is enough* **then**
            Quit the loop and output $L^{opt, scID}$ as a result;
        else
            $L^{init,itr}$ ← Add new light source candidates around the candidates that have values to $L^{init, itr-1}$;

$L^{opt, *}$ ← Eliminate phantoms by computing 'and operation' to $L^{opt, 1}, \ldots, L^{opt, numberSkeletonCubes}$;

Consequently, the algorithm for one frame of the captured sequence is shown in the following:

1. Configure a scene.
2. Estimate 3-D distribution of light sources:

   (a) Directional search with phantom elimination.
   (b) Depth identification.
   (c) Distribute light source candidates around the light sources that are estimated by the above.

3. Estimate radiant intensities with the estimated distribution.

For an image sequence, we estimate a light source distribution of the first image with the full processing of the above, and then use the result as an initial distribution for the succeeding images. We distribute light source candidates around the initial distribution and go to step 3 in the above procedure.

5 Performance Evaluation

We evaluate the performance of our method with a real scene under the following configuration (Fig. 14).

Skeleton Cubes   Two different sizes of cubes. The small cube is hanged with a fine thread inside the large cube. The sizes are 100 mm and 300 mm, respectively. We assume the reflectance as Lambertian, and ignore the interreflection.

3-D Lighting Environment Estimation with Shading and Shadows

Fig. 14 Configuration for experiments

Lighting environment	Four candles in a dark room and reflected lighting from the walls. We moved the candles around the cubes during the capture.
Camera	Sony PMW-EX3, 1920 × 1080 pixels, 29.97 fps. We rectified the captured image with easycalib [22] and estimate the camera parameters with the Skeleton Cubes.
Directional search	We utilize an initial hemisphere that has 301 vertices, and specify $R_{min} = 400$ and $R_{max} = 1600$ with spacing 200, and the number of iterations to 4.
Depth identification	We place 80 light sources for each detected direction in the directional search from 400 from the center of the cubes with spacing 15.

The figures in the left of Fig. 15 show the captured images. We find that the position, shape and radiant intensity vary with time. The figures in the right of Fig. 15 show the estimated light sources that represent the candles in the lighting environment. We find that the time-varying 3-D volume of the candles are represented, but the estimation of the distance is not so accurate as the estimation of the direction. This is because that the shape of the shadow is affected by the direction of light sources much more than their distance, and thus there is uncertainty for the distance estimation.

Although the estimation of the distance is not quite accurate, we have realistically relit images of the skeleton cubes with the estimated lighting environment as shown in the left of Fig. 16. The figures in the middle of Fig. 16 show that the difference of the radiance on the surfaces of the skeleton cubes in the captured images and the relit images, blue denote low and red denotes high values (Numerically, see Table 1). While most of regions of the images show low errors (The average errors are at most 2% of the maximum intensity, 255.), we can find some high errors in the regions where the edges of the shadows and concave portions due to the interreflection.

Finally, the figures in the right of Fig. 16 show that the superimposed virtual objects under the estimated lighting environment. In the figures, the candles to the left are real and the all other objects to the right are virtual. We can find that the objects are illuminated photo-realistically with the flicker of the candles and the soft-shadows on the objects represented.

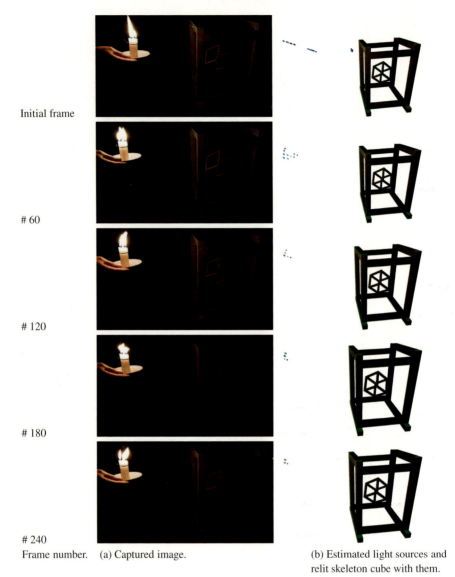

Frame number. (a) Captured image. (b) Estimated light sources and relit skeleton cube with them.

Fig. 15 Experimental results of real scene (I). The color of the points in the right of (**b**) denotes the value of the estimated radiant intensity, i.e. *blue* is low and *red* is high. The *green regions* on the skeleton cube in (**b**) denote areas we do not use sampling intensities, such as edges of the cubes and surfaces with screws

3-D Lighting Environment Estimation with Shading and Shadows

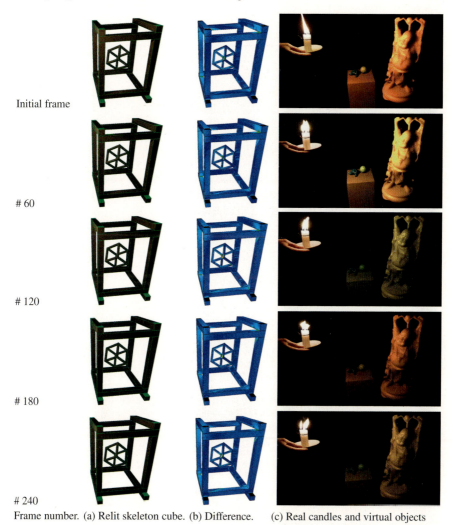

Frame number. (a) Relit skeleton cube. (b) Difference. (c) Real candles and virtual objects under estimated lighting environment.

Fig. 16 Experimental results of real scene (II). The *green regions* in (**a**) denote areas we do not use sampling intensities as described in Fig. 15. The color in (**b**) denote the value of the difference, *blue* is low and *red* is high. Figures in (**c**) show superimposed virtual objects in the real scene, where the candles are real, and the objects in the right are virtual ones under the estimated lighting environment

6 Conclusion

We have presented a method for estimating complex 3-D lighting environment with a combination of skeleton cubes. Taking into account the geometric structure of the real world enables us to solve what simple linear algebra cannot solve. Experimental

Table 1 Experimental results of real scene

	Initial frame	# 60	# 120	# 180	# 240
Max difference	64.3	55.8	49.4	38.7	50.0
Average of differences	4.66	4.31	3.84	2.67	3.71
Variance of differences	18.4	15.1	11.1	5.91	10.7

results have demonstrated that our method can estimate a complex 3-D lighting environment by localizing a distribution of 3-D volumetric light sources. As future work, we point out the following issues; a mathematical proof of why our method can eliminate phantoms, the optimal shape of the reference object, precise depth identification, and multiple camera system to observe the blind side of the reference objects so as to stabilize the distribution estimation.

Acknowledgements This work was supported by the "Foundation of Technology Supporting the Creation of Digital Media Contents" project (CREST, JST) and Ministry of Education, Culture Sports, Science and Technology under the Leading Project, "Development of High Fidelity Digitization Software for Large-Scale and Intangible Cultural Assets."

References

1. Basri, R., Jacobs, D.W.: Lambertian reflectance and linear subspaces. IEEE Trans. Pattern Anal. Mach. Intell. **25**(2), 218–233 (2003)
2. Debevec, P.: Rendering synthetic objects into real scenes: Bridging traditional and image-based graphics with global illumination and high dynamic range photography. In: SIGGRAPH '98: Procedings of the 25th annual conference on Computer graphics and interactive techniques, pp. 189–198 (1998)
3. Hara, K., Nishino, K., Ikeuchi, K.: Determining reflectance and light position from a single image without distant illumination assumption. In: IEEE International Conference on Computer Vision, vol. 2, pp. 560–567 (2003)
4. Ihrke, I., Magnor, M.: Image-based tomographic reconstruction of flames. In: Proc. ACM/EG Symposium on Animation (SCA'04), pp. 367–375. Grenoble, France (2004)
5. Ikeuchi, K., Sato, K.: Determining reflectance properties of an object using range and brightness images. IEEE Trans. Pattern Anal. Mach. Intell. **13**(11), 1139–1153 (1991)
6. Lawson, C.L., Hanson, R.J.: Solving Least Squares Problems. Society for Industrial Mathematics, Philadelphia (1987)
7. Marschner, S.R., Greenberg, D.P.: Inverse lighting for photography. In: *Fifth Color Imaging Conference*, pp. 262–265 (1997)
8. Matsuyama, T., Wu, X., Takai, T., Nobuhara, S.: Real-Time 3D Shape Reconstruction, Dynamic 3D Mesh Deformation, and High Fidelity Visualization for 3D Video. Comput. Vis. Image Understand. **96**(3), 393–434 (2004)
9. Nillius, P., Eklundh, J.O.: Automatic estimation of the projected light source direction. In: IEEE Computer Society Conference on Computer Vision and Pattern Recognition, vol. I, pp. 1076–1083 (2001)
10. Pentland, A.P.: Finding the illuminant direction. J. Opt. Soc. Am. **72**(4), 448–455 (1982)
11. Phong, B.T.: Illumination for computer generated pictures. Comm. ACM **18**(6), 311–317 (1975)
12. Powell, M.W., Sarkar, S., Goldgof, D.: A simple strategy for calibrating the geometry of light sources. IEEE Trans. Pattern Anal. Mach. Intell. **23**(9), 1022–1027 (2001)

13. Ramamoorthi, R., Hanrahan, P.: A signal-processing framework for inverse rendering. In: ACM SIGGRAPH, pp. 117–128 (2001)
14. Sato, I., Sato, Y., Ikeuchi, K.: Acquiring a radiance distribution to superimpose virtual objects onto a real scene. IEEE Trans. Visual. Comput. Graph. **5**(1), 1–12 (1999)
15. Sato, I., Sato, Y., Ikeuchi, K.: Illumination from shadows. IEEE Trans. Pattern Anal. Mach. Intell. **25**(3), 290–300 (2003)
16. Takai, T., Maki, A., Matsuyama, T.: Self shadows and cast shadows in estimating illumination distribution. In: Fourth European Conference on Visual Media Production (2007)
17. Torrance, K.E., Sparrow, E.M.: Theory for off-specular reflection from roughened surfaces. J. Opt. Soc. Am. **57**(9), 1105–1114 (1967)
18. Unger, J., Wenger, A., Hawkings, T., Gardner, A., Debevec, P.: Capturing and rendering with incident light fields. In: EGRW '03: Proceedings of the 14th Eurographics workshop on Rendering, pp. 141–149 (2003)
19. Wang, Y., Samaras, D.: Estimation of multiple directional illuminants from a single image. Image Vis. Comput. **26**(9), 1179–1195 (2008)
20. Yang, Y., Yuille, A.: Sources from shading. In: IEEE Computer Society Conference on Computer Vision and Pattern Recognition, pp. 534–539 (1991)
21. Zhang, Y., Yang, Y.-H.: Multiple illuminant direction detection with application to image system. IEEE Trans. Pattern Anal. Mach. Intell. **232**(8), 915–920 (2001)
22. Zhang, Z.: A flexible new technique for camera calibration. IEEE Trans. Pattern Anal. Mach. Intell. **22**(11), 1330–1334 (2000)
23. Zheng, Q., Chellappa, R.: Estimation of illuminant direction, albedo, and shape from shading. IEEE Trans. Pattern Anal. Mach. Intell. **13**(7), 680–702 (1991)
24. Zhou, W., Kambhamettu, R.: Estimation of illuminant direction and intensity of multiple light sources. In: European Conference on Computer Vision, pp. 206–220 (2002)

3-D Cinematography with Approximate or No Geometry

Martin Eisemann, Timo Stich, and Marcus Magnor

Abstract 3-D cinematography is a new step towards full immersive video, allowing complete control of the viewpoint during playback both in space and time. One major challenge towards this goal is precise scene reconstruction, either implicit or explicit. While some approaches exist which are able to generate a convincing geometry proxy, they are bound to many constraints, e.g., accurate camera calibration and synchronized cameras. This chapter is about methods to remedy some of these constraints.

1 Introduction

As humans we perceive most of our surroundings through our eyes and visual stimuli affect all of our senses, drive emotion, arouse memories and much more. That is one of the reasons why we like to look at pictures. A major revolution occurred with the introduction of moving images, or videos. The dimension of time was suddenly added, which gave incredible freedom to film- and movie makers to tell their story to the audience.

With more powerful hardware, computation power and clever algorithms we are now able to add a new dimension to videos, namely the third spatial dimension. This will give users or producers the possibility to change the camera viewpoint on the fly.

The standard 2-D video production pipeline can be split into two parts, acquisition and display. Both are well understood. 3-D cinematography adds a new intermediate step, the scene reconstruction. This could be a 3-D reconstruction of the scene combined with retexturing for displaying a new viewpoint or a direct estimation of a plausible output image for a new viewpoint without the need for reconstruction. Both methods have their benefits and drawbacks and will be further discussed in the next sections. Purely image-based approaches as Lightfields [36]

M. Eisemann (✉), M. Magnor, and T. Stich
TU Braunschweig, Mühlenpfordtstr. 23, 38102 Braunschweig, Germany
e-mail: eisemann@cg.cs.tu-bs.de, magnor@cg.cs.tu-bs.de, stich@cg.cs.tu-bs.de

R. Ronfard and G. Taubin (eds.), *Image and Geometry Processing for 3-D Cinematography*, Geometry and Computing 5, DOI 10.1007/978-3-642-12392-4_12,
© Springer-Verlag Berlin Heidelberg 2010

or the Lumigraph [27] might not need any sophisticated reconstruction as the sheer amount of images allows for smooth interpolation. But more practical systems need to rely on some clever reconstruction as the number of cameras is seldom larger than a dozen. These are the systems we are interested in this chapter.

2 3-D Reconstruction

Despite the acquired image or video footage one of the most important factors for 3-D cinematography is arguably a 3-D model of the scene. Depending on the task only a 3-D model of the foreground or a complete scene reconstruction is needed. Most methods aim at the reconstruction of the foreground in controlled environments, e.g. the reconstruction of an actor. For proper reconstruction the camera parameters need to be known in advance. These can be acquired by several methods and the choice depends on the task [29, 53, 58].

Usually the most simple camera, a frontal pinhole camera, is assumed. With this model every point $\mathbf{x}_w$ in a 3-D scene can be simply projected to its image space position $\mathbf{x}_{ip}$ using a projection matrix $\mathbf{P}$. Given this dependency between the 3-D world and its 2-D image equivalent reconstruction of the scene geometry is then possible. There are basically three established methods that can compute 3-D information from images alone. These are parameterized template model matching, shape-from-silhouette and depth-from-stereo.

2.1 Model Reconstruction

In Hollywood movies it is quite common to replace actors with hand-made, rendered 3-D models, which are crafted by some designers using modelling tools such as Maya, 3DS Max, Blender, Cinema4D, etc. Using marker-based motion capture methods the model animation can be brought into congruence with the motion of the actor. Nevertheless this is very laborious work and only suitable for studios that are willing to spend a tremendous amount of money on skilled artists and designers, additionally in some situations, like sports, or public events it might be impossible to place markers on the object of interest.

The Free-Viewpoint Video System of Carranza et al. [15] combines motion capture and 3-D reconstruction by using a single template model. In a first step the silhouettes of the object of interest are extracted in all input images. A generic human body model consisting of several segments, i.e. submeshes, and a corresponding bone system is then adapted to resemble the human actor and fitted to the silhouettes of each video frame by an analysis-by-synthesis approach.

A single parameterized template model can seldom represent all possibilities of human shapes sufficiently, therefore the result can be improved by identifying multi-view photo-inconsistent regions and fine-tuning the mesh in these regions by enforcing a color-consistency criterion [3].

Small details usually cannot be sufficiently recovered by these methods, as the underlying mesh is quite coarse, usually only up to 21k triangles [15]. An improvement can be achieved by acquiring a detailed mesh beforehand. One way to do this would be to use a structured light approach [33, 46, 50]. Anguelov et al. [7] make use of detailed laser scans of an actor in different poses, from which they learn a pose deformation model and a model of variation for the body shape in order to simulate realistic muscle behaviour on the model. Using a set of markers on the human, their system can use the motion capture data to animate the model. De Aguiar et al. also make use of detailed laser scans of the actor which they deform in order to maximize the congruence with the multi-view recordings [5]. Their system is not aiming at muscle behaviour but is more focussed on arbitrary inputs, as e.g. humans wearing different kinds of apparel, and markerless tracking, which is less intrusive. Similar to Carranza et al. [15] a template model is fitted to the videos first. In a next step the laser scan is deformed to fit the template model by specifying correspondence points between the two meshes.

To capture non-rigid surface deformations occurring from wearing wide apparel for example, optical flow-based 3-D correspondences are estimated and the laser scanned model is deformed using a Laplacian mesh deformation such that it follows the motion of the actor over time [4]. Fusing of efficient volume- and surface-based deformation schemes, a multi-view analysis-through-synthesis procedure, and a multi-view stereo approach, cf. Sect. 2.3, can lead to an even higher correspondence match of the mesh with the input video [2]. This allows to capture performances of people wearing a wide variety of everyday apparel and performing extremely fast and energetic motions.

While this approach delivers an evidentiary high quality it is not so well suited for situations in which a high-quality laser scan of the actor cannot be acquired beforehand. For such situations more general methods are needed that are described in the following sections.

2.2 Shape-From-Silhouettes

The shape-from-silhouettes approach by Laurentini et al. [34] uses the extracted silhouettes of the object to define its visual hull. In 2-D the visual hull is equivalent to the convex hull, in 3-D the visual hull is a subset of the convex hull including hyperbolic regions. It is the maximal volume constructed from reprojecting the silhouettes cones of each input image back into 3-D space and computing their intersection. As the number of input images is limited, only an approximation of the visual hull, sometimes called inferred visual hull, can be reconstructed, ref. Fig. 1. As this method rather conservatively estimates the real geometry, results can become relatively coarse. On the other hand this algorithm can be easily implemented on graphics hardware to achieve real-time frame rates, e.g. [40], and can even be calculated in image-space rather than 3-D space [41]. An improvement can be achieved by adding color constraints in order to detect concavities as well [32, 52], also known as space-carving, or to employ an optimization process, as it is done by

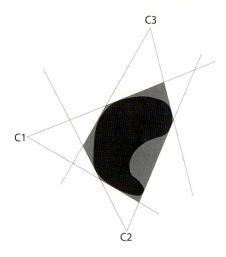

Fig. 1 The inferred visual hull (*light gray*) of an object (*dark gray*) is estimated by reprojecting each silhouette cone and computing the intersection

Starck et al. [54]. Their approach combines cues from the visual hull and stereo-correspondences, cf. Sect. 2.3, in an optimization framework for reconstruction. The visual hull is used as an upper bound on the reconstructed geometry. Contours on the underlying surface are extracted using a wide-baseline feature matching, which serve as further constraints. In a final step an optimization framework based on graph cuts reconstructs the smooth surface within the visual hull that passes through the features while reproducing the silhouette images and maximizing the consistency in appearance between views.

2.3 Depth-From-Stereo

Sometimes a whole scene has to be reconstructed, in which case all previously mentioned methods fail (except for [32]), as they are based on silhouettes, which can no longer be extracted. In this case depth-from-stereo systems perform better, as they extract a depth map for each input image, which can then be used for 3-D rendering. The basic principle of depth-from-stereo is triangulation. Given two corresponding points in two images and the camera parameters, the exact position of this point in 3-D can be reconstructed, see Fig. 2. Finding these correspondences can be arbitrarily hard and sometimes even ambiguous. To relax the problem of doing an exhaustive search for similarity over the whole image, one usually makes use of the epipolar constraint. By employing epipolar constraints the search can be reduced to a 1D line search along the conjugate epipolar lines, Fig. 2. Usually a rectification precedes the line search so that the search can be performed along the same scanline, i.e. the input images are projected onto a plane parallel to the line between the optical centers of the input cameras. The correspondence finding can be performed for example by a window-based matching, i.e. the matching is performed on a $n \times n$ window instead of a single pixel. If further knowledge about the scene is given or scene characteristics are assumed, as for example local smoothness, more

3-D Cinematography with Approximate or No Geometry

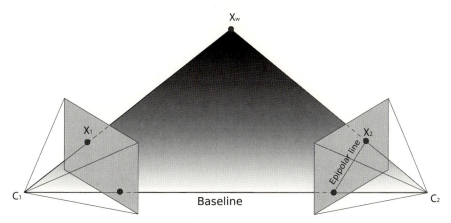

Fig. 2 Using epipolar constraints and triangulation the 3-D position of any scene point can be reconstructed

sophisticated methods based on energy minimization can be employed, e.g. [10, 11]. If more than two images should be used for depth estimation a plane sweep algorithm can be used [18]. In this approach a plane is placed at different depths, the input images are projected onto it and the plane is rendered from the virtual viewpoint. The color variation at every fragment serves as a quality estimate for this depth value. This approach is especially appealing in real-time acquisition systems as it can be computed very efficiently on graphics hardware [26, 37, 62]. Even dedicated hardware is nowadays available for multi-view stereo reconstruction and has already been successfully applied in an image-based rendering system [43].

One of the first systems to achieve high quality interpolation with a relatively sparse camera setup was the approach by Zitnick et al. [64]. Instead of matching single-pixels or windows of pixels, they match segments of similar color. As they assume that all pixels inside a segment have similar disparities an oversegmentation of the image is needed. The segments are then matched and the estimated disparities are further smoothed to remove outliers and create smooth interpolation between connected segments belonging to the same object.

Methods based on this matching are commonly applied only for dense stereo, i.e. the distance between the cameras are rather small, only up to a few dozen inches. For larger distances, or fewer cameras, additional information is needed for reconstruction. Waschbüsch et al. [60] use video bricks, which consist of a color camera for texture acquisition, and two calibrated grayscale cameras that are used together with a projector to estimate depth in the scene using structured light. The benefit of these bricks is that depth ambiguities are resolved in textureless areas. These depth estimations are used as initialization for a geometry filtering, based on bilateral filtering, for generation of time-coherent models with removed quantization noise and calibration errors.

There are many other 3-D reconstruction methods, like Shape-from-Texture, Shape-from-Shading, etc., but these are commonly not used for multi-view stereo reconstruction for 3-D cinematography and therefore we refer the interested reader to the appropriate literature.

3 Texturing with Imprecise Geometry and Sparse Camera Setups

In the previous section several approaches have been presented to reconstruct 3-D geometry from a collection of images. All these reconstruction approaches have usually one thing in common: they are imprecise (especially the faster ones, which are essential for real-time acquisition). While this may not be such a big problem when looking at the mesh alone, it will become rather obvious when the mesh is to be textured again using only a set of input images. One way to circumvent this problem is to use a large enough number of input cameras or a camera array [13, 25, 27, 36, 42] and then simply interpolate between the images in ray space. This is of course arguably the way to achieve the best rendering quality, but rather impractical as thousands of images might be needed for sufficient quality and freedom of camera movement [36]. The number can be effectively reduced by making use of a more precise geometry proxy [27]. But reconstruction of scene geometry is seldom perfect. Another possibility is to make use of dynamic textures [17]. Here a coarse geometry is used to capture large scale variations in the scene, while the residual statistical variability in texture space is captured using a PCA basis of spatial filters. It can be shown that this is equivalent to the analytical basis. New poses and views can then be reconstructed by first synthesizing the texture by modulating the texture basis and then warping it back onto the geometry. However for a good estimation of the basis quite a few input images are needed. Therefore the challenge is generating a perceptually plausible rendering with only a sparse setup of cameras and a possibly imperfect geometry proxy.

Commonly in image-based rendering (IBR) the full bidirectional reflectance distribution function (i.e. how a point on a surface appears depending on the viewpoint and lighting) is approximated by projective texture mapping [51] and image blending. Assuming the camera parameters of the input cameras are given the recorded images can be reprojected onto the geometry. If more than one camera is used the corresponding projected color values must be somehow combined for the final result. Usually the influence of a cameras' projected color value to the result is based on either its angular deviation to the normal vector at the corresponding mesh position [15] or its angular deviation to the view vector [13, 19]. The first approach is suitable for Lambertian surfaces but can result in either cracks in the texture or, even worse, a complete loss of the 3-D impression, this would, e.g., happen to a light-field [36] where the geometry proxy is only a simple quad. The second approach is more general but has to deal with ghosting artifacts if the textures do not match on the surface. This is the case if too few input images are available or the geometry is too imprecise, see Fig. 3 for an example. Some approaches prevent the ghosting artifacts by smoothing the input images at the cost of more blurriness.

Fig. 3 Comparison of standard linear interpolation using the Unstructured Lumigraph weighting scheme [13] (*left*) and the Floating Textures approach [9] (*right*) for similar input images and visual hull geometry proxy. Ghosting along the collar and blurring of the shirt's front, noticeable in linear interpolation, are eliminated by the non-linear approach

The bandlimiting approach, discussed by Stewart et al. in [55], chooses the amount of blur based on the disparity. For light field rendering they propose to add back high frequency details from a wide aperture filtered image, but this approach is only suitable to two-plane light field rendering and not for general IBR or even sparse camera setups. Eisemann et al. [21] vary the amount of smoothness depending on the current viewpoint, but the constant change of blurriness in the rendered images can cause distractions. Let us take a more detailed look at the underlying problem before dealing with more convincing solutions.

In a slightly simplified version, the plenoptic function $P(x, y, z, \theta, \phi)$ describes radiance as a function of 3-D position in space (x, y, z) and direction (θ, ϕ) [1]. The notion of IBR now is to approximate the plenoptic function with a finite number of discrete samples of P for various (x, y, z, θ, ϕ) and to efficiently re-create novel views from this representation by making use of some sort of object geometry proxy.

Any object surface that one chooses to display can be described as a function $\mathbf{G} : (x, y, z, \theta, \phi) \rightarrow (x_o, y_o, z_o)$, i.e., by a mapping of viewing rays (x, y, z, θ, ϕ) to 3-D coordinates (x_o, y_o, z_o) on the object's surface. Of course, the function $\mathbf{G}$ is only defined for rays hitting the object, but this is not crucial since one can simply discard the computation for all other viewing rays. Let $\mathbf{G_O}$ denote the function of the true surface of the object, and $\mathbf{G_A}$ denote a function that only approximates this surface, Fig. 4.

Next, a variety of camera calibration techniques exist to establish the projection mapping $\mathbf{P}_i : (x, y, z) \rightarrow (s, t)$ which describes how any 3-D point (x, y, z) is mapped to its corresponding 2-D position (s, t) in the i-th image [29, 58]. From its projected position (s, t) in image $\mathbf{I}_i$, the 3-D point's color value (r, g, b) can be read out, $\mathbf{I}_i : (s, t) \rightarrow (r, g, b)$. Then, any novel view $\mathbf{I}_{\text{Linear}}^V$ from a virtual viewpoint V synthesized by a weighted linear interpolation scheme, as employed by most IBR systems, can be formulated as

$$\mathbf{I}_{\text{Linear}}^V(x, y, z, \theta, \phi) = \sum_i \mathbf{I}_i^V(x, y, z, \theta, \phi)\, \omega_i \qquad (1)$$

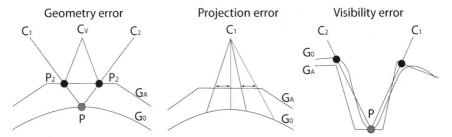

Fig. 4 *Left*: Geometry inaccuracies cause ghosting artifacts. Point P on the original surface $\mathbf{G_O}$ is erroneously projected to 3-D-position P_1 from camera C_1 and to 3-D-position P_2 from camera C_2 if only the approximate geometry $\mathbf{G_A}$ is available. *Middle*: Small imprecisions in camera calibration can lead to false pixel projections (*dark gray lines*, compared to correct projections displayed as *light gray lines*). This leads to a texture shift on the object surface and subsequently to ghosting artifacts. *Right*: Visibility errors. Given only approximate geometry $\mathbf{G_A}$, point P is classified as being visible from C_2 and not visible from camera C_1. Given correct geometry $\mathbf{G_O}$, it is actually the reverse, resulting in false projections

with

$$\mathbf{I}_i^V(x,y,z,\theta,\phi) = \mathbf{I}_i(\mathbf{P}_i(\mathbf{G_A}(x,y,z,\theta,\phi))) \tag{2}$$
$$\omega_i = \delta_i(\mathbf{G_A}(x,y,z,\theta,\phi))\, w_i(x,y,z,\theta,\phi) \tag{3}$$

and $\sum_i \omega_i = 1$. The notation $\mathbf{I}_i^V$ is used to denote the image rendered for a viewpoint V by projecting the input image $\mathbf{I}_i$ as texture onto $\mathbf{G_A}$. δ_i is a visibility factor which is one if a point on the approximate surface $\mathbf{G_A}$ is visible by camera i, and zero otherwise. w_i is the weighting function which determines the influence of camera i for every viewing ray, also called weight map.

Note that (1) is the attempt to represent the plenoptic function as a linear combination of re-projected images. For several reasons, weighted linear interpolation cannot be relied on to reconstruct the correct values of the plenoptic function:

1. Typically, $\mathbf{G_O} \neq \mathbf{G_A}$ almost everywhere, so the input to (2) is already incorrect in most places, Fig. 4 left.
2. Due to calibration errors, $\mathbf{P}_i$ is not exact, leading to projection deviations and, subsequently, erroneous color values, Fig. 4 middle.
3. In any case, only visibility calculations based on the original geometry $\mathbf{G_O}$ can provide correct results. If only approximate geometry is available, visibility errors are bound to occur, Fig. 4 right.

In summary, in the presence of even small geometric inaccuracies or camera calibration imprecisions, a linear approach is not able to correctly interpolate from discrete image samples.

4 Floating Textures

As pointed out in the last section an adaptive, non-linear approach is necessary for high quality texturing in the presence of imprecise geometry and undersampled input data to reduce blurring and ghosting artifacts. Assuming that a set of input images, the corresponding, possibly imprecise, calibration data and a geometry proxy are given (cf. to Sect. 2 for common 3-D reconstruction methods), the task is to find a way to texture this proxy again without noticeable artifacts and shadowing the imprecision of the underlying geometry.

Without occlusion, any novel viewpoint can, in theory, be rendered directly from the input images by warping, i.e. by simply deforming the images, so that the following property holds:

$$\mathbf{I}_j = \mathbf{W}_{\mathbf{I}_i \to \mathbf{I}_j} \circ \mathbf{I}_i \quad , \tag{4}$$

where $\mathbf{W}_{\mathbf{I}_i \to \mathbf{I}_j} \circ \mathbf{I}_i$ warps an image $\mathbf{I}_i$ towards $\mathbf{I}_j$ according to the warp field $\mathbf{W}_{\mathbf{I}_i \to \mathbf{I}_j}$. The problem of determining the warp field $\mathbf{W}_{\mathbf{I}_i \to \mathbf{I}_j}$ between two images $\mathbf{I}_i, \mathbf{I}_j$ is a heavily researched area in computer graphics and vision and several techniques exist which try to solve this problem, the most famous known are optical flow estimations, e.g. [31, 38]. If pixel distances between corresponding image features are not too large, algorithms to robustly estimate per-pixel optical flow are available [12, 45]. The issue here is that in most cases these distances will be too large.

In order to relax the correspondence finding problem, the problem can literally be projected into another space, namely the output image domain. By first projecting the photographs from cameras C_i onto the approximate geometry surface $\mathbf{G_A}$ and rendering the scene from the desired viewpoint V, creating the intermediate images $\mathbf{I}_i^V$, the corresponding image features are brought much closer together than they have been in the original input images, Fig. 5. This opens up the possibility of using well-established techniques like optical flow estimation to the intermediate images $\mathbf{I}_i^V$ to robustly determine the pairwise flow fields $\mathbf{W}_{\mathbf{I}_i^V \to \mathbf{I}_j^V}$, i.e. to find the corresponding features in both images. To compensate for more than two input images, a linear combination of the flow fields according to (6) can be applied to all intermediate images $\mathbf{I}_i^V$, which can then be blended together to obtain the final rendering result $\mathbf{I}_{\text{Float}}^V$ [9]. To reduce computational cost, instead of establishing for n input photos $(n-1)n$ flow fields, it often suffices to consider only the three closest input images to the current viewpoint. If more than three input images are needed, the quadratic effort can be reduced to linear complexity by using intermediate results.

It is important to use an angular weighting scheme as proposed in [13,19] because it provides smooth changes of the camera influences and therefore prevents snapping problems which could otherwise occur.

The processing steps are summarized in the following functions and visualized in Fig. 5:

$$\mathbf{I}_{\text{Float}}^V = \sum_{i=1}^{n} (\mathbf{W}_{\mathbf{I}_i^V} \circ \mathbf{I}_i^V) \omega_i \tag{5}$$

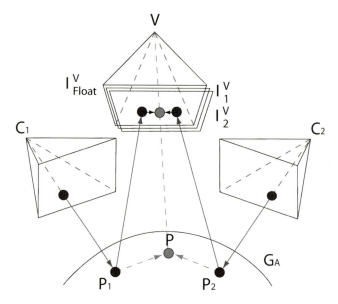

Fig. 5 Rendering with Floating Textures [9]. The input photos are projected from camera positions C_i onto the approximate geometry $\mathbf{G_A}$ and onto the desired image plane of viewpoint V. The resulting intermediate images $\mathbf{I}_i^V$ exhibit mismatch which is compensated by warping all $\mathbf{I}_i^V$ based on the optical flow to obtain the final image $\mathbf{I}_{\text{Float}}^V$

$$\mathbf{W}_{\mathbf{I}_i^V} = \sum_{j=1}^{n} \omega_j \mathbf{W}_{\mathbf{I}_i^V \to \mathbf{I}_j^V} \qquad (6)$$

$\mathbf{W}_{\mathbf{I}_i^V}$ is the combined flow field which is used for warping image $\mathbf{I}_i^V$. Equation (5) is therefore an extension of (1) by additionally solving for the non-linear part in P.

4.1 Soft Visibility

Up to now only occlusion-free situations can be precisely handled, which is seldom the case in real-world scenarios. Simple projection of imprecisely calibrated photos onto an approximate 3-D geometry model typically causes unsatisfactory results in the vicinity of occlusion boundaries, Fig. 6 top left. Texture information from occluding parts of the mesh project incorrectly onto other geometry parts. With respect to Floating Textures, this not only affects rendering quality but also the reliability of flow field estimation.

A common approach to handle the occlusion problem is to establish a binary visibility map for each camera, multiply it with the weight map, and normalize the weights afterwards so they sum up to one. This efficiently discards occluded pixels in the input cameras for texture generation. In [15] the camera is slightly displaced

3-D Cinematography with Approximate or No Geometry

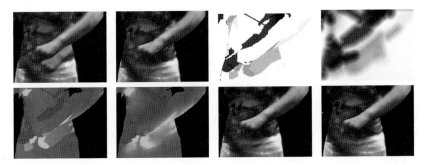

Fig. 6 *Top row, left*: Projection errors occur if occlusion is ignored. *Middle left*: Optical flow estimation goes astray if occluded image regions are not properly filled. *Middle right*: Visualization of a binary visibility map for three input cameras. *Right*: Visualization of a soft visibility map for three input cameras. *Second row, left*: Weight map multiplied with the binary visibility map. *Middle left*: Weight map multiplied with the soft visibility map; note that no sudden jumps of camera weights occur anymore between adjacent pixels. *Middle right*: Final result after texture projection using a weight map with binary visibility. *Right*: Final result after texture projection using a weight map with soft visibility. Note that most visible seams and false projections have been effectively removed

several times in order to reliably detect occluded pixels. Lensch et al. [35] discard samples which are close to large depth changes, as they cannot be relied on. One drawback of this approach is that it must be assumed that the underlying geometry is precise, and cameras are precisely calibrated. In the presence of coarse geometry, the usage of such binary visibility maps can create occlusion boundary artifacts at pixels where the value of the visibility map suddenly changes, Fig. 6 bottom row, middle right.

To counter these effects, a "soft" visibility map Ω for the current viewpoint and every input camera can be generated using a distance filter on the binary map:

$$\Omega(x, y) = \begin{cases} 0 & \text{if } \delta(x, y) = 0 \\ \frac{occDist(x,y)}{r} & \text{if } occDist(x, y) \leq r \\ 1 & \text{else} \end{cases} \quad (7)$$

Here r is a user-defined radius, and $occDist(x, y)$ is the distance to the next occluded pixel. If Ω is multiplied with the weight map, (7) makes sure that occluded regions stay occluded, while hard edges in the final weight map are removed. Using this "soft" visibility map the above mentioned occlusion artifacts effectively disappear, Fig. 6 bottom right.

To improve optical flow estimation, occluded areas in the projected input images I_i^V need to be filled with the corresponding color values from that camera whose weight ω for this pixel is highest, as the probability that this camera provides the correct color is the highest. Otherwise, the erroneously projected part could seriously influence the result of the Floating Texture output as wrong correspondences could be established, Fig. 6 top row, middle left. Applying the described filling procedure noticeably improves the quality of the flow calculation, Fig. 6 bottom right.

4.2 GPU Implementation

The non-linear optimization before the blending step is computationally very intensive and cannot be sufficiently calculated in advance. Therefore for immediate feedback it is important to compute the whole rendering part on-the-fly exploiting the power of modern graphics hardware. A block diagram is given in Fig. 7. The geometry representation can be of almost arbitrary type, e.g., a triangle mesh, a voxel representation, or a depth map (even though correct occlusion handling with a single depth map is not always possible due to the 2.5D scene representation).

First, given a novel viewpoint, the closest camera positions are queried. For sparse camera arrangements, typically the two or three closest input images are chosen. The geometry model is rendered from the cameras' viewpoints into different depth buffers. These depth maps are then used to establish for each camera a binary visibility map for the current viewpoint. These visibility maps are used as input to the soft visibility shader which can be efficiently implemented in a two-pass fragment shader. Next, a weight map is established by calculating the camera weights per output pixel, based on the Unstructured Lumigraph weighting scheme [13]. The final camera weights for each pixel in the output image are obtained by multiplying the weight map with the visibility map and normalizing the result.

To create the input images for the flow field calculation, the geometry proxy is rendered from the desired viewpoint several times into multiple render targets in turn, projecting each input photo onto the geometry. If the weight for a specific camera is 0 for a pixel, the color from the input camera with the highest weight at this position is used instead.

To compute the optical flow between two images efficient GPU implementations are needed [9, 45]. Even though this processing step is computationally expensive and takes approximately 90% of the rendering time, interactive to real-time speedups are possible with modern GPUs. Once all needed computations have been carried out, the results can be combined in a final render pass, which warps and blends the projected images according to the weight map and flow fields. The benefits of the Floating Textures approach are best visible in the images in Fig. 8, where a comparison of different image-based rendering approaches is given.

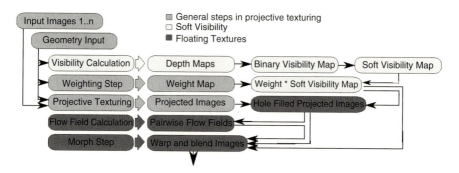

Fig. 7 Complete overview of the Floating Textures algorithm on GPU. See text for details

3-D Cinematography with Approximate or No Geometry

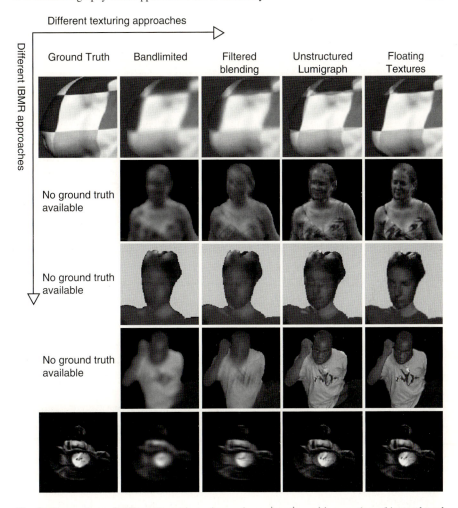

Fig. 8 Comparison of different texturing schemes in conjunction with a number of image-based modelling and rendering (IBMR) approaches. *From left to right*: Ground truth image (where available), bandlimited reconstruction [16], Filtered Blending [21], Unstructured Lumigraph Rendering [13], and Floating Textures. The different IBMR methods are (*from top to bottom*): Synthetic data set, Polyhedral Visual Hull Rendering [24], Free-Viewpoint Video [15], SurfCap [54], and Light Field Rendering [36]

4.3 Static Correspondence Finding

Under some circumstances it might be important to prewarp the textures, not for each viewpoint but once for each time step. One application in this direction would be the estimation of the BRDF of the model. Therefore reflectance information is needed for every point on a surface throughout the whole sequence.

Ahmed et al. [6] incorporate BRDF estimation into the free-viewpoint video system [15]. They specifically solve two problems. First, due to the underlying parameterized model a consistent image to surface correspondence for each frame must be found. This is done by reprojecting the input images onto the geometry and back into the views of the other cameras. Then they optimize the projected image for each camera to create a multi-view video texture. For every point on the surface they estimate the camera for which the surface point to the camera deviates the least from the normal vector at that position and use this projected color as reference value. They then warp the input image so that it most resembles this view.

A second registration problem is the model change over time. A parameterized model cannot directly cope with changes of the recorded object, as e.g. shifting clothes. This would invalidate the assumption that a constant set of BRDF parameters could be assigned to each location on the object. To deal with this the texture is transformed into a square domain, similar to geometry images [28] and frame to frame correspondences are computed to handle the shift.

During acquisition two recording passes are usually needed, one pass to acquire the reflectance information, where the actor needs to slowly turn himself around and one recording for the actual motion that one wants to capture. For both, calibrated light sources need to be used.

Assigning constant texture information through warping for each vertex is only possible if a mesh with consistent vertex topology is given. In many reconstruction approaches, cf. Sect. 2, this is not provided. Furthermore assigning constant texture coordinates to each vertex even per frame may lead to wrong results on coarse geometry. This is due to the assumption, that at least one camera projects the correct color value onto the mesh is not always true and the amount of warping must be based on the current viewpoint [9]. That means the correspondences can still be precomputed but the amount of warping during rendering must be scaled depending on the viewpoint to theoretically generate an artifact free image. In our experience the dynamic approach from Sect. 4 reveals more realistic results and should be preferred if no complete BRDF estimation is needed.

5 View and Time Interpolation in Image Space

Up to now we considered the case where at least an approximate geometry could be reconstructed. In some cases however it is beneficial not to reconstruct any geometry at all, but instead work solely in image space. In some sense reconstructing geometry imposes an implicit quality degradation by creating a 3-D scene from a 2-D video, for the purpose creating a 2-D video out of the 3-D scene again.

While sophisticated methods are still able to create high quality, in controlled studio environments, cf. Sect. 3, these methods also pose several constraints on the acquisition setup. First of all, many methods only reconstruct foreground objects, which can be easily segmented from the rest of the image. Second, the scene to be reconstructed must be either static or the recording cameras must be synchronized, so that frames are captured at exactly the same time instance, otherwise reconstruction will fail for fast moving parts. Even though it is possible to trigger synchronized capturing for modern state-of-the-art cameras, it still poses a problem in outdoor environments or for moving cameras, due to the amount of cables and connectors. Third, if automatic reconstruction fails, laborious modelling by hand might be necessary. Additionally sometimes even this approach seems infeasible due to fine, complicated structures in the image like, e.g., hair.

Working in image-space directly can solve or at least ease most of the aforementioned problems for 3-D cinematography, as the problem is altered from a 3-D reconstruction problem to a 2-D correspondence problem. If perfect correspondences are found between every pixel of two or more images, morphing techniques can create the impression of a real moving camera to the human observer, plus time and space can be treated equally in a common framework. While this enforces some constraints, as e.g. limiting the possible camera movement to the camera hull,[1] see Fig. 9, it also opens up new possibilities as e.g. easier acquisition and rendering of much more complex scenes. In addition rendering quality is better in many cases.

Computing the true motion field from the images alone, however, is a formidable task that, in general, is hard to solve due to inherent ambiguities. For example the aperture problem and insufficient gradient strength can make it impossible to compute the correct motion field using e.g. optical flow. However, the true motion field is not needed if the goal is to generate perceptually convincing image interpolations. Because a perceptually plausible motion is interpreted as a physically correct motion by a human observer, we can rely on the capabilities of the human visual system to understand the visual input correctly in spite of all ambiguities. It is thus sufficient to focus on the aspects that are important to human motion perception to solve the interpolation problem. Or in other words:

> The human eye does not care about optimal solutions in a least squares sense, as long as it looks good.

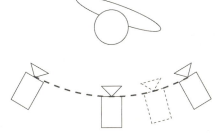

Fig. 9 Image-based interpolation techniques can create the impression of a moving camera along the space spanned by the input cameras (here depicted by the *dashed line*)

---

[1] This is not completely true. Extrapolation techniques could be used to go beyond this limitation, but quality will quickly prevail.

5.1 Image Morphing and Spatial Transformations

Image morphing aims at creating smooth transitions between pairs or arbitrary numbers of images. For simplicity of explanation we will stick to two images first. The basic procedure is to warp, i.e. to deform, the input images I_1 and I_2 towards each other depending on some warp functions $W_{I_1 \to I_2}$, $W_{I_2 \to I_1}$ and a time step α, with $\alpha \in [0, 1]$ so that $\alpha W_{I_1 \to I_2} \circ I_1 = (1-\alpha) W_{I_2 \to I_1} \circ I_2$ and vice versa in the best case. This optimal warp function can usually only be approximated, so to achieve more convincing results when warping image I_1 towards I_2, one usually also computes the corresponding warp from I_2 towards I_1 and blends the results together. More mathematically formulated we can write

$$I_{1,2}(\alpha) = B((\alpha \mathbf{W}_{\mathbf{I_1} \to \mathbf{I_2}}) \circ I_1, ((1 - \alpha) \mathbf{W}_{\mathbf{I_2} \to \mathbf{I_1}}) \circ I_2, \alpha) \tag{8}$$

where the blending function B is usually a simple linear cross-dissolve. We will have a more detailed look on how to implement a sophisticated warping function in Sect. 5.4.

5.2 Image Deformation Model for Time and View Interpolation

Analyzing properties of the human visual system shows that it is sensitive to three main aspects [30, 47, 48, 59]. These are:

1. Edges
2. Coherent motion for parts belonging to the same object
3. Motion discontinuities at object borders

It is therefore important to pay special attention to these aspects for high-quality interpolation.

Observing our surroundings we might notice that objects in the real world are seldom completely flat, even though many man-made objects are quite flat. However they can be approximated quite well by flat structures, like planes or triangles, as long as these are small enough. Usually this limit is given by the amount of detail the eye can actually perceive. In computer graphics it is usually set by the screen resolution (you may try as hard as you wish, but details smaller than a pixel are simply not visible).

If it is assumed that the world consists of such planes, then the relation between two projections of such a 3-D plane can be directly described via a homography in image space. Such homographies for example describe the relation between a 3-D plane seen from two different cameras, the 3-D rigid motion of a plane between two points in time seen from a single camera or a combination of both. Thus, the interpolation between images depicting a dynamic 3-D plane can be achieved by a per pixel deformation according to the homography directly in image space without the need to reconstruct the underlying 3-D plane, motion and camera parameters explicitly.

Fig. 10 An image (*upper left*) and its decomposition into its homogeneous regions (*upper right*). Since the transformation estimation is based on the matched edglets, only superpixels that contain actual edglets (*lower left*) are of interest. Stich et al. [56, 57] merge superpixels with insufficient edglets with their neighbors (*lower right*)

Only the assumption that natural images can be decomposed into regions, for which the deformation of each element is sufficiently well described by a homography has to be made, which is surprisingly often the case. Stich et al. [56, 57] introduced translets which are homographies that are spatially restricted. Therefore a translet is described by a 3×3 matrix H and a corresponding image segment. To obtain a dense deformation, they enforce that the set of all translets is a complete partitioning of the image and thus each pixel is part of exactly one translet, an example can be seen in Fig. 10 on the bottom right. Since the deformation model is defined piecewise, it can well describe motion discontinuities as for example resulting from occlusions.

The first step in estimating the parameters of the deformation model is to find a set of point correspondences between the images from which the translet transformation can be derived. This may sound contradictive as we stated earlier that this is the overall goal. However at this stage we are not yet interested in a complete correspondence field for every pixel. Rather we are looking for a subset for which the transformation can be more reliably established and which convey already most of the important information concerning the apparent motion in the image. As it turns out classic point features such as edges and corners, which have a long history of research in computer vision, are best suited for this task. This is in accordance to the human vision, which measures edge- and corner-features early on.

Using the Compass operator [49], a set of edge pixels, called edglets, is obtained.[2] Depending on the scene, between 2,000 and 2,0000 pixels are edglets. Having extracted these edges in both images, the task is now to find for each edglet in image

---

[2] Other edge detectors could be used for this step as well, as the Canny operator [14].

I_1 a corresponding edglet in image I_2 and this matching should be an as complete as possible one-to-one matching. This problem can be posed as a maximum weighted bipartite graph matching problem, or in other words, one does not simply assign the best match to each edglet, but instead tries to minimize an energy function to find the best overall solution. Therefore descriptors for each edglet need to be established. The shape context descriptor [9] has been shown to perform very well at capturing the spatial context C_{shape} of edglets and is robust against the expected deformations. To reduce computational effort and increase robustness for the matching process only the k nearest neighbor edglets are considered as potential matches for each edglet. Also one can assume that edglets will not move from one end of the image I_1 to the other in image I_2 as considerable overlap is always needed to establish a reliable matching. Therefore an additional distance term C_{dist} can be added. One prerequisite for the reformulation is that for each edglet in the first set a match in the second set exists, otherwise the completeness cannot be achieved. While this is true for most edglets, some will not have a correspondence in the other set due to occlusion or small instabilities of the edge detector at faint edges. However, this is easily addressed by inserting virtual occluder edglets for each edglet in the first edglet set. The graph for the matching problem is then build as depicted in Fig. 11. Each edge pixel of the first image is connected by a weighted edge to its possibly corresponding edge pixels in the second image and additionally to its virtual occluder edglet. The weight or cost function for edglet $\mathbf{e}_i$ in I_1 and $\mathbf{e}'_j$ in I_2 is then defined as

$$C(\mathbf{e_i}, \mathbf{e}'_j) = C_{dist} + C_{shape} \tag{9}$$

where the cost for the shape is the χ^2-test between the two shape contexts and the cost for the distance is defined as

$$C_{dist}(\mathbf{e_i}, \mathbf{e}'_j) = \frac{a}{(1 + e^{-b \, \|\mathbf{e}_i - \mathbf{e}'_j\|})} \tag{10}$$

with $a, b > 0$ such that the maximal cost for the euclidean distance is limited by a. The cost $C_{occluded}$ to assign an edglet to its occluder edglet is user defined and controls how aggressively the algorithm tries to find a match with an edglet of the second image. The lower $C_{occluded}$ the more conservative the resulting matching will be, as more edges will be matched to their virtual occluder edglets.

Now that the first reliable matches have been found this information can be used to find good homographies for the translets of both images. But first the spatial support for these translets need to be established, i.e. the image needs to be segmented into coherent, disjoint regions. From Gestalt theory [61] it is known that for natural scenes, these regions share not only a common motion but in general also share other properties such as similar color and texture. Felzenszwalb and Huttenlocher's superpixel segmentation [22] can be exploited to find an initial partitioning of the image into regions to become translets, based on neighboring pixel similarities. Then from the matching between the edge pixels of the input images, local homographies for each set of edge pixels in the source image that are within one superpixel

3-D Cinematography with Approximate or No Geometry

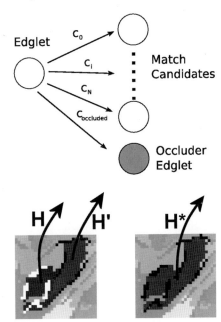

Fig. 11 Subgraph of the weighted bipartite graph matching problem for a single edglet. Each edglet has an edge to its possible match candidates and an additional edge to its virtual occluder edglet

Fig. 12 During optimization, similar transformed neighboring translets are merged into a single translet. After merging, the resulting translet consists of the combined spatial support of both initial translets (*mid and dark gray*) and their edglets (*black and white*)

are estimated. In order to do this four reliable point correspondences need to be found to compute the homography. Since the least-squares estimation based on all matched edglets of a translet is sensitive to outliers and often more than the minimal number of four matched edge pixels is available, a RANSAC approach to obtain a robust solution and filter match outliers is preferred instead [23]. Usually still between 20% and 40% of the computed matches are outliers and thus some translets will have wrongly estimated transformations. Using a greedy iterative approach, the most similar transformed neighboring translets are merged into one, as depicted in Fig. 12, until the ratio of outliers to inliers is lower than a user defined threshold. When two translets are merged, the resulting translet then contains both edglet sets and has the combined spatial support. The homographies are re-estimated based on the new edglet set and the influence of the outliers is again reduced by the RANSAC filtering.

Basically in this last step a transformation for each pixel in the input images towards the other image was established. Assuming linear motion only, the deformation vector $d(\mathbf{x})$ for a pixel $\mathbf{x}$ is thus computed as

$$d(\mathbf{x}) = H_t \mathbf{x} - \mathbf{x}. \tag{11}$$

H_t is the homography matrix of the translet t with $\mathbf{x}$ being part of the spatial support of t. However, when only a part of a translet boundary is at a true motion discontinuity, noticeably incorrect discontinuities still produce artifacts along the rest of the boundary. Imagine for example the motion of an arm in front of the body. It is discontinuous along the silhouette of the arm, while the motion at the shoulder changes continuously. We can then resolve the per pixel smoothing by an anisotropic diffusion [44] on this vector field using the diffusion equation

$$\delta I/dt = div(\ g(min(|\nabla d|, |\nabla I|)\ \nabla I) \tag{12}$$

which is dependent on the image gradient ∇I and the gradient of the deformation vector field ∇d. The function g is a simple mapping function as defined in [44]. Thus, the deformation vector field is smoothed in regions that have similar color or similar deformation, while discontinuities that are both present in the color image and the vector field are preserved. During the anisotropic diffusion, edglets that have an inlier match, meaning they are only slightly deviating from the planar model, are considered as boundary conditions of the diffusion process. This results in exact edge transformations handling also non-linear deformations for each translet and significantly improves the achieved quality.

5.3 Optimizing the Image Deformation Model

There are three ways to further optimize the image deformation model from the previous section:

1. Using motion priors
2. Using coarse-to-fine translet estimation
3. Using a scale-space hierarchy

Since the matching energy function (9) is based on spatial proximity and local geometric similarity, a motion prior can be introduced by pre-warping the edglets with a given deformation field. The estimated dense correspondences described above can be used as such a prior. So the algorithm described in Sect. 5.2 can be iterated using the result from the i-th iteration as the input to the $(i + 1)$-th iteration.

To overcome local matching minima a coarse to fine iterative approach on the translets can be applied. In the first iteration, the number of translets is reduced until the coarsest possible deformation model with only one translet is obtained. Thus the underlying motion is approximated by a single perspective transformation. During consecutive iterations, the threshold is decreased to allow for more accurate deformations as the number of final translets increases.

Additionally, solving on different image resolutions similar to scale-space [63] further improves robustness. Thus a first matching solution is found on the coarse resolution images and is then propagated to higher resolutions. Using previous solutions as motion prior significantly reduces the risk to getting stuck in local matching minima, cf. Fig. 13.

In rare cases, some scenes still cannot be matched automatically sufficiently well. For example, when similar structures appear multiple times in the images the matching can get ambiguous and can only be addressed by high level reasoning. To resolve this, a fallback on manual intervention is necessary. Regions can be selected in both

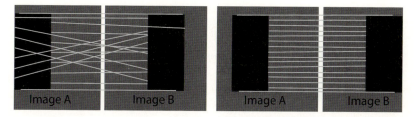

Fig. 13 Local matching minima (*left*) can be avoided by multiple iterations. In a coarse to fine manner, in each iterations the number of translets increases avoiding local matching minima by using the previous result as prior (*right*)

images by the user and the automatic matching is computed again only for the so selected subset of edglets. Due to this restriction of the matching, the correct match is found and used to correct the solution.

5.4 Rendering

Given the pixel-wise displacements from the previous sections the rendering can then be efficiently implemented on graphics hardware to allow for real-time image interpolation. Therefore a regular triangle mesh is placed over the image plane, so that each pixel in the image is represented by two triangles with appropriate texture coordinates. A basic morphing scheme algorithm as presented in (8) would be straight-forward to implement by just displacing the vertices in the vertex shader by the scaled amount of the corresponding displacement vector according to the α-value chosen. However, two problems arise with forward warping at motion discontinuities: Fold-overs and missing regions. Fold-overs occur when two or more pixels in the image end up in the same position during warping. This is the case when the foreground occludes parts of the background. Consistent with motion parallax it is assumed that the faster moving pixel in x-direction is closer to the viewpoint to resolve this conflict. When on the other hand regions get disoccluded during warping the information of these regions is missing in the image and must be filled in from the other image. Mark et al. [39] proposed to use a connectedness criterion evaluated on a per-pixel basis after warping. This measure can be computed directly from the divergence of the deformation vector field such that

$$c_{I_1} = 1 - div(d_{I_1 \to I_2})^2. \tag{13}$$

with c_{I_1} being the connectedness and $d_{I_1 \to I_2}$ is the vector field between the images I_1 and I_2 (cf. Fig. 14). The connectedness is computed on the GPU during blending to adaptively reduce the alpha values of pixels with low connectedness. Thus, in missing regions only the image which has the local information has an influence on the rendering result.

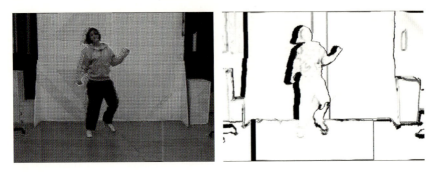

Fig. 14 *Left*: Per-vertex mesh deformation is used to compute the forward warping of the image, where each pixel corresponds to a vertex in the mesh. The depicted mesh is at a coarser resolution for visualization purposes. *Right*: The connectedness of each pixel that is used during blending to avoid a possibly incorrect influence of missing regions

Opposed to recordings with cameras, rendered pixels at motion boundaries are no longer a mixture of background and foreground color but are either foreground or background color. In a second rendering pass, the color mixing of foreground and background at boundaries can be modelled using a small selective low-pass filter applied only to the detected motion boundary pixels. This effectively removes the artifacts with a minimal impact on rendering speed and without affecting rendering quality in the non-discontinuous regions.

The complete interpolation between two images I_1 and I_2 can then be described as

$$I(\alpha) = \frac{c_{I_1}(1-\alpha)(\alpha d_{I_1 \to I_2} \circ I_1) + c_{I_2}(\alpha)((1-\alpha) d_{I_2 \to I_1} \circ I_2)}{c_{I_1}(1-\alpha) + c_{I_2}(\alpha)} \quad (14)$$

where $c_X(\phi)$ is the locally varying influence of each image on the final result which is modulated by the connectedness

$$c_X(\alpha) = c_X \cdot \alpha \quad (15)$$

Thus, the (possibly incorrect) influence of pixels with low connectedness on the final result is reduced.

The interpolation is not restricted to two images. Interpolating between multiple images is achieved by iteratively repeating the warping and blending as described in (14), where I takes over the role of one of the warped images in the equation. To stay inside the image manifold that is spanned by the images the interpolation factors must sum to one, $\sum_i \alpha_i = 1$.

As can be seen in Table 1 the proposed algorithm produces high-quality results, e.g. using the Middlebury examples [8].

The results have been obtained without user interaction. As can be seen the approach is best when looking at the interpolation errors and best or up to par in the sense of the normalized interpolation error. It is important to point out that from a perception point of view the normalized error is less expressive than the

3-D Cinematography with Approximate or No Geometry

Table 1 Interpolation, normalized interpolation and angular errors computed on the Middlebury optical flow examples by comparison to ground truth with results obtained by our method and by other methods taken from Baker et al. [8]

Venus	Interp.	Norm. Interp.	Ang.
Stich et al.	**2.88**	**0.55**	**16.24**
Pyramid LK	3.67	0.64	14.61
Bruhn et al.	3.73	0.63	8.73
Black and Anandan	3.93	0.64	7.64
Mediaplayer	4.54	0.74	15.48
Zitnick et al.	5.33	0.76	11.42
Dimetrodon	Interp.	Norm. Interp.	Ang.
Stich et al.	**1.78**	**0.62**	**26.36**
Pyramid LK	2.49	0.62	10.27
Bruhn et al.	2.59	0.63	10.99
Black and Anandan	2.56	0.62	9.26
Mediaplayer	2.68	0.63	15.82
Zitnick et al.	3.06	0.67	30.10
Hydrangea	Interp.	Norm. Interp.	Ang.
Stich et al.	**2.57**	**0.48**	**12.39**
RubberWhale	Interp.	Norm. Interp.	Ang.
Stich et al.	**1.59**	**0.40**	**23.58**

unnormalized error since discrepancies at edges in the image (e.g. large gradients) are dampened. Interestingly, relatively large angular errors are observed with the presented method emphasizing that the requirements of optical flow estimation and image interpolation are different.

Acknowledgements We would like to thank Jonathan Starck for providing us with the Surf-Cap test data (http://www.ee.surrey.ac.uk/CVSSP/VMRG/surfcap.htm) and the Stanford Computer Graphics lab for the buddha light field data set.

References

1. Adelson, E.H., Bergen, J.R.: The plenoptic function and the elements of early vision. In: M. Landy, J.A. Movshon (eds.) Computational Models of Visual Processing, pp. 3–20. MIT Press, Cambridge, MA (1991)
2. de Aguiar, E., Stoll, C., Theobalt, C., Ahmed, N., Seidel, H.P., Thrun, S.: Performance capture from sparse multi-view video. ACM Trans. Graph. **27**(3), 1–10 (2008)
3. de Aguiar, E., Theobalt, C., Magnor, M., Seidel, H.P.: Reconstructing human shape and motion from multi-view video. In: European Conference on Visual Media Production, pp. 42–49 (2005)
4. de Aguiar, E., Theobalt, C., Stoll, C., Seidel, H.P.: Marker-less deformable mesh tracking for human shape and motion capture. In: International Conference on Computer Vision and Pattern Recognition, pp. 1–8 (2007)

5. de Aguiar, E., Theobalt, C., Stoll, C., Seidel, H.P.: Rapid animation of laser-scanned humans. In: Virtual Reality, pp. 223–226 (2007)
6. Ahmed, N., Theobalt, C., Magnor, M.A., Seidel, H.P.: Spatio-temporal registration techniques for relightable 3D video. In: International Conference on Image Processing, pp. 501–504 (2007)
7. Anguelov, D., Srinivasan, P., Koller, D., Thrun, S., Rodgers, J., Davis, J.: SCAPE: shape completion and animation of people. ACM Trans. Graph. **24**(3), 408–416 (2005)
8. Baker, S., Scharstein, D., Lewis, J., Roth, S., Black, M., Szeliski, R.: A database and evaluation methodology for optical flow. In: International Conference on Computer Vision, pp. 1–8 (2007)
9. Belongie, S., Malik, J., Puzicha, J.: Matching shapes. In: International Conference on Computer Vision, pp. 454 – 461 (2001)
10. Bhat, P., Zitnick, C.L., Snavely, N., Agarwala, A., Agrawala, M., Curless, B., Cohen, M., Kang, S.B.: Using photographs to enhance videos of a static scene. In: J. Kautz, S. Pattanaik (eds.) Eurographics Symposium on Rendering, pp. 327–338. Eurographics (2007)
11. Boykov, Y., Veksler, O., Zabih, R.: Fast approximate energy minimization via graph cuts. Trans. Pattern Anal. Mach. Intell. **23**(11), 1222–1239 (2001)
12. Brox, T., Bruhn, A., Papenberg, N., Weickert, J.: High accuracy optical flow estimation based on a theory for warping. In: European Conference on Computer Vision, pp. 25–36 (2004)
13. Buehler, C., Bosse, M., McMillan, L., Gortler, S., Cohen, M.: Unstructured lumigraph rendering. ACM Trans. Graph. **20**(3), 425–432 (2001)
14. Canny, J.: A Computational approach to edge detection. Trans. Pattern Anal. Mach. Intell. **8**, 679–714 (1986)
15. Carranza, J., Theobalt, C., Magnor, M., Seidel, H.P.: Free-viewpoint video of human actors. ACM Trans. Graph. **22**(3), 569–577 (2003)
16. Chai, J.X., Chan, S.C., Shum, H.Y., Tong, X.: Plenoptic sampling. ACM Trans. Graph. **19**(3), 307–318 (2000)
17. Cobzaş, D., Yerex, K., Jägersand, M.: Dynamic textures for image-based rendering of fine-scale 3D structure and animation of non-rigid motion. Comput. Graph. Forum **21**(3), 493–502 (2002)
18. Collins, R.T.: A space-sweep approach to true multi-image matching. In: Conference on Computer Vision and Pattern Recognition, pp. 358–363 (1996)
19. Debevec, P.E., Taylor, C.J., Malik, J.: Modeling and rendering architecture from photographs: a hybrid geometry- and image-based approach. ACM Trans. Graph. **15**(3), 11–20 (1996)
9. Eisemann, M., Decker, B.D., Magnor, M., Bekaert, P., de Aguiar, E., Ahmed, N., Theobalt, C., Sellent, A.: Floating textures. Comput. Graph. Forum **27**(2), 409–418 (2008)
21. Eisemann, M., Sellent, A., Magnor, M.: Filtered blending: a new, minimal reconstruction filter for ghosting-free projective texturing with multiple images. In: Vision, Modeling, and Visualization pp. 119–126 (2007)
22. Felzenszwalb, P., Huttenlocher, D.: Efficient graph-based image segmentation. Int. J. Comput. Vis. **59**, 167–181 (2004)
23. Fischler, M., Bolles, R.: Random sample consensus: a paradigm for model fitting with applications to image analysis and automated cartography. Commun. ACM **24**(6), 381–395 (1981)
24. Franco, J.S., Boyer, E.: Exact polyhedral visual hulls. In: British Machine Vision Conference, pp. 329–338 (2003). Norwich, UK
25. Fujii, T., Tanimoto, M.: Free viewpoint TV system based on ray-space representation. In: SPIE, vol. 4864, pp. 175–189. SPIE (2002)
26. Gallup, D., Frahm, J.M., Mordohai, P., Yang, Q., Pollefeys, M.: Real-time plane-sweeping stereo with multiple sweeping directions. In: Computer Vision and Pattern Recognition, pp. 1–8 (2007)
27. Gortler, S.J., Grzeszczuk, R., Szeliski, R., Cohen, M.F.: The lumigraph. ACM Trans. Graph. **15**(3), 43–54 (1996)
28. Gu, X., Gortler, S., Hoppe, H.: Geometry images. ACM Trans. Graph. **21**(3), 355–361 (2002)
29. Hartley, R., Zisserman, A.: Multiple View Geometry in Computer Vision, 2nd edn. Cambridge University Press, Cambridge (2003)

30. Heeger, D., Boynton, G., Demb, J., Seidemann, E., Newsome, W.: Motion opponency in visual cortex. J. Neurosci. **19**, 7162–7174 (1999)
31. Horn, B., Schunck, B.: Determining optical flow. Artif. Intell. **17**, 185–203 (1981)
32. Kutulakos, K.N., Seitz, S.M.: A Theory of shape by space carving. Int. J. Comput. Vis. **38**(3), 199–218 (2000)
33. Lanman, D., Crispell, D., Taubin, G.: Surround structured lighting for full object scanning. In: International Conference on 3-D Digital Imaging and Modeling, pp. 107–116 (2007)
34. Laurentini, A.: The visual hull concept for silhouette-based image understanding. Trans. Pattern Anal. Mach. Intell. **16**(2), 150–162 (1994)
35. Lensch, H.P.A., Kautz, J., Goesele, M., Heidrich, W., Seidel, H.P.: Image-based reconstruction of spatial appearance and geometric detail. ACM Trans. Graph. **22**(2), 234–257 (2003)
36. Levoy, M., Hanrahan, P.: Light field rendering. ACM Trans. Graph. **15**(3), 31–42 (1996)
37. Li, M., Magnor, M., Seidel, H.P.: Hardware-accelerated rendering of photo hulls. Comput. Graph. Forum **23**(3), 635–642 (2004)
38. Lucas, B., Kanade, T.: An iterative image registration technique with an application to stereo vision. In: International Joint Conference on Artificial Intelligence, pp. 674–679 (1981)
39. Mark, W., McMillan, L., Bishop, G.: Post-rendering 3D warping. In: Symposium on Interactive 3D Graphics, pp. 7–16 (1997)
40. Matsuyama, T., Wu, X., Takai, T., Nobuhara, S.: Real-time 3D shape reconstruction, dynamic 3D mesh deformation, and high fidelity visualization for 3D video. Comput. Vis. Image Underst. **96**(3), 393–434 (2004)
41. Matusik, W., Buehler, C., Raskar, R., Gortler, S.J., Mcmillan, L.: Image-based visual hulls. ACM Trans. Graph. **19**(3), 369–374 (2000)
42. Matusik, W., Pfister, H.: 3D TV: A scalable system for real-time acquisition, transmission, and autostereoscopic display of dynamic scenes. ACM Trans. Graph. **19**(3), 814–824 (2004)
43. Naemura, T., Tago, J., Harashima, H.: Real-time video-based modeling and rendering of 3D scenes. Comput. Graph. Appl. **22**(2), 66–73 (2002)
44. Perona, P., Malik, J.: Scale-space and edge detection using anisotropic diffusion. Trans. Pattern Anal. Mach. Intell. **12**(7), 629–639 (1990)
45. Pock, T., Urschler, M., Zach, C., Beichel, R., Bischof, H.: A duality based algorithm for tv-l1-optical-flow image registration. In: International Conference on Medical Image Computing and Computer Assisted Intervention, pp. 511–518 (2007)
46. Posdamer, J., Altschuler, M.: Surface measurement by space-encoded projected beam systems. Comput. Graph. Image Process. **18**(1), 1–17 (1982)
47. Qian, N., Andersen, R.: A physiological model for motion-stereo integration and a unified explanation of Pulfrich-like phenomena. Vis. Res. **37**, 1683–1698 (1997)
48. Reichardt, W.: Autocorrelation, a principle for the evaluation of sensory information by the central nervous system. In: W. Rosenblith (ed.) Sensory Communication, pp. 303–317. MIT Press-Willey, New York (1961)
49. Ruzon, M., Tomasi, C.: Color edge detection with the compass operator. In: Conference on Computer Vision and Pattern Recognition, pp. 160–166 (1999)
50. Salvi, J., Pagés, J., Batlle, J.: Pattern codification strategies in structured light systems. Pattern Recognit. **37**, 827–849 (2004)
51. Segal, M., Korobkin, C., van Widenfelt, R., Foran, J., Haeberli, P.: Fast shadows and lighting effects using texture mapping. ACM Trans. Graph. **11**(3), 249–252 (1992)
52. Seitz, S.M., Dyer, C.R.: Photorealistic scene reconstruction by voxel coloring. Int. J. Comput. Vis. **35**, 1067–1073 (1997)
53. Snavely, N., Seitz, S., Szeliski, R.: Photo tourism: exploring photo collections in 3D. ACM Trans. Graph. **25**(3), 835–846 (2006)
54. Starck, J., Hilton, A.: Surface capture for performance based animation. Comput. Graph. Appl. **27**(3), 21–31 (2007)
55. Stewart, J., Yu, J., Gortler, S.J., McMillan, L.: A new reconstruction filter for undersampled light fields. In: Eurographics Workshop on Rendering, pp. 150–156 (2003)
56. Stich, T., Linz, C., Albuquerque, G., Magnor, M.: View and time interpolation in image space. Comput. Graph. Forum **27**(7), 1781–1787 (2008)

57. Stich, T., Linz, C., Wallraven, C., Cunningham, D., Magnor, M.: Perception-motivated interpolation of image sequences. In: Symposium on Applied Perception in Graphics and Visualization, pp. 97–106 (2008)
58. Tsai, R.: An efficient and accurate camera calibration technique for 3D machine vision. In: Conference on Computer Vision and Pattern Recognition, pp. 364–374 (1986)
59. Wallach, H.: Über visuell wahrgenommene Bewegungsrichtung. Psychol. Forsch. **20**, 325–380 (1935)
60. Waschbüsch, M., Würmlin, S., Gross, M.: 3D video billboard clouds. Comput. Graph. Forum **26**(3), 561–569 (2007)
61. Wertheimer, M.: Laws of organization in perceptual forms. In: W. Ellis (ed.) A Source Book of Gestalt Psychology, pp. 71–88. Kegan Paul, Trench, Trubner & Co., London (1938)
62. Yang, R., Pollefeys, M.: Multi-resolution real-time stereo on commodity graphics hardware. In: Conference on Computer Vision and Pattern Recognition, pp. 211–217 (2003)
63. Yuille, A.L., Poggio, T.A.: Scaling theorems for zero crossings. Trans. Pattern Anal. Mach. Intell. **8**(1), 15–25 (1986)
64. Zitnick, C., Kang, S., Uyttendaele, M., Winder, S., Szeliski, R.: High-quality video view interpolation using a layered representation. ACM Trans. Graph. **23**(3), 600–608 (2004)

View Dependent Texturing Using a Linear Basis

Martin Jagersand, Neil Birkbeck, and Dana Cobzas

Abstract We present a texturing approach for image-based modeling and rendering, where instead of texturing from one (or a blend of a few) sample images, new view-dependent textures are synthesized by modulating a differential texture basis. The texture basis contains image derivatives, and it models the first order intensity variation due to image projection errors, parallax and illumination variation. We derive an analytic form for this basis and show how to obtain it from images. Experimentally, we compare rendered views to ground truth real images and quantify how the texture basis can generate a more accurate rendering compared to conventional view-dependent textures. In a hardware accelerated implementation, we achieve frame rate on regular PCs and consumer graphics cards.

1 Introduction

Texture normally refers to fine-scale visual or tactile properties of a surface. The word is related to textile. Originally it was used to describe the particular surface created by the the interwoven threads in a fabric. In computer graphics, texturing is the process of endowing a surface with fine-scale properties. Often this is used to make the visualization richer and more natural than if only the 3D geometry had been rendered. There are a wide range of computer graphics texturing approaches. Early texturing involved replicating a small texture element over the surface of an object to enrich its appearance. Commonly this is done by warping a small, artificially generated 2D intensity image onto the structure, but other variations exist including 3D texturing. Texture can also be used to model light and reflections by texturing a model with a specular highlight texture.

The focus of this chapter is on texturing from photographic images. We will study how to compose a texture image suitable for photo-realistic image-based ren-

---

M. Jagersand (✉), N. Birkbeck, and D. Cobzas
University of Alberta, Canada, http://www.cs.ualberta.ca/~vis,
e-mail: jag@cs.ualberta.ca, birkbeck@cs.ualberta.ca, dana@cs.ualberta.ca

R. Ronfard and G. Taubin (eds.), *Image and Geometry Processing for 3-D Cinematography*, Geometry and Computing 5, DOI 10.1007/978-3-642-12392-4_13,
© Springer-Verlag Berlin Heidelberg 2010

dering. In this case the texturing element is a comparably large image, and unlike in the above mentioned techniques, the texture image is not repeated over the surface. Hence, we are to some extent transcending out of the original domain of texturing by now not only representing a repetitive fine scale structure, but also potentially medium and large scale changes in texture, including light and geometry aspects not included in the geometric 3-D model. Here we will focus on aspects that are specific to image based modeling and rendering. We will not cover standard issues such as implementation of 2-D warps, filtering and multi-resolution texturing. The background on basic texturing is covered in the literature [16] and recent text books [20].

One purpose of the texture representation is to capture intensity variation on the pixel level. This includes the view dependency of potentially complex light surface interactions. Another purpose, for models captured from images, is to compensate for discrepancies between the approximate captured geometry and the true object surface. The first purpose is similar to that of Bi-directional Texture Function (BTF) representations [6] and the second similar to view-dependent texture mapping (VDTM) [7].

In a parallel line of research Freeman, Adelson and Heeger noted that small image motions could be modulated using a fixed spatial image basis [11]. Using more basis vectors extracted by PCA from example images whole motion sequences could be synthesized [17]. A later work used the same idea to animate stochastic motion [8]. The texture basis can also be decomposed into a multi-linear form, where two or more variations (e.g. light and viewpoint) are represented separately [27]. The above works all represent intensity variation on the 2D image plane, but others realized that it is more efficient to represent the intensity variation on the surface facets of a 3D triangulated model [5, 12]. Both in spirit and in actual implementation all these representations are quite similar in their use of a set of basis textures/images to modulate a new texture.

The work presented in this chapter falls between *relief textures* and *lightfield* representations in its approach to photorealistic rendering. Relief textures provide an explicit geometric solution for adding 3D structure to a planar texture [21]. However, relief textures require a detailed a-priori depth map of the texture element. This is normally not available in image-based modeling if only 2-D camera video is used. Thus, relief textures have been mostly used with a-priori graphics models. The floating textures approach is similar to relief textures and performs a geometric perturbation of the pixels at render time, but instead of a depth map it uses a 2D motion vector field to drive the perturbation [9].

While initially lightfield (e.g. [13, 18]) and geometry-plus-texture approaches to image-based rendering were disparate fields, recent work attempts to close this gap. Our work is in the intersection of the two, using a relatively dense image sampling from real-time video, and representing appearance on an explicit geometric model. Work in the lightfield area using geometric proxies is closely related. Here the lightfield can be represented on a geometry that closely envelops an object [4]. However, lightfield rendering directly interpolates input images, unlike in our approach which uses an intermediate basis with well defined geometric and photometric interpolation capabilities. In surface light fields, instead of a geometric proxy, an accurate object geometry is used. Wood et al. [29] studies how to efficiently parameterize these light fields to capture complex reflectance. They do not address the issues arising from geometric misalignments, but instead rely on range scanned precise models being accurately hand-registered with calibrated imagery. On the other hand, our approach explicitly addresses the misalignment issues and the necessary correction is built into the texturing process.

The main contributions of the chapter cover theoretical and practical issues in texturing models from images, and experimentally evaluates these.

1. Theoretically, we derive the analytical form of texture variation under a full perspective camera, where previous formulations have been either image-plane based or used simplified linear camera models. The derived variation to captures misalignments between the geometric model and the texture images, parallax arising when planar model facets approximate non-planar scenes, and light variation occurring naturally in camera-based texture capture.
2. Practically, we show how the actually occurring variability in a particular texture-image sequence can be estimated. We show how the the above mentioned analytically derived forms can be identified in real data. To show that our method is practical, we present an implementation allowing real-time rendering on consumer grade PC's and graphics cards.
3. Experimentally, we compare rendering results from our model to static textures, to traditional view-dependent texturing, and to lumigraph ray-based rendering. For the experimental comparison, we use image sequences from four increasingly difficult objects.

Our texturing method combined with standard geometry acquisition using Structure-From-Motion (SFM) or Shape-from-Silhouette (SFS) is particularly suited for the consumer market. Anyone with any video camera, from a $100 web cam to a high quality digital camera, can capture image sequences of scenes and objects, build his or her own image-based scene models, and then generate reasonable quality renderings. To stimulate use by others we provide a downloadable capture and modeling system and a renderer [2].

2   Background: Image Geometry

Given images of a scene or object, the 3D geometry can be recovered in a variety of ways. Classic photogrammetry recovers 3D from 2D point correspondences using calibrated cameras. In the past decade, much work was devoted to 3D recovery from uncalibrated images [15]. Despite this, no system can recover accurate dense geometry robustly and reliably from general scenes. One of the few publicly accessible systems is KU Leuven's 3D Webservice [28], for which one can upload image sets of scenes and get back 3D reconstructions. It sequences Structure-From-Motion (SFM), auto-calibration, and dense stereo. The procedure is computationally demanding and runs in parallel on a computer network. Reconstructions often take hours to complete. Practically, care must be taken in selecting both scenes and viewpoints for the system to work well. Nonetheless, it is a representative of the state of the art in SFM based systems.

Shape-From-Silhouette (SFS) [25], on the other hand is a very robust method that obtains a visual hull geometry. It only requires the object silhouette and the calibration of the cameras, and it is quite robust to silhouette or calibration errors. In our system, we implement an efficient algorithm for silhouette carving that uses an orthogonal ray set and the Marching Intersections [26] algorithm. This decreases storage cost and improves geometric precision (by recording silhouette intersections exactly on the rays) compared to the conventional discrete voxel representation.

For the examples in this chapter, a rough 3D geometry has been obtained using either SFM or SFS as indicated. Independent of how the geometry was obtained, but central to image-based modeling, is that this 3D structure can be reprojected into a new virtual camera and thus novel views can be rendered. In mathematical terms we assume that starting with a set of m images $I_1 \ldots I_m$ from different views of a scene, a structure of n physical scene points $X_1 \ldots X_n$, and m view projections $P_1 \ldots P_m$ has been computed. These project onto image points $x_{j,i}$ as

$$x_{j,i} = P_j X_i \quad i \in 1 \ldots n, j \in 1 \ldots m \tag{1}$$

The structure is divided into Q planar facets (triangles or quadrilaterals are used in our experiments) with the points $x_{j,i}$ as node points. For texture mapping, each one of the model facets are related by a planar projective homography to a texture image; see Fig. 1.

3 Texture Basis

In conventional texture mapping, one or more of the real images are used as a source to extract texture patches from. These patches are then warped onto the re-projected structure in the new view.

Instead of using one image as a source texture, here we study how to relate and unify all the input sample images into a texture basis. Let $x_{T,i}$ be a set of texture coordinates in one-to-one correspondence to each model point X_i and thus also for each view j with the image points $x_{j,i}$ above. A texture warp function $\mathcal{W}$ translates the model-vertex to texture correspondences into a pixel-based re-arrangement (or warp) between the texture space T to screen image coordinates I.

$$T(\mathbf{x}) = I(\mathcal{W}(\mathbf{x}; \mu)) \tag{2}$$

where μ are the warp parameters and $\mathbf{x}$ the image pixel coordinates. For notational clarity in later derivations, we let the warp function act on the parameter space, as is usual in the computer vision literature, but perhaps less often seen in graphics.

View Dependent Texturing Using a Linear Basis

Common texture warp functions are affine, bi-linear and projective warps. The warp function $\mathcal{W}$ acts by translating, rotating and stretching the parameter space of the image, and hence for discrete images a re-sampling and filtering step is needed between the image and texture spaces. Details of these practicalities can be found in Moller and Haines [20].

Now if for two sample views j and k, we warp the real images I_j and I_k from image to texture coordinates into a texture images T_j and T_k, we would find that in general the two texture images are not identical, $T_j \neq T_k$, $j \neq k$ as illustrated in Fig. 1. Typically, the closer view j is to k, the smaller the difference is between T_j and T_k. This is the rationale for view-dependent texturing, where a new view is textured from the closest sample image, or by blending the three closest sample images [7].

In this chapter we will develop a more principled approach, where we seek a texture basis B such that for each sample view:

$$\mathbf{T}_j = B\mathbf{y}_j, \ j \in 1\ldots m. \qquad (3)$$

Here, assuming gray-scale images[1], $\mathbf{T}$ is a column vector representation of the image, i.e. the $q \times q$ texture image flattened into a $q^2 \times 1$ column vector. B is a $q^2 \times r$

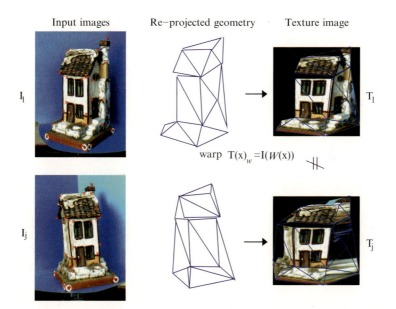

Fig. 1 Textures generated from two different images using are usually different. In the figure this is exemplified using a very coarse geometry that is only approximately aligned with the images. Common texturing problems are misalignment of texture coordinates, as visible on the right house edge, and parallax as visible on windows and door

---

[1] For color images we have a similar equation for each color channel.

matrix, where normally $r \ll m$, and $\mathbf{y}$ is a modulation vector. The texture basis B needs to capture geometric and photometric texture variation over the sample sequence, and to correctly interpolate new in-between views. We first derive a first-order geometric model and then add the photometric variation. For clarity, we derive these for a single texture warp (as in Fig. 7), whereas in practical applications a scene will be composed by texturing several model facets (as in Fig. 1).

3.1 Geometric Texture Variation

The starting point for developing a spatial texture basis to represent small geometric variations is the well known optic flow constraint. For small image plane translations it relates texture intensity change $\Delta \mathbf{T} = T_j - T_k$ to spatial derivatives $\frac{\partial}{\partial u} T, \frac{\partial}{\partial v} T$ with respect to texture coordinates $\mathbf{x} = [u, v]^T$ under an image constancy assumption [14].

$$\Delta \mathbf{T} = \frac{\partial T}{\partial u} \Delta u + \frac{\partial T}{\partial v} \Delta v \tag{4}$$

Note that given one reference texture T_0 we can now build a basis for small image plane translations $B = [T_0, \frac{\partial T}{\partial u}, \frac{\partial T}{\partial v}]$ and from this generate any slightly translated texture $T(\Delta u, \Delta v) = B[1, \Delta u, \Delta v]^T = B\mathbf{y}$.

In a real situation, a texture patch deforms in a more complex way than just translation. Recall that the texture warp, (2), has re-arranged pixel coordinates from the image into the texture coordinate space, $T = I(\mathcal{W}(\mathbf{x}; \mu))$. This warp may not be exact; image variability introduced by the imperfect stabilization can be captured by an additional warp, $\mathcal{W}(\mathbf{x}; \Delta \mu)$, occurring in the texture coordinates.[2] We study the residual image variability, $\Delta \mathbf{T} = T(\mathcal{W}(\mathbf{x}; \Delta \mu)) - T$, introduced by this imperfect perturbed warp. Rewriting as an approximation of image variability to the first order, we have the following:

$$\begin{aligned} \Delta \mathbf{T} &= T(\mathcal{W}(\mathbf{x}; \Delta \mu)) - T \\ &\approx T(\mathcal{W}(\mathbf{x}; \mathbf{0})) + \nabla T \frac{\partial \mathcal{W}}{\partial \mu} \Delta \mu - T \\ &= \nabla T \frac{\partial \mathcal{W}}{\partial \mu} \Delta \mu \\ &= \left[\frac{\partial T}{\partial u}, \frac{\partial T}{\partial v} \right] \begin{bmatrix} \frac{\partial u}{\partial \mu_1} & \cdots & \frac{\partial u}{\partial \mu_k} \\ \frac{\partial v}{\partial \mu_1} & \cdots & \frac{\partial v}{\partial \mu_k} \end{bmatrix} \Delta [\mu_1 \ldots \mu_k]^T \end{aligned} \tag{5}$$

The above equation expresses an optic flow type constraint in an abstract formulation without committing to a particular form or parameterization of $\mathcal{W}(\mathbf{x}; \mu)$, except that $\mu = 0$ gives the identity warp: $\mathbf{x} = \mathcal{W}(\mathbf{x}; \mathbf{0})$. The main purpose in the following

---

[2] An alternative to this texture-based formulation is to consider the perturbed warp as a composition with the warp from image space, e.g., $T(\mathcal{W}(\mathbf{x}; \Delta \mu)) = I(\mathcal{W}(\mathbf{x}; \mu \circ \Delta \mu))$, where $\mathcal{W}(\mathbf{x}; \mu \circ \Delta \mu) = \mathcal{W}(\mathcal{W}(\mathbf{x}; \Delta \mu); \mu))$. The texture-based representation is more convenient for our argument.

is that (5) lets us express (small) texture perturbation due to a geometric shift, $\Delta\mu$, without the explicit pixel shifting used in [9, 21], but rather using a basis of derivative images, namely the spatial derivatives of the image texture ΔT multiplied by the Jacobian of the warp function $\frac{\partial W}{\partial\mu}$. In practice, the function W is usually discretized using e.g., triangular or quadrilateral mesh elements. Next we give examples of how to concretely express image variability from these discrete representations.

For image-based modeling and rendering we warp real source images into new views given an estimated scene structure. Errors between the estimated and true scene geometry cause these warps to generate imperfect renderings. We divide these up into two categories, *image plane* and *out of plane* errors. The planar errors cause the texture to be sourced with an incorrect warp.[3] The out of plane errors arise when piecewise planar facets in the model are not true planes in the scene. Rewarping into new views under a false planarity assumption will not correctly represent parallax. This can be due to the geometry being coarse (common when using SFM) and/or inaccurate (e.g. a visual hull from SFS).

Planar texture variability. First we will consider geometric errors in the texture image plane. In most IBR (as well as conventional rendering) textures are warped onto the rendered view from a source texture $\mathbf{T}$ by means of a projective homography.

$$\begin{bmatrix} u' \\ v' \end{bmatrix} = W_h(\mathbf{x}_h; \mathbf{h}) = \frac{1}{1 + h_7 u + h_8 v} \begin{bmatrix} (1 + h_1)u + h_3 v + h_5 \\ h_2 u + (1 + h_4)v + h_6 \end{bmatrix} \tag{6}$$

We rewrite (5) with the partial derivatives of W_h for the parameters $h_1 \ldots h_8$ into a Jacobian matrix. Let $c_1 = 1 + h_7 u + h_8 v$, $c_2 = (1 + h_1)u + h_3 v + h_5$, and $c_3 = h_2 u + (1 + h_4)v + h_6$. The resulting texture image variability due to variations in the estimated homography is (to the first order) spanned by the following spatial basis:

$$\Delta\mathbf{T}_h(u, v)$$

$$= \frac{1}{c_1} \begin{bmatrix} \frac{\partial\mathbf{T}}{\partial u}, & \frac{\partial\mathbf{T}}{\partial v} \end{bmatrix} \begin{bmatrix} u\ 0\ v\ 0\ 1\ 0 & -\frac{uc_2}{c_1} & -\frac{vc_2}{c_1} \\ 0\ u\ 0\ v\ 0\ 1 & -\frac{uc_3}{c_1} & -\frac{vc_3}{c_1} \end{bmatrix} \begin{bmatrix} \Delta h_1 \\ \vdots \\ \Delta h_8 \end{bmatrix} \tag{7}$$

$$= [\mathbf{B}_1 \ldots \mathbf{B}_8][y_1, \ldots, y_8]^T = B_h \mathbf{y}_h$$

Where here and throughout the paper $\mathbf{y}$ is used for the texture modulation coefficients. Examples of the $\mathbf{B}_1 \ldots \mathbf{B}_8$ derivative images can be seen in Fig. 3. Similar expressions can be derived for other warps. For example, dropping the two last columns of the above Jacobian gives the variability for the affine warp.

---

[3] Errors in tracking and point correspondences when computing an SFM geometry, as well as projection errors due to differences between the estimated camera model and real camera (SFM and SFS) both cause model points to be reprojected incorrectly in images.

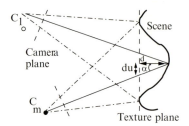

Fig. 2 Texture parallax between two views when using planar facets. True scene points project to different points on the model facet (*dashed line*). A texture representation needs to account for this

Non-planar parallax variation. In image-based modeling, a scene is represented as piecewise planar model facets, but the real-world scene is seldom perfectly represented by and aligned with these model planes. In rendering this gives rise to parallax errors. Figure 2 illustrates how the texture plane image T changes for different scene camera centers C. Given a depth map $d(u, v)$ representing the offset between the scene and texture plane, relief texturing [21] can be used to compute the rearrangement (pre-warp) of the texture plane before the final homography renders the new view. In image-based methods, an accurate depth map is seldom available. However, we can still develop the analytic form of the texture intensity variation as above. For a point on the model facet, let $[\alpha, \beta]$ be the angles between the facet normal vector and the ray pointing to the camera center C_j along the u and v axis (i.e. $\mathbf{v} = |C_j - P| = (v_x, v_y, v_z)^T$, $\alpha = \tan^{-1}(\frac{v_x}{v_z})$, and $\beta = \tan^{-1}(\frac{v_y}{v_z})$). The pre-warp rearrangement needed on the texture plane to correctly render this scene using a standard homography warp is then:

$$\begin{bmatrix} \delta u \\ \delta v \end{bmatrix} = \mathcal{W}_p(\mathbf{x}; \mathbf{d}) = d(u, v) \begin{bmatrix} \tan \alpha \\ \tan \beta \end{bmatrix} \qquad (8)$$

As before, taking the derivatives of the warp function with respect to a camera angle change and inserting into (5), we get:

$$\Delta \mathbf{T}_p(u, v) = d(u, v) \begin{bmatrix} \frac{\partial \mathbf{T}}{\partial u}, \frac{\partial \mathbf{T}}{\partial v} \end{bmatrix} \begin{bmatrix} \frac{1}{\cos^2 \alpha} & 0 \\ 0 & \frac{1}{\cos^2 \beta} \end{bmatrix} \begin{bmatrix} \Delta \alpha \\ \Delta \beta \end{bmatrix} = B_p \mathbf{y}_p \qquad (9)$$

3.2 Photometric Variation

In image-based rendering real images are re-warped into new views, and hence the composite of both reflectance and lighting is used. If the light conditions are the same for all sample images, there is no additional intensity variability introduced. However, commonly the light will vary at least somewhat. In the past decade, both empirical studies and theoretical motivations have shown that a low dimensional intensity subspace of dimension 5–9 is sufficient for representing the light variation of most natural scenes. Recently, Barsi and Jacobs [3] and Ramamoorthi and Hanrahan [23] have independently derived an analytic formula for irradiance (and reflected radiance from a convex Lambertian object) under distant illumination, explicitly considering attached shadows. They shown that the irradiance can be regarded as a convolution of the incident illumination with the Lambertian reflectance function and express the irradiance in terms of coefficients for a spherical harmonics function basis of the illumination. An important result of their work is that Lambertian reflections act as a low-pass filter, so actual irradiance lies very close to a 9-D subspace.

Let (α, β) be the spherical coordinates of the distant light source and $T(\alpha, \beta, \theta, \phi)$ the intensity of the image at a point with surface normal whose spherical coordinates are (θ, ϕ). Assuming a Lambertian surface and ignoring albedo, $T(\alpha, \beta, \theta, \phi)$ can be thought as the irradiance at orientation (θ, ϕ) due to a unit directional source at (α, β). The analytical formula for T is then [3]:

$$T(\alpha, \beta, \theta, \phi) = \sum_{l=0}^{\infty} \sum_{k=-l}^{l} A_l L_{lk}(\alpha, \beta) Y_{lk}(\theta, \phi) \tag{10}$$

$$\approx \sum_{l=0}^{2} \sum_{k=-l}^{l} A_l L_{lk}(\alpha, \beta) Y_{lk}(\theta, \phi) \tag{11}$$

where $Y_{lk}(\theta, \phi)$ are the spherical harmonics, A_l is a constant that vanishes for odd $l > 1$ and $L_{lk}(\alpha, \beta)$ are the spherical harmonic coefficients of the incident illumination.

The first nine spherical harmonics and constants:

$$(x, y, z) = (\cos\theta \sin\phi, \sin\theta \sin\phi, \cos\phi)$$

$$Y_{00}(\theta, \phi) = \sqrt{(4\pi)^{-1}},$$

$$Y_{1-1}(\theta, \phi) = \sqrt{\frac{3}{4\pi}} y, \quad Y_{10}(\theta, \phi) = \sqrt{\frac{3}{4\pi}} z, \quad Y_{11}(\theta, \phi) = \sqrt{\frac{3}{4\pi}} x,$$

$$Y_{2-2}(\theta, \phi) = \sqrt{\frac{15}{4\pi}} xy, \quad Y_{2-1}(\theta, \phi) = \sqrt{\frac{15}{4\pi}} yz, \quad Y_{21}(\theta, \phi) = \sqrt{\frac{15}{4\pi}} xz,$$

$$Y_{20}(\theta, \phi) = \sqrt{\frac{5}{16\pi}} (3z^2 - 1), \quad Y_{22}(\theta, \phi) = \sqrt{\frac{15}{16\pi}} (x^2 - y^2)$$

$$A_0 = \pi, \quad A_1 = \frac{2\pi}{3}, \quad A_2 = \frac{\pi}{4}$$

Using single index notation

$$Y_1, Y_2, Y_3, Y_4 = Y_{00}, Y_{1-1}, Y_{10}, Y_{11}$$

$$Y_5, Y_6, Y_7, Y_8, Y_9 = Y_{2-2}, Y_{2-1}, Y_{20}, Y_{21}, Y_{22}$$

$$\hat{A}_j = \begin{cases} A_0 & \text{if } j = 1 \\ A_1 & \text{if } j \in 2, 3, 4 \\ A_2 & \text{if } j \in 5 \dots 9 \end{cases}$$

and defining $B_j(\theta, \phi) = \hat{A}_j Y_j(\theta, \phi)$, $j = 1 \ldots 9$, we can rewrite (11) for all the pixels in the images as:

$$\mathbf{T} = [\mathbf{B}_1 \ldots \mathbf{B}_9][L_1 \ldots L_9]^T \tag{12}$$

The image difference caused by light change can be then expressed as:

$$\Delta \mathbf{T}_l = [\mathbf{B}_1 \ldots \mathbf{B}_9][y_1 \ldots y_9]^T = B_l \mathbf{y}_l \tag{13}$$

3.3 Estimating Composite Variability

In textures sampled from a real scene using an estimated geometric structure we expect that the observed texture variability is the composition of the above derived planar, parallax and light variation plus other unmodeled errors. Hence we can write the texture for any sample view k, and find a corresponding texture modulation vector $\mathbf{y}_k$:

$$\mathbf{T}_k = [\mathbf{T}_0, B_h, B_p, B_l][1, y_1, \ldots, y_{19}] = B\mathbf{y}_k \tag{14}$$

where $\mathbf{T}_0$ is the reference texture. Textures for new views are synthesized by interpolating the modulation vectors from the nearest sample views into a new $\mathbf{y}$, and computing the new texture $\mathbf{T}_{new} = B\mathbf{y}$.

Since this basis was derived as a first order representation it is valid for (reasonably) small changes only. In practical image-based modeling the geometric point misalignments and parallax errors are typically within 3–5 pixel, which is small enough.

Often in IBR, neither dense depth maps nor light is available. Hence B_p, and B_l cannot be directly analytically computed using (9) and (11). Instead the only available source of information are the sample images $I_1 \ldots I_m$ from different views of the scene, and from these, the computed corresponding textures $\mathbf{T}_1 \ldots \mathbf{T}_m$.

However, from the above derivation we expect that the effective rank of the sample texture set is the same as of the texture basis B, i.e. $rank[\mathbf{T}_1, \ldots, \mathbf{T}_m] \approx 20$. Hence, from $m \gg 20$ (typically 100–200) sample images, we can estimate the best fit (under some criterion) rank-20 subspace using e.g. PCA, SVD, or ICA. This yields an estimated texture basis $\hat{B}$ and corresponding space of modulation vectors $\hat{\mathbf{y}}_1, \ldots \hat{\mathbf{y}}_m$ in one-to-one correspondence with the m sample views. From the derivation of the basis vectors in B, we know this variation will be present and dominating in the sampled real images. Hence, the analytical B and the estimate $\hat{B}$ span the same space and just as before, new view dependent textures can now be modulated from the estimated basis by interpolating the $\hat{\mathbf{y}}$ from $\hat{\mathbf{y}}_j, \hat{\mathbf{y}}_k, \hat{\mathbf{y}}_l$ corresponding to the closest sample views j, k, l and modulating a new texture $\mathbf{T} = \hat{B}\hat{\mathbf{y}}$. We call this a dynamic texture (DynTex) as it is continuously changing with viewpoint variation.

3.4 Experimental Comparison for Analytical and PCA Basis

To validate the equivalence between the analytical formulation of the texture basis (Sects. 3.1 and 3.2) and the statistically estimated one (Sect. 3.3), we performed several experiments where we isolated different types of texture variability (planar, parallax, light). In the following sections we show that the analytical basis is represented in the estimated PCA subspace.

Planar texture variation. The planar variation is usually caused by tracking inaccuracies that are about 1–5 pixels. To replicate this variability in a controlled way, a planar region from a toy house (Fig. 3a was selected and warped using a homography warp to a 128×128 texture (Fig. 3b). The corners were then randomly perturbed with 1–5 pixels to generated 200 textures. We calculated the analytical texture basis using (7) with the initial texture (see Fig. 3 (c1, d1, e1)) and the PCA basis for the perturbed textures. We projected the analytical basis onto a subspace of the PCA texture basis. Figure 3 (c2, d2, e2) illustrates the recovered analytical textures from the PCA subspace. The average intensity pixel error between the original and recovered basis was 0.5%. Hence the PCA basis spans the analytically derived texture variability quite well.

Non-planar texture variation (parallax). The parallax variability is caused by a planar facet in the geometric model representing a non-planar scene. To simulate this variability 90 images from different pan angles were captired of a non-planar wall from the toy house used above while tracking four corners of a quadrilateral region.

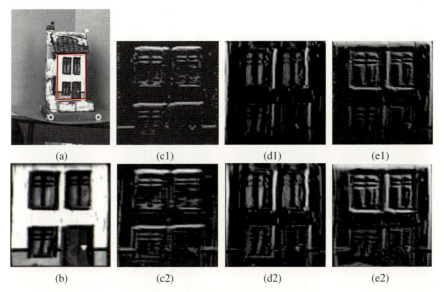

Fig. 3 Comparison of analytical and PCA basis for planar variability. (**a**) Original image; (**b**) warped texture quadrilateral; (c1), (d1), (e1) analytical basis (1st, 4th, 7th from (7)); (c2), (d2), (e2) corresponding recovered basis using PCA subspace

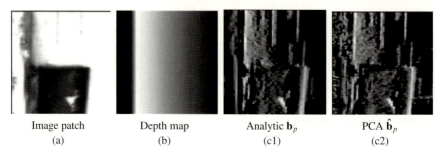

Image patch	Depth map	Analytic $\mathbf{b}_p$	PCA $\hat{\mathbf{b}}_p$
(a)	(b)	(c1)	(c2)

Fig. 4 Comparison of analytical form (9) of $\mathbf{b}_l$ and the estimated PCA basis $\hat{\mathbf{b}}_l$ for parallax variability. Image patch is from the right side wall of the house, see Fig. 1 *top row*

The corners are used to warp each quad into a standard shape for generating the texture images. Choosing a reference texture (see Fig. 4a), we manually inputted the depth map (see Fig. 4b) and calculated the analytical texture basis using (9). From the other sample textures we estimated a PCA subspace and projected the analytical basis into this space. Figure 4 (c1) shows the original analytical basis (B1) and Fig. 4 (c2) shows the recovered basis from the PCA subspace.

Photometric texture variation. Photometric variation is caused by changing light conditions or by object rotation relative to the light source. We simulated this variability by moving a toy house on a pivot rig relative to incoming sunlight. Other forms of variation were avoided by attaching the camera to the pivot rig (i.e., the projection of the house in the sequence is fixed). A laser-scanned model was then aligned to the image sets, and the spherical harmonic functions were computed for this geometry. Figure 5 shows the first few of these analytical spherical harmonic basis functions and their reconstruction from a PCA basis that was computed from the image sequence. The similarity of the harmonics reconstructed from the PCA basis to the analytic harmonic functions illustrates that the empirical basis sufficiently encodes light variation.

4 Model and Texture Capture System Implementation

Computing the texture basis involves reprojecting input images using the object or scene geometry. The geometry is usually computed from the same images, but could be obtained in some other way. We developed a software integrating the steps from images to model. The whole procedure of making a textured model takes only a few minutes in most cases, see Video 1 [1] . The software is downloadable; see [2]. To quickly capture views from all sides of an object we use a rotating platform (Radio Shack TV stand). Our software can take live video from an IEEE1394 camera, (we use a Unibrain web cam or a PtGrey Scorpion 20SO in the experiments to follow) or it can import digital image files from a still camera. Camera calibration is obtained with a pattern, and object silhouettes through bluescreening. Figure 6 shows the

View Dependent Texturing Using a Linear Basis

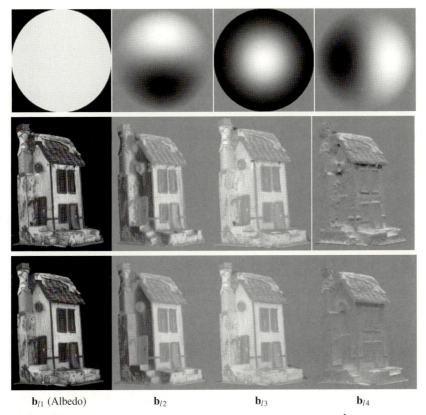

$\mathbf{b}_{l1}$ (Albedo)　　　　　$\mathbf{b}_{l2}$　　　　　$\mathbf{b}_{l3}$　　　　　$\mathbf{b}_{l4}$

Fig. 5 Comparison of analytical form (11) of $\mathbf{b}_l$ and the estimated PCA basis $\hat{\mathbf{b}}_l$ for light variability. *Top row*: Angular map of the spherical harmonics. Middle: Analytic spherical harmonic basis. *Bottom*: Corresponding light basis computed by PCA

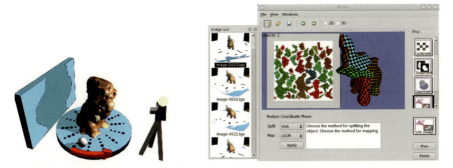

Fig. 6 *Left*: experimental capture-setup. *Right*: GUI for our capture system. The screenshot shows the texture coordinate step

capture setup. An object is rotated in front of the camera to capture different viewpoints. Hence light variation is implicitly captured with viewpoint direction into a lit texture. Alternatively light direction can be separately parameterized using the image of the specular ping-pong ball put on the rotating platform. The geometry is then computed using SFS as in Sect. 2. Alternatively, a separately obtained geometry can be imported. To stay with the camera-based paradigm we have used KU Leuven's 3D Webservice [28].

Texture coordinates. While in computer vision it is common to texture directly from images, in graphics applications a unified texture space is desired and often necessary. To automatically compute texture coordinates, the object geometry is first split along high curvature regions. Then each region is flattened using a conformal mapping [19] and packed into an OpenGL texture square (the GUI screenshot in Fig. 6 illustrates an example of this mapping). In the PCA texture basis computation, all input images are transformed into and processed in this space.

Texture basis generation. The projection of the estimated structure into the sample images, x_j, is divided into planar facets (triangles). Practically, using HW accelerated OpenGL each frame I_j is loaded into texture memory and warped to a standard shape texture T_j based on the texture coordinate atlas. We next estimate a texture basis $\hat{B}$ and a set of texture coefficients $\hat{\mathbf{y}}_j$ by performing PCA on the set of zero-mean textures $[T_j - \overline{T}]$, $j \in 1 \ldots m$.

Final model. The PCA/DynTex basis is the largest component of a model. However it compresses well using jpeg, and model storage size of 50 kB–5 MB are typical. (The storage is proportional to texture image dimension and number of basis vectors.) The complete model consisting of the 3-D geometry and texture basis can be exported, either for inclusion in Maya or Blender, for which we have written a dynamic texture rendering plugin, or direct real-time rendering by a stand-alone program downloadable from our web site. An example of several objects and people captured separately using our capture system and incorporated into a scene from Edmonton can be seen in Fig. 8, and Video 1 [1]. A cylindrical panorama from a location near the Muttart conservatory is used as a city backdrop.

Rendering. A desired view is given by the projection matrix P with the camera direction $\mathbf{v}$. For calculating the texture blending $\mathbf{y}$ we first apply 2-dimensional Delaunay triangulation over the camera viewing directions in the training set. Then we determine which simplex the new camera direction is contained in, and estimate the new texture modulation coefficients by linearly interpolating the coefficients associated with the corner points of the containing simplex. The new texture is generated from the basis textures, and then the geometric model is rendered into the desired pose. The most computationally demanding part of rendering is blending the texture basis. Hardware accelerated blending helps to achieve real time rendering. Depending on the graphics hardware capabilities, one of several methods are choosen. On old, modest graphics cards, multipass rendering is used to blend the basis textures. On newer graphics hardware, a shader program is used to directly blend the textures. If graphics hardware acceleration is unavailable, a SIMD MMX routine performs the texture blending. Rendering our textures on midrange HW, with shader programs, single objects render at well over 100 Hz using $20\ 512 \times 512$ resolution basis textures per object. Ten dynamically textured objects in a scene still render at over 30 Hz.

Example renderings. The first example illustrates the difference between a modulated texture and standard image texturing. A wreath made of natural straw and flowers was captured and processed into a texture basis. In Fig. 7, a rotating quadrilateral is textured with the image of the wreath. Using only one image, the texture appears unnatural from all but the capture direction, as illustrated in the top row. On the other hand, by modulating the view dependent texture, the fine scale variability from the wreath physical geometric texture as well as its photometric properties is realistically reproduced (bottom row).

As mentioned, we are not limited to small objects. We can import geometries from 3D Webservice [28]. In Video 3 [1] and Fig. 9, we show a preacher's chair captured in situ from the Seefeld church in Germany.

5   Experiments

Rendering quality of textures can be judged subjectively by viewers and evaluated numerically by comparing to ground truth images. Unlike comparisons of geometry alone, numeric errors are not indicative of perceptual quality. Furthermore, a static image does not show how light and specularities move. Therefore we rely mainly on the video renderings to argue photo-realistic results. As far as we know there is no commonly accepted standard for a perceptually relevant numerical measure. We use just the mean pixel intensity difference between the rendered model and a real image (from a pose not used in the capture data to compute the model). For each experiment, a set of input images were acquired using the turntable setup. Half were used to compute the model, and the other half (from different viewpoints) were used as reference in the comparison videos and intensity error computation. For the three sequences below captured in our lab (house, elephant, and wreath) a PtGrey Scorpion camera at 800×600 resolution was used. Due to the calibration pattern taking up image space, the effective object texture resolution is however closer to web-cam VGA (640×480) resolution.

Four algorithms compared. To evaluate the subspace-based dynamic texture (DynTex), we compared it to several other popular texturing methods in the literature. As a base case we use standard single texturing with the pixel values of the single texture computed to minimize the reprojection error in all training views. Next we choose the popular view dependent texturing (VDTM) [7], and our final comparison is against ray-based "Unstructured Lumigraph" rendering [4]. While more methods have been published, many are variations or combinations of the four we compare. To put the methods on an equal footing we use 20 basis vectors in our dynamic texture. The VDTM texturing blends textures sourced from 20 input images. In the lumigraph, for each texturespace pixel a list of view rays and their corresponding

Fig. 7 Texturing a rotating quadrilateral with a wreath. *Top*: by warping a flat texture image. *Bottom*: by modulating the texture basis B and generating a continuously varying texture which is then warped onto the same quad

Fig. 8 Several objects and persons composed in Blender, **Video 1**

View Dependent Texturing Using a Linear Basis

Fig. 9 Seefeld Kanzel: input image, geometry, static and dynamic texture rendering

colors are stored. The lumigraph is then computed on the geometry by picking the 20 rays per texture pixel that minimize the reprojection error over all input images. (Note: Unlike the VDTM and DynTex basis, jpeg compression does not work for the ray indices yielding in practice a larger data representation).

Selection of four test data sets. Depending on the complexity of the scene or object to be captured and rendered, different texturing methods can be used. In the following evaluation we choose four data sets of increasing complexity to challenge the texturing methods.

For a comparison to existing literature, we start with the downloadable temple scene from [24] (Fig. 10, I.) A close approximation to the true geometry is computed using SFS by our system, with 90% reconstructed within 1.7 mm of ground truth, a further geometry refinement improves this to 1.1 mm. Our geometry is not quite as good as Hernandez et al. (0.5 mm) [10], but comparing texture renderings for the initial SFS model with those of the refined model, there is next to no perceptual difference. Likewise for this simple BRDF we find little perceptual or numerical error difference between using just a conventional static single texture or any of the view dependent textures (see Video 5 [1]).

Our second data set is of a house, with wood, bark and moss materials, and a more complex structure. For the house there is a significant difference between the SFS visual hull geometry and the underlying true geometry (particularly in the middle inside corner). Also can be seen in Video 6 [1] and Fig. 11 now the static texture compares badly to the view dependent ones, which there is little difference between.

Third we try an elephant carved in jade. This object has a complex reflectance with both specularities and subsurface scattering. Here a single texture gives a dull

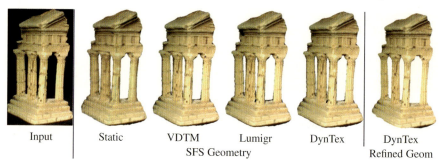

Fig. 10 Renderings of the temple from Middlebury multiview stereo image set. Texture is simple, and renders well with any texturing method, even on the initial SFS geometry. Texturing a refined model (*right*) gives almost no perceptual improvement

Fig. 11 Renderings of a textured SFS house model. The static texture is blurred due to averaging colors on different rays, while the other textures are sharp with indistinguishable quality differences

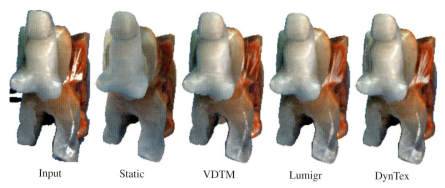

Fig. 12 A Jade elephant with complex reflectance. Static and VDTM textures are dull and completely miss the specularity on the ear. DynTex and Lumigraph capture the light and material more faithfully

flat appearance. VDTM is perceptually better, but a close analysis shows that some specularities are missing (e.g. on ears in Fig.12), and others have incorrect gradients. The DynTex and unstructured lumigraph show better results both visually and numerically for difficult (particularly specular) views, with a max intensity error of 6% compared to 10% for the standard view dependent texture and 19% for a static texture, Fig. 12 (Video 7 [1]).

Finally, we show an example of a straw wreath, where obtaining a good geometry is very difficult (Fig. 13 IV, Video 8 [1]). Here, a purely image-based method

| Input Image | static | VDTM | Lumi | DynTex |

Fig. 13 Detail crops showing results for the Wreath with complex micro-geometry rendered on a rough proxy geometry

Table 1 Numerical texture intensity errors and (variance). %-scale

Error (variance)	Temple	House	Elephant	Wreath
Static texture	10.8 (1.5)	11.8 (1.2)	19.0 (1.4)	28.4 (2.8)
VDTM	8.3 (1.9)	9.8 (1.3)	10.1 (1.9)	21.4 (3.5)
Lumigraph	10.8 (2.5)	9.8 (1.2)	5.9 (0.7)	14.3 (1.3)
DynTex	7.3 (1.0)	9.4 (1.0)	6.6 (0.7)	13.4 (1.2)

can represent a dense sample of the rayset, but at a huge storage (gigabytes) cost. We used a rough visual hull proxy geometry. The static texture is blurred out. The VDTM looks sharper because input images are used directly, but a close inspection shows somewhat jumpy transitions in the video, and during these transitions two input images are blended on top, creating a wreath with more straws. Both the DynTex and Unstructured Lumigraph code view dependency in texture space in different ways. These instead blur detail somewhat but give an overall lower error as explained below.

Summarizing the experiments we find that for simple reflectance and geometry, any texturing method works well, while for more complex cases, view-dependent appearance modeling helps, and for the two most complex cases the DynTex has a better performance than VDTM. Both of these can be rendered in hardware using simple texture blending. The unstructured lumigraph has similar performance to the DynTex, but at a much higher storage cost, and would require a complicated plugin to render in Maya or Blender. The maximum image errors and error variance are summarized in Table 1. The variance indicates smoothness of texture modulation over viewpoint changes. Perceptually a high value manifests itself as a jumpy appearance change. An example of viewpoint error variation can be seen in Fig. 14. The jumpy appearance of the VDTM is due to it working better when close to an image in the reference set.

6 Discussion

We have presented a texturing method where for each new view a unique view-dependent texture is modulated from a texture basis. The basis is designed so that it encodes a texture intensity spatial derivatives with respect to warp and parallax

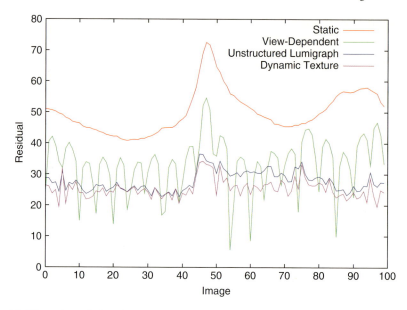

Fig. 14 Viewpoint variation of rendering error for the wreath

parameters in a set of basis textures. In a rendered sequence the texture modulation plays a small movie on each model facet, which correctly represents the underlying true scene structure to a first order. This effectively compensates for small (up to a few pixels) geometric errors between the true scene structure and captured model.

The strength of our method lies in its ability to capture and render scenes with reasonable quality from images alone. Hence, neither a-priori models, expensive laser scanners nor tedious manual modeling is required. Only a PC computer and a camera is needed. This can potentially enable applications of modeling from images such as virtualized and augmented reality in the consumer market. Experiments show better performance than static or conventional view dependent textures, and equal performance to more cumbersome and slower unstructured Lumigraph methods.

References

1. Movies of the experiments are available on http://www.cs.ualberta.ca/~vis/scalesmod
2. Software is downloadable from http://www.cs.ualberta.ca/~vis/ibmr
3. Barsi, R., Jacobs, D.: Lambertian refrectance and linear subspace. In: IEEE International Conference on Computer Vision, pp. 383–390 (2001)
4. Buehler, C., Bosse, M., McMillan, L., Gortler, S.J., Cohen, M.: Unstructured lumigraph rendering. In: Computer Graphics (SIGGRAPH), pp. 43–54 (2001)
5. Cobzas, D., Yerex, K., Jagersand, M.: Dynamic textures for image-based rendering of fine-scale 3D structure and animation of non-rigid motion. In: Eurographics (2002)

6. Dana, K.J., van Ginneken, B., Nayar, S.K., Koenderink, J.J.: Reflectance and texture of real-world surfaces. ACM Trans. Graph. **18**(1), 1–34 (1999)
7. Debevec, P.E., Taylor, C.J., Malik, J.: Modeling and rendering architecture from photographs: a hybrid geometry- and image-based approach. In: SIGGRAPH (1996)
8. Doretto, G., Chiuso, A., Wu, Y.N., Soatto, S.: Dynamic textures. Int. J. Comput. Vis. **51**(2), 91–109 (2003)
9. Eisemann, M., Decker, B.D., Magnor, M., Bekaert, P., de Aguiar, E., Ahmed, N., Theobalt, C., Sellent, A.: Floating textures. Comput. Graph. Forum (Proc. Eurographics EG'08) **27**(2), 409–418 (2008)
10. Esteban, C.H., Schmitt, F.: Silhouette and stereo fusion for 3D object modeling. Comput. Vis. Image Understand. **96**(3), 367–392 (2004)
11. Freeman, W.T., Adelson, E.H., Heeger, D.J.: Motion without movement. In: SIGGRAPH (1991)
12. Furukawa, R., Kawasaki, H., Ikeuchi, K., Sakauchi, M.: Appearance based object modeling using texture database: acquisition, compression and rendering. In: Proceedings of the 13th Eurographics Workshop on Rendering. Eurographics Association, Aire-la-Ville, Switzerland, Switzerland, pp. 257–266 (2002)
13. Gortler, S.J., Grzeszczuk, R., Szeliski, R.: The lumigraph. In: Computer Graphics (SIGGRAPH'96), pp. 43–54 (1996)
14. Hager, G.D., Belhumeur, P.N.: Efficient region tracking with parametric models of geometry and illumination. IEEE Trans. Pattern Anal. Mach. Intell. **20**(10), 1025–1039 (1998)
15. Hartley, R.I., Zisserman, A.: Multiple View Geometry in Computer Vision. Cambridge University Press (2000)
16. Heckbert, P.: Fundamentals of Texture Mapping. Msc thesis. Technical Report No. UCB/CSD 89/516, University of California, Berkeley (1989)
17. Jagersand, M.: Image based view synthesis of articulated agents. In: CVPR (1997)
18. Levoy, M., Hanrahan, P.: Light field rendering. In: Computer Graphics (SIGGRAPH'96), pp. 31–42 (1996)
19. Levy, B., Petitjean, S., Ray, N., Maillot, J.: Least squares conformal maps for automatic texture atlas generation. ACM Trans. Graph., 362–371 (2002)
20. Möller, T., Haines, E.: Real-Time Rendering. A.K. Peterson, Natick (2002)
21. Oliveira, M.M., Bishop, G., McAllister, D.: Relief texture mapping. In: Computer Graphics (SIGGRAPH'00) (2000)
22. Ramamoorthi, R.: Analytical pca construction for theoretical analysis of light variability in a single image of a lambertian object. IEEE Trans. Pattern Anal. Mach. Intell. **24**(10) (2002)
23. Ramamoorthi, R., Hanarahan, P.: On the relationship between radiance and irradiance: determining the illumination from images of a convex lambertian object. J. Optical Soc. Am. A **18**(10), 2448–2459 (2001)
24. Seitz, Curless, Diebel, Scharstein, Szeliski: A comparison of multiview stereo reconstruction algorithms. In: CVPR (2006)
25. Slabaugh, G.G., Culbertson, W.B., Malzbender, T., Schafer, R.W.: A survey of methods for volumetric scene reconstruction from photographs. In: International Workshop on Volume Graphics (2001)
26. Tarini, M., Callieri, M., Montani, C., Rocchini, C.: Marching intersections: an efficient approach to shape from silhouette. In: Proceedings of VMV 2002 (2002)
27. Vasilescu, M.A.O., Terzopoulos, D.: Tensortextures: multilinear image-based rendering. ACM Trans. Graph. **23**(3), 336–342 (2004)
28. Vergauwen, M., Gool, L.V.: Web-based 3D reconstruction service. Mach. Vis. Appl. **17**, 411–426 (2006)
29. Wood, D.N., Azuma, D.I., Aldinger, K., Curless, B., Duchamp, T., Salesin, D.H., Stuetzle, W.: Surface light fields for 3D photography. In: Computer Graphics (SIGGRAPH'00) (2000)

Printing: Ten Brink, Meppel, The Netherlands
Binding: Stürtz, Würzburg, Germany